How to Think Impossibly

HOW TO

IMPOSSI

The University of Chicago Press CHICAGO AND LONDON

JEFFREY J. KRIPAL

THINK BLY

About Souls, UFOs,
Time, Belief, and
Everything Else

The University of Chicago Press, Chicago 60637
The University of Chicago Press, Ltd., London
© 2024 by Jeffrey J. Kripal
Published 2024
Printed in the United States of America

33 32 31 30 29 28 27 26 25 24   1 2 3 4 5

ISBN-13: 978-0-226-83368-2 (cloth)
ISBN-13: 978-0-226-83369-9 (e-book)
DOI: https://doi.org/10.7208/chicago/9780226833699.001.0001

Library of Congress Cataloging-in-Publication Data

Names: Kripal, Jeffrey J. (Jeffrey John), 1962- author.
Title: How to think impossibly : about souls, UFOs, time, belief,
    and everything else / Jeffrey J. Kripal.
Description: Chicago ; London : The University of Chicago Press, 2024. |
    Includes bibliographical references and index.
Identifiers: LCCN 2023053669 | ISBN 9780226833682 (cloth) |
    ISBN 9780226833699 (ebook)
Subjects: LCSH: Parapsychology—Philosophy. | Extrasensory
    perception—Philosophy. | Curiosities and wonders—Philosophy. |
    Altered states of consciousness—Philosophy. | Mysticism—Philosophy. |
    Metaphysics. | Monism. | Dualism. | BISAC: PHILOSOPHY / Metaphysics |
    SCIENCE / Philosophy & Social Aspects
Classification: LCC BF1040 .K75 2024 | DDC 133.8—dc23/eng/20231218
LC record available at https://lccn.loc.gov/2023053669

♾ This paper meets the requirements of ANSI/NISO z39.48-1992
(Permanence of Paper).

That systematization of pseudo-data is approximation to realness or final awakening.

CHARLES FORT, The Book of the Damned

Any phenomenon that is widely and deeply studied becomes central to its field, while puzzling "fringe" phenomena may be ignored for so long that their very existence is forgotten. Yet it is often those same phenomena, when someone finally takes note of them, that force a paradigm shift.

BARBARA NEWMAN, The Permeable Self: Five Medieval Relationships

To say that a whole new dimension opened up for me on the spot at that instant, is to understate the magnitude of the experience. . . . It's hard to explain what actually happened, really, but somehow, that very rational moment of clarity immediately sparked some kind of cognitive firestorm that I can't describe in precise terms. . . . I had been trained to *never, ever* take the supernatural or miraculous as facts in any way, under any circumstances.

CARLOS EIRE, They Flew: A History of the Impossible

Contents

Prologue. Knowledge before Its Time   ix

Introduction. The Fantastic Foundations of Reality   1

Part One. When the Impossible Happens

One. Words Are Experiences: Evolutionary Origins
    and the World of the Dead   23
Two. Why They Don't Land: Mantis, Mystical
    Theology, and Social Criticism   52
Three. "That They Are Not Human":
    Thinking on the Autistic Spectrum   92

Part Two. Making the Impossible Possible

Four. The Timeswerve: Theorizing in a Block Universe   123
Five. The World Is One, and the Human Is Two:
    Some Tentative Conclusions   152
Six. We Are God (and the Devil): Further Thoughts
    and Moral Objections   179
Conclusion. How to Think Impossibly   217

Epilogue. The Three Bars   239

ACKNOWLEDGMENTS. A SOCIOLOGY OF THE IMPOSSIBLE   247
NOTES   251
INDEX   289

Knowledge before Its Time

So there must be something wrong with a child that questions the gods of the pigeon-holes. I was always asking and making myself a crow in a pigeon's nest. It was hard on my family and surroundings, and they in turn were hard on me. I did not know then, as I know now, that people are prone to build a statue of the kind of person that it pleases them to be. And few people want to be forced to ask themselves, "What if there is no me like my statue?"

ZORA NEALE HURSTON, *Dust Tracks on a Road*

I want to begin with an early scene in Zora Neale Hurston's memoir. Hurston was a student of the anthropologist Franz Boas at Barnard College, engaged in important ethnographic work in Haiti and Jamaica, and was a major voice in the Harlem Renaissance.[1] I want to begin with Hurston's story because it enacts quite obviously what I mean by the impossible, and because my encounter with her memoir first took place within my teaching experience, which is itself filled with impossible moments.[2]

Hurston's story implicitly forces a series of questions that I want to ask explicitly and repeatedly in the following pages about consciousness, time, and the fantastic foundations of reality. What do we do with the *empirical* or *physical* implications of impossible phenomena? And how do these same physical-mental phenomena challenge and change our conceptions of the human, of consciousness, of embodiment, and, perhaps most of all, of the relationship of the human being to space-time and the physical cosmos itself? Actually, how do they change *everything*?

I mean it. Consider precognitive phenomena. I have long thought of these as the most well-documented and philosophically important of all impossible events. As such, they carry immense potential for influencing everything intellectuals and scientists do. If taken as real (by which I simply mean, "they happen"), such experiences and events (and they are *both*) would transform the entire order of knowledge upon which our present culture depends, the sciences included. For a start, they could tell us something stunning about the practice of history (time goes both ways), the history of religions (divination is globally distributed because it is based on an actual, if unreliable, human ability), the philosophy of mind (consciousness and cognition are not stuck in the present skull cavity or in this temporal slice of a body), and even something as abstract as causality itself (agency can act from the future). If we want to begin to learn how to think impossibly, precognitive phenomena are going to be a key to any such new order of knowledge.

Back to Zora Neale Hurston. Her early chapter on the scene I want to foreground is entitled "The Inside Search." It is probably sometime in the late 1890s, just before the turn of the century, just south of Orlando, Florida, in the only incorporated Black town in the United States, a place called Eatonville. Her father was John Hurston, the three-time mayor of the town, a Baptist preacher, and a light-skinned person of mixed white and Black ancestry, "with whom," as Maya Angelou diplomatically puts it, Zora Neale Hurston "rarely shared a peaceful hour."[3] At the time of this event, Hurston is around seven years of age. The little girl is on a neighbor's porch, hiding because of a childhood prank with a hen:

> I sat down, and soon I was asleep in a strange way. Like clearcut stereopticon slides, I saw twelve scenes flash before me, each one held until I had seen it well in every detail, and then be replaced by another. There was no continuity as in an average dream. Just disconnected scene after scene with blank spaces in between. I knew that they were all true, a preview of things to come. . . . So when I left the porch, I left a great deal behind me. I was weighed down with a power I did not want. I had knowledge before its time.[4]

"Knowledge before its time." There is a double meaning. These detailed "unhappy" images of future history would return to Hurston throughout her life up to the point where they stopped and her life began to flourish.[5] She kept quiet about them, but she knew perfectly well that she was not like everyone else—actually, she was like *no one else*. Secrecy was a kind of social protection. "I never told anyone around me about these strange things. It was too different. . . . I had a feeling of difference from my fellow men, and I did not want it to be found out. . . . I was told inside myself that there was no one. . . . A cosmic loneliness was my shadow. Nothing and nobody around me really touched me. It is one of the blessings of this world that few people see visions and dream dreams."[6]

Hurston does not let her readers escape into easy explanations about this difference. She is clear that she was not sleeping when the visions began: "I did not wake up when the last one flickered and vanished, I merely sat up and saw the Methodist Church, the line of moss-draped oaks, and our strawberry patch stretching out to the left."[7] The precognitive visions all came about. "Time was to prove the truth of my visions, for one by one they came to pass. As soon as one was fulfilled, it ceased to come." Zora Neale Hurston was being led through a life that had already been lived in a future that already existed. This was her "fate."[8]

Hurston does something very interesting with this alternative rationality. She locates it in professional writers; in her case, the New York City newspaper columnist O. O. McIntyre and the British Indian author Rudyard Kipling. But we could name many others in this context, including Mark Twain, Samuel Taylor Coleridge, Mary Shelley, J. R. R. Tolkien, Franz Kafka, Virginia Woolf, Stanislaw Lem, and Philip K. Dick.[9] Precognition and literary creativity, it turns out, are often deeply intertwined. Indeed, a professional writer like Twain openly and confessedly connected them. Hurston takes comfort in this occult-professional connection. She reads McIntyre and Kipling as "fellow pilgrims." It turns out that she is not alone.

Hurston also writes of a "psychic bond" with Charlotte Osgood Mason, a patron of the Harlem Renaissance who supported writers like Langston Hughes, and whose husband, Rufus Osgood Mason, was an early psychical researcher. Indeed, it was meeting this very woman that constituted the end of Hurston's prophetic visions and their spiritual guidance: "It was decreed in the beginning of things that I should meet Mrs. R. Osgood Mason. She had been in the last of my prophetic visions from the first coming of them. I could not know that until I met her. But the moment I walked in the room I knew that was the end." Hurston knew this instantly because of the minute details she saw around her (this is a common feature of precognitive visions and dreams), like the "strange flowers" and the physical posture of the second woman in the room, "as I had seen it hundreds of times." The conclusion of this last of time loops transcended race and class: "Born so widely apart in every way, the key to certain phases of my life had been placed in her hand. I had been sent to her to get it. I owe her and owe her and owe her! Not only for material help, but for spiritual guidance."[10]

There was more. Hurston, it turns out, could read Mrs. R. Osgood Mason's thoughts, and Mason could in turn read Hurston's mind, despite whatever spatial distance there was between them—for example, while Hurston was "in Alabama, or Florida, or in the Bahama Islands." "Laugh if you will,"

Hurston writes.[11] In a similar spirit, Hurston describes a series of hoodoo initiations in New Orleans, including one in which she had a realistic dream in which she "strode across the heavens with lightning flashes from under [her] feet, and grumbling thunder following in [her] wake." A "symbol of lightning" was painted on her back. The spiritual message was clear: she was "to walk with the storm and hold [her] power, and get [her] answers to life and things in storms."[12]

Unsurprisingly, Hurston asserts a kind of spiritual elitism. She is simply and obviously special, very special. She has a theory about impossibility, which is also my own. It goes like this. There are certain human capacities, like precognition, that most people will never know—*can* never know— not because this is a piece of data that they haven't been told but because they would not believe such a thing *even if they were told*. Why? Because they do not have this ability. Accordingly, they *cannot* know this truth. It is simply not them.

Hurston has "ears to hear." But the public, the "millions," as she calls them, do not have a clue and never will.[13] Secrecy here is no social stratagem or political gambit. It is a necessary function of the nature of the impossible itself. The moral conclusion follows: if we are going to think impossibly, we should not assume that this thought will be widely accepted. In fact, it will not be. It cannot be.

Not that it has not been heard, experienced, and translated into folklore and mythology. Unsurprisingly, Hurston loved Norse mythology—Thor with his hammer and lightning bolts in particular. She wanted to be like Hercules, not the "gentle little girl who gave up her heart to Christ and good works. Almost always they died from it, preaching as they passed."[14] Forget that nonsense. This is Storm of the X-Men. I mean, really, Hurston had a lightning bolt painted on her back, and she gets guiding wisdom and her power from, well, *storms*.

This was Zora Neale Hurston. She was nothing like the statue that her village shaped her to become. She was different. "My soul was with the gods and my body in the village."[15] As for the Bible, she loved David, who was always killing someone, and the book of Leviticus, for its great detail about all the naughty things one was not supposed to do.[16]

I think Zora Neale Hurston was telling us the truth of her actual life experience, and I think that we need to change our conceptions of the human and the nature of nature, accordingly. That is, I think that both the humanities and the sciences are deeply implicated here. If we are going to think-with Hurston, I see few, if any, conventional interpretive options of the visions on the porch, of her perfectly empirical knowledge before its time. The twelve historical scenes that were clearly seen years before they happened are not presented as some rhetorical device for the sake of the memoir. They cannot be adequately treated as literary metaphors, psychological hallucinations, misremembered exaggerations, cultural representations, discursive strategies, or any other at-arms-length category that we want to invoke to salvage our own sense of reality. *They simply happened.*

And so we begin there—on the Florida porch, with the little precog girl who did not fit in, who was not like the others, who would eventually become "Zora Neale Hurston."

The Fantastic Foundations of Reality

What if you slept, and what if in your sleep you dreamed, and what if in your dream you went to heaven and there plucked a strange and beautiful flower, and what if when you awoke you had the flower in your hand? Ah, what then?

SAMUEL TAYLOR COLERIDGE

And the third rose? he heard himself thinking. *Where do you want me to put it? Lay the album down, and show me where you want the rose put. The third rose....* Very carefully and with much excitement he opened the album. A freshly picked rose, maybe such as he had seen but once before, was there in the middle of the page. Happily, he picked it up.

MIRCEA ELIADE, Youth without Youth

I remember well the event. I had invited a group of colleagues to the Esalen Institute in Big Sur, California.[1] I asked them to come with two things: a story about the most unusual thing that ever happened to them and a theory of the imagination that could explain or at least make sense of this story. Every one of them came with a whopper of a story. No one came with a theory of the imagination.

It was obvious: these things happen to us, as they have always happened to others around the world and throughout human history. They obviously have everything to do with "religion," and yet we mostly dismiss these events, ignore them, do not speak about them, as if we somehow cannot. We certainly do not integrate them into our thinking and public culture. Basically, we keep them secret because they violate pretty much everything that we are supposed to think about ourselves and the physical world. They are not supposed to happen.

But they do.

I never forgot that week. Because of those stories and those silences, I kept saying that we need a new theory of the imagination to put these

common extraordinary experiences back on our academic and public tables. If I have learned anything over the decades, however, it is that when I
say something like that, what it really means is, "I should do that."

So I did it. Here it is. This book constitutes my theory of the imagination,
particularly as this theory is intuited and foretold in the furthest reaches
of the imagination—that is, in the drop-your-jaw stories like those I heard
in Big Sur. As you will quickly realize, I think those furthest reaches—and
they are *way* out there—are the key. I want to consider what happens to
our thought after we dream about a flower and awake to find that it is so.
This book is an answer to Coleridge's question: "Ah, what then?" It is following up on that question—not letting it go, trying to answer it, refusing
to consider it as just a metaphor—that constitutes impossible thinking.

So, *what then?*

To think impossibly is, first and foremost, to think-with individuals and
their experiences, however fantastic these experiences become or, better,
precisely *because* they become so fantastic.

That sentence is loaded. For one thing, I added the hyphen. I needed
a neologism to signal something much more than the ordinary sense of
thinking with another person. This book explores what is at stake in a
kind of thinking-with that takes absolutely seriously the altered state of
consciousness and embodiment of the other as a potential form of real
knowledge about one's own self and the material world. I also intend the
repeated *fantastic* of the sentence in a somewhat shocking sense, signaled
again by Coleridge's dreamed flower that becomes actual or, closer to home
(my home, anyway), Eliade's rose that materializes in the middle of the
page for his central character, the aging academic and the author's spiritual
doppelgänger, Dominic Matei.[2]

Double, indeed. Eliade was himself an author of *la littérature fantas-
tique*, short stories and novels that were themselves literary transformations of his own paranormal openings around the transcendence of space
and time that went at least as far back as the 1920s. One such literary work
was his last novel, *Youth without Youth* (1976), which was beautifully recreated by Francis Ford Coppola in a 2007 film by the same title after the book
was recommended to the director by my own mentor, Wendy Doniger, who
happened to be Coppola's high school friend.

The title of the present introduction and one of the earliest lines of the
prologue are both allusions to this line from Coppola: "I was happy to re-
discover in Eliade's short-story the key concepts I want to understand better: time, consciousness, and the fantastic foundations of reality."[3] So do I.

My focus is not on Eliade, however, but on those living human beings he might have well called mutants. Dominic Matei himself was such a mutant. So, I would argue, was Mircea Eliade. Listen to this self-description by Dominic, as he wakes from a dream and reflects on his electrocution (he had been struck by lightning) and the paranormal powers that it had actualized in him, including something he calls *anamnesis*, basically a kind of instant remembrance about what a particular text encodes:

> His mental activity anticipated somewhat the condition men will attain some tens of thousands of years hence. The principal characteristic of the new humanity will be the structure of its psycho-mental life: all that has ever been thought or done by men, expressed orally or in writing, will be recoverable through a certain exercise of concentration. In fact, education then will consist in the learning of this method under the direction of instructors. In short, I'm a "mutant," he said to himself on awakening. I anticipate the posthistoric existence of man. Like in a science-fiction novel, he added, smiling with amusement.[4]

A voice immediately explains to Dominic that the difference between his own evolving condition and the traditional figures of science fiction is that he has the freedom "to accept or reject this new condition."[5]

So do we. Perhaps that is why I address so many of my colleagues, correspondents, and students as mutants, "smiling with amusement." I am perfectly serious, but I am also playful, reflexive, and self-critical. I do not mean the description in any final or perfect sense (I am not even convinced that all altered states are expressions of evolution, or time). I simply mean to open a conversation in the only remotely adequate mythical terms that our culture provides at the moment. The science-fiction language, after all, encodes a particular (Nietzschean) evolutionary esotericism that is still working its way out, still being explored, in that sci-fi genre, for sure, but also in so many modern spiritual-intellectual lives—the human is becoming the superhuman.

Such individuals are free, of course, to accept or reject my descriptions. They almost always accept them with a sense of overwhelming relief and gratitude, no doubt because they can see something of themselves in the mirrors of these literary, cinematic, spiritual, and intellectual lineages. They feel *seen*.

If you read far enough and turn the last page of the present book, this superhuman vision will end in a kind of future of the past that is not an ending at all. Here is an ending that is another beginning, one that is still being explored, still being debated, challenged, and nuanced, *as it should be.*[6]

But that is the last page. This is among the first, and there is much to do before we end, including and especially developing a theory of the fantastic, which will gradually appear as this same book, like the dreamed flower on the bed or the materialized rose on the page. The fantastic so conceived is the beginning of impossible thinking. It, too, is a kind of mutation.[7]

By the fantastic, I do not mean an exaggerated event in a dream or fantasy that is functioning as a wish-fulfillment or symptomatic expression, as we have the term in psychoanalysis. Much less do I mean the fantastic in the common Marxist sense—some kind of unreachable utopia or religious opiate that takes one's attention to some nonexistent place and so away from the hard work of transforming economic labor and capitalist society in the here and now, a form of "false consciousness." Nor do I mean (although this is closer) the fantastic in the literary sense, as a particular genre of literature that puts the reader in a state of suspense, never really knowing if this event happened or not, if the ghost really appeared or not, perhaps toward some larger conception of what is possible. No. The fantastic in these pages is shockingly much more than these standard psychoanalytic, Marxist, and literary categories.[8]

By the fantastic, I mean *something that is exotic, outlandish, or shocking (apparently, it is trying to get our attention), something that actually happened in* both *the material* and *mental domains, which are in turn being mediated by the imagination.* I will define and nuance this opening definition extensively in what follows. Still, it is worth italicizing immediately.

By the fantastic, I mean something that is often very physical, empirical, and sometimes even witnessed by many people. I mean something that possesses its own agency and purpose and often changes history, be it of a single person or an entire civilization. I do not mean something that is "good," neat, and clear. I mean something, or someone, that is known as inherently elusive, apparently *by nature and intent,* trickster-like.[9] I mean something that bends back on the human experiencer to reflect and refract that human being in shocking, confusing, paradoxical, and sometimes morally troubling ways. I mean something is imagined, yes, but something imagined that is mediating, translating, and expressing something super-real, which is what our modern *surreal* originally meant.

This, anyway, is where I am going to push, and push hard. In this understanding of the fantastic, human beings (saints, witches, mediums, and the possessed) literally float off the ground in ecstatic rapture or terrifying possession, and other people watch them, witness this in amazement and fear—which is to say, *they find them as impossible as we do*. People appear in two places at once. They "bilocate." Things (honey jars, rocks, flowers, real-life versions of the flower and the rose of our epigraphs) "teleport" across a room or materialize out of nowhere and fall from the ceiling or out of thin air. Authors meet their own fictional characters in real life. Human beings know, instantly, that a loved one has died hundreds, even millions of miles away. Distant historical or even future events in precise banal detail appear in dreams and visions.

These events may or may not fulfill unconscious wishes, posit other dimensions beyond the socioeconomic world, or put their witnesses in a state of metaphysical suspense (I have nothing against these other theories of the fantastic). But they are nevertheless reported to happen as historical facts. There are literally *thousands* of testimonies, as numerous and as reliable as any other historical event that no one ever questions.

Why is that?

The historian of early modern Europe Carlos Eire provocatively puts the matter in this most blunt of ways with respect to the seventeenth-century Italian Franciscan friar Joseph of Cupertino, who did things (before thousands of witnesses) like fly up into olive trees or churches and perch on branches that could not possibly hold his weight:

> What kind of nonsense is this? Who is this liar? Human beings can't fly, or kneel on slender tree limbs like little birds, and they have never, ever done so. Such a feat is absolutely impossible, and everyone can agree on this, for sure. Or at least everyone nowadays who doesn't want to be taken for a fool or an unhinged eccentric. So, how is it that in the sixteenth and seventeenth centuries—the very era that gave birth to aggressive skepticism and empirical science—countless people could swear that they had witnessed some such event? And how is it that some of these sworn testimonies are legal records, archived alongside lawsuits and murder trials, and that the testimonies come from all sorts of people, not just illiterate mud-caked peasants, but also some elites at the apex of the social, intellectual, and political hierarchy? What sense are we to make of this? How does any historian deal with such accounts? How does one write a history of what could never have happened, a history of the impossible?[10]

Eire is adamant. What he calls "freakish folk" are not tangential to the history of religions. They are the "main event." That is why fringe phenomena are so very important today. They are only fringe *because we have made them so*. Such appearances are in fact trying desperately to get our attention and knock us out of our foolish normalcy. Radical skepticism is not what we are being told. Not even close. Here is the real intellectual life:

> The history of the impossible is all about questioning, about being evenhand-edly skeptical—that is, being as skeptical about strictly materialist interpre-tations of seemingly impossible events as about the actual occurrence of the event itself. Counterintuitive as this might seem—given that the impos-sibility of certain events is deemed unquestionable in our dominant culture and that dogmatic materialists tend to think of themselves as the only truly objective skeptics—this sort of nonconformist skepticism is necessary if one is to claim any kind of genuine objectivity.[11]

Much depends, of course, upon *which* historical texts one chooses to study and—more to my present point—*with whom* one is thinking. In the following pages, I will focus on some very specific historical texts that do not look away from these histories of the impossible, indeed that express them in some very colorful ways. I will also think-with some especially gifted human beings. Some would call them freaks. They are not like us. They are certainly not like me. I call them superhumans, because that is what they are, which is not at all to say that they are not also humans.

As the book develops, the self-reported experiences of these superhu-mans become the building blocks of new thought and theory, which I will perform and explain in the last chapters and conclusion. This new thought must be built up slowly and carefully, as it were, but the material—if you can call it that—is already there, already so. And it is not me. It is them. It is the freakish folk, the mutants, the superhumans.

But there is more to it than that. I think-with these superhumans with the understanding that their fantastic experiences are trying to speak, to communicate, to say something important to all of us. Such events are basi-cally shouting in the only language they have, which is often not a language at all. The meanings of these events, in other words, often have to be drawn out, thought through, translated, and finally integrated into public culture through conversations, texts, images, and books, like this one. Such experi-ences require our attention and intention, our cocreation. In a word, they must be *interpreted*.

Even that is too easy, though, and much too safe. In the conventional understanding of the academy, I am engaging these individuals as cocre-ators of knowledge. I am certainly not some distant expert commenting on them as objects of study. We are very much doing this together (hence the hyphen in thinking-with). Except that these people are not just cocreators of knowledge. More like destroyers of knowledge. They abolish me. They annihilate my assumptions. Part of this abolishment is the fact that they describe their experiences as a *discovering* of new knowledge, not some construction or creation. *Something is given.*

I confess that I am more than sympathetic to this shocking sense of discovery, perhaps because I am a historian of religions, and this is how revelation has long worked—as a more or less passive receiving, as a show-ing or gift that elicits awe, indeed a sense of wonder *so* great that it often ends up transforming the course of human history. We can question the content of that revelation (I certainly do) and explore its cultural pre-cedents and obvious shaping (I certainly do), but I do not see how we can question the passivity of the revelation itself or the cultural and historical effects that it has.

So, too, here, with my superhumans. Things appear. These individuals do not "make up" what they see or know. I strongly suspect that what they see and know is shown to them by some other form of mind or intelligence that is *also* them in some paradoxical sense. Such a form of mind uses the cultural bits and pieces that are lying around, *whatever* is lying around, to express itself. We are forever confusing those bits and pieces with the su-perintelligence that is ordering them, shuffling them around into meaning and message. This book is a plea: "Please stop doing that."

Which is all to say that I am also more than sympathetic to, but also suspicious of, the secular conviction that everything about human experi-ence has a history. That is *mostly* true. What appears to the experiencer, after all, is a part of culture. Indeed, one of my deepest intuitions that I will play out in these pages is that *the paranormal is culture working in hereto-fore unimagined ways—expressing itself, yes, but also overcoming, surpass-ing itself, pointing in new directions and to new futures.* The deep speaks, shows itself to the surface. But it also largely determines the behavior of that surface. And it is both that deep (the extreme experiencer is correct) and that surface (the historian is correct). It is all one sea. There is no final hierarchy or difference.

Part of how to think impossibly, then, is holding these two levels to-gether: recognizing the passive, given, or shown nature of revelation and also acknowledging and pursuing its culturally or historically constructed expression or appearance. It is that both-and paradox that I am after in these pages.

In any case, you picked up this book. Prepare to be abolished.

Blurry Fuzziness

There is another recurring pattern of impossible thinking. It explodes between and beyond what we have come to call the humanities and the sciences. More philosophically, the impossible emerges from a knowing place
before and beyond the subject/object split. This is one fundamental reason,
I suspect, its many appearances are so often fuzzy or blurry. They are in
between. The presence is not quite humanistic, subjective, psychological,
or social, but neither is it entirely material, physical, or empirical. It is both.
And it is neither. It is before these. And it is beyond these. Creatures are
coming through the walls. And they are coming for us. What we think of as
the objective world is at stake. So are we.

This shocking doubleness—at once humanistic and scientific—is the
most basic reason why I will alternate between "experience" and "event" in
what follows. Subjective states sometimes have objective correspondences
in the shared physical environment, and objective states in the environment
sometimes have subjective correspondences. And I do not mean that safely,
as in "the physical causes the mental."

Let me lay my cards on the table. I personally think that this is all so
because *there is no final distinction between subjective and objective states,
between consciousness and cosmos.* Moreover, and this sounds even more
outrageous, the past, present, and future are all one "thing," and they are
constantly interacting outside or to the side of us. These impossible things
happen precisely because of this fundamental unity of space, time, and
mind. I will say much more about my own ontological thought experiment
later, no doubt inadequately so.

Alas, it is *always* inadequate. It is never enough. It cannot be enough,
but for reasons that I think we can come to understand. If we cannot think
impossibly in any full sense, at least we can come to understand why.

The Humanist's Secret

We already know why it can never be enough. What we have come to realize in the humanities is that our assumed sensibilities around the nature
of mind, self and society, and secularism and science are all socially constructed. They are not innocent, natural, or universal. Even the conditioned
sensorium of the body-brain, what we experience to be true with our senses
and thoughts, is informed by centuries of affirmations and denials, religious
fights, and too many colonial, imperial, and racist projects to name. We are

who we are because of what our ancestors did and did not do, because of who won and who lost the cultural wars, because of who killed whom, who enslaved whom.[12]

I wish I could say that some culture is genuinely innocent here, but no such community is. It just depends on how far we go back in time and, more to my point, what genetic line, or fractal branch, we choose to follow—genetically speaking, we are not single genetic lines at all but biological branches or bushes that all go back to a shared root and, far enough back, to a single organism. At some point, some branch of people you think of as your ancestors most likely killed, overwhelmed, or colonized another people, who were also your ancestors from another line or branch. And countless peoples have moved, migrated, and eradicated entire species in the process. That is "you" too.

And yet there is a bright flipside to this genetically informed humanist vision of global history: none of these constructions of who we are—these religious, political, ethnic, and racial identities—are permanent, natural, or necessary. It is *all* constructed, relative, "historical," as we say. And so reality is always punching holes in our assumptions and constructions, including in the assumptions and constructions of those ancestors who won the wars.

Much of this hole punching, moreover, displays unbelievable correspondences, even identities, between the human being and the larger environment. Accordingly, one begins to suspect—I do, anyway—that the quick and common divisions between the inside and the outside, between the subject and the object, and *everything* that this split produces in human minds (which includes the entire present order of knowledge of the modern university), may not finally be adequate to our actual cosmic condition. This is the deeper secret that I wish to whisper to the reader in these pages. Reality only appears to be divided into mind and matter, into consciousness and cosmos, or into a past, present, and future because *you* experience it as such. *You* are the splitter of the real and the creator of linear time.

I should add that I am not at all certain that many are up to this secret. Perhaps this is why it still remains a whisper. The secret, after all, suggests that the social ego's constructions of reality are not finally so. That is another reason that no explanation of this conscious cosmos is ever adequate—or can be adequate. The social ego does not want to hear it. It actually *cannot* hear it, since this human subject is precisely what is splitting reality into a mental and a material dimension, neurologically creating the arrow of temporality, and suppressing or denying the deeper unity of everything and everyone.

We cannot hear this secret, then, because this psychophysical split and these social constructions, enacted over centuries and in multiple cultural zones, have been incredibly effective. They created the most intimate subjectivities—the negotiations, compromises, dreams, fears, bigotries, violences, and paradoxes that are so many of us in the modern world. They have also been devastatingly myopic. This repression renders most of us basically inept, incapable of understanding, much less accepting, common human experiences that violate our socially assumed binaries. We think these events are "impossible." But they are only impossible within our historically constructed frameworks, which means that they are not impossible at all.

Impossible things certainly happen in numerous cultural zones and through specific ritual practices on a fairly routine basis. This is why a comparative practice is so necessary, so liberating and healing, but also so devastating to our local assumptions. Trained intellectuals who experience the impossible are commonly shocked and often see and say more or less the same thing: "It is not what we thought."[13]

That is the very first lesson of thinking impossibly: recognizing that the impossible is a function not of reality but of our own present social constructions and subsequent perceptions and cognitions. That is also the humanist's secret taken to its furthest point until it bends back on itself and consumes its own historical perspective. That is the deconstruction of deconstruction. That is the snake biting its own tail.

Lightning Strikes in the Nightscape

The idea for *How to Think Impossibly* came in the wake of an international conference at my home university, entitled "Opening the Archives of the Impossible," that attracted over 1,700 registrants and, within a single month, had resulted in over 150,000 views on YouTube (at the time of this writing, it is well over a quarter million). This book, moreover, was finished two days after the second such conference, in May 2023.

Both events were overwhelming. It was like a series of lightning strikes temporarily illuminating a nightscape from this and that direction—from the histories of the Cold War, slavery, lynching, and the UFO abduction experience to New Testament criticism, medieval female mystical theology, secret US government research, poltergeists, and pedagogy to psychedelics, altered states, film, Chicana Mexican American "mutation," and European, Hindu, and Buddhist philosophy. We could see something of the landscape from the flashes of the lightning strikes in the darkness, but we saw different things with every new thunderclap. I think the impressions with which many came away were something close to my own: we exist in a largely hidden ecology that can be understood from a plurality of perspectives, and we appear to be in the midst of an intergenerational transformation around all of this.

Just as telling, I saw a kind of genuine awe in the stunned eyes and trembling voices of some of the professional staff who ran the event and with whom I work every day—a kind of, "Wait . . . you do *this*?" They were gobsmacked.

This book builds on those intellectual, emotional, and spiritual responses and the responses of hundreds of colleagues these events have involved in one way or the other. I assure you: there is a kind of global renaissance going on around the impossible. Different colleagues have advanced a "history of the impossible," which we will see in chapter 6, to an "anthropology of the impossible," which is what Hussein Ali Agrama has called for.[14] The ideas are simply everywhere in the humanities and the social sciences, even in the sciences, which are by no means immune from these factual influences. A new order of knowledge is on the horizon. The present book works in front of the curtain. It tries to practice what I have seen up close in public but also what I have heard in private behind the curtain. In essence, this is a kind of intellectual manual on how to think impossibly in full view, in public.[15]

The Physics of Mystics: A Thought Experiment Yet to Be Fully Realized

Contemporary secular thought is based largely on an unconscious way of understanding reality that is very useful for particular projects but essentially incorrect in its exclusions. This is why the chapters in this volume are about *the impossible*—bluntly, phenomena like precognitive dreams (Hurston again), mystical experiences of cosmic unity, clairvoyant visions, near-death experiences, and UFO encounters that are not supposed to happen but do all the time. I am most interested in these events and experiences not because I understand them, nor because I believe them in the terms in which they appear, but because they show every sign of issuing from another order of knowledge and reality outside what we currently think of as "the human." They promise, if never quite deliver, a future human, a new world.

They cannot deliver such a future human or new world because it is not up to them to deliver such things. It is up to us. Impossible phenomena are urgings, nudges, inspirations, premonitions, and pointers, not the full deal. The fantastic is the *beginning* of impossible thinking—its early realization not its mature practice or public authorization.

There are other reasons that such impossible phenomena cannot *ever* be understood within our present conventional academic and scientific categories. Let me speak in the metaphors of the physical sciences. One of the topics that I want to explore in what follows is the likelihood that these experiences and events are meaningful expressions of a deeper unthinkable or unimaginable reality that is quantum in nature but that has historically—which is to say, in the history of religions—been known as "groundless," "nondual," "empty," "infinite," or, perhaps most offensive to the modern rational ear, "mystical." Regardless of such offense, the adjective *mystical* works especially well here, as it literally translates from the Greek as *secret*—unspeakable, indescribable, unknowable in *any* historical, cognitive, rational, or scientific sense. You just cannot think this. Much less can you speak, language, or measure it.

There are some very good philosophical reasons for this unspeakable or unthinkable nature. No historical subject can experience such a realm, since there are no historical subjects in such a realm (nor are there any objects, causality, space, or time). Technically, then, there can be no "experience," either, not at least as we ordinarily understand that expression in the sense of a subject experiencing an object. To speak in the terms of philosophy, there is no subject or object, no intentionality, no "aboutness."

Still, human beings have consistently reported this empty or infinite nonground, which they have known, or better still, *become*. Actually, *they are already it.* To speak in the metaphors of physics, everything and everyone are made of the exact same physical stuff, or nonstuff, all the way down in a quantum world, which is our physical world right here, right now. That should come as no surprise, but it always does. But what these human beings are reporting, or realizing, is something more fundamental still, something that is nothing, something that is not the physical *or* the mental, something that is both and neither, something deeper or higher.

And this is the ultimate reason that quantum mechanics and mystical literature correspond so powerfully: because they are finally speaking of different aspects or expressions of the same deeper reality, one "from the outside" (the physics), the other "from the inside" (mystical experience). Indeed, if quantum physics and comparative mystical literature are really pointing to or emerging from the same deeper reality, they *must* correspond. And that is why they do. This is certainly what the founders of quantum mechanics intuited. Physicists like Werner Heisenberg, Niels Bohr, Erwin Schrödinger, and Wolfgang Pauli all turned to comparative mystical literature to make human sense of what the quantum mechanics implied. They knew, immediately, where to go—to the furthest reaches of the history of religions.

I do not invoke such a quantum deep in a dilettantish *or* a professional scientific way. I am not a quantum physicist and will not pretend to be one. But neither am I a loose cannon.[16] Moreover, and most importantly, I have spent large chunks of the last quarter of a century sitting with, speaking to, and carefully reading colleagues who *are* professional quantum physicists and historians of science who want to talk to philosophers and historians of religions like me.

Foremost among these colleagues is the Swiss-German quantum theorist Harald Atmanspacher, whose dual-aspect monism I adopted some time ago as my own working ontology and whose recent writings on the "deep structure of meaning" (meaning is real and an "objective" feature of the physical world) with the historian and philosopher of physics Dean Rickles is very close to where I want to go in these pages. Indeed, this book is organized around this dual-aspect monism and this deep structure of meaning and begins to end with a penultimate chapter that expands on an essay I originally wrote for Harald.

I am doing this partly to answer my own call that the study of religion desperately needs new ontologies to imagine new futures. I also do so to heed the call of another intellectual—in this case, the science writer Philip Ball, who has called on artists and cultural creators to translate the new quantum real into public forms so that people can hear, process, and, above all, appreciate what is, in truth, a shockingly new worldview.

Ball may or may not agree with what I do. That is not my point. My point is that it is important to try, with the proviso that my own project is *philosophical* in nature *not* quantum mechanical, mathematical, or scientific, which is precisely why I can adopt, understand, and use dual-aspect monism. This is a philosophical framework that helps me think about the empirical data of my historical materials that would otherwise go unnoticed or be immediately dismissed. Put a bit differently, I am not doing quantum physics here. I am thinking-with quantum physicists, who are in turn thinking-with me.

It will simply no longer do to avoid the philosophical implications of quantum physics and focus exclusively on the mathematics and the equations (mathematics, by the way, is probably the clearest sign that the inner and the outer worlds of human experience correspond). "Shut up and calculate" was the norm of physics for much of the twentieth century. That kind of snarky attitude got people jobs and gave us much of our modern technology, including the nuclear bomb and modern computer communication, but it also prevented any effective cultural integration of what we have learned about the physical world with our own human experience. Making fun of those who have tried to integrate the two dimensions is no answer. It is just another form of cultural schizophrenia, which intellectuals like the neuroscientist of perception Donald Hoffman have called out as preposterous and unthinking.

Hoffman, in fact, lands pretty much where I land, with a "case against reality" that looks like something straight out of *The Matrix*. In Hoffman's view, physical objects are not objectively "out there." They are icons on the screen of our virtual reality game of perception that allow us to maneuver through the world, survive, reproduce, and so "win" the game. They are practical, not actual. As Hoffman puts it in a Darwinian soundbite, "fitness beats truth."[17] In short, we have evolved to win the game of life not to perceive reality as it really is.

Still, some of us *do* perceive reality as it is—become it, actually.

Not unlike Hoffman, Ball thinks that we should be profoundly unsettled by the results of quantum experiments, experiments that "make possible what sounds as though it should be impossible."[18] Such experimental results imply or make possible things like an unobserved photon being both a wave and a particle at the same time (Niels Bohr's complementarity principle); a quantum particle as a smeared-out set of statistical possibilities before observation, when the "quantum wave function" appears to "collapse" (a particular interpretation of quantum mechanics suggesting the causal role of human observation in physical reality); a quantum event being determined by a *future* intention (John Wheeler's delayed-choice thought experiment); and entangled particles responding to one another across great reaches of cosmic space and time, clearly outside of any possible speed-of-light communication.

Some of this is *highly* contentious and by no means settled, particularly what is often called the "Copenhagen interpretation" (Bohr worked in Copenhagen) about the causal role of observation in the collapse of the wave function. But some of it has been experimentally confirmed, over and over again, and is not in serious dispute, although no one has any settled idea about what to make of the bizarrerie that follows. Quantum nonlocality or acausal "entanglement," for example, has been experimentally confirmed repeatedly, including on a cosmic scale involving ancient quasars, massive telescopes, and literally billions of years of traveling light.[19]

Ball has written that if we are serious about the cultural project that follows such discoveries and confirmations, "we're going to need some philosophy."[20] Which is another way of saying we are going to need the humanities. What I will add here is that we are going to need a long, hard look at extreme anomalous experiences, which, if I am correct, may well be quantum effects just staring us in the proverbial face. They certainly *look* quantum. But they are also generally mediated by the imagination, and so they also appear as fantastic.

I am not the first to see or say such things. The historian of science David Kaiser points out that the "hippie physicists" of the 1960s, '70s, and '80s kept the quantum theory of entanglement alive until it could be experimentally established, even as they also linked quantum physics to things like comparative mystical literature and parapsychology.[21] Such comparisons were not new either. In historical fact, the physics of the nineteenth century, especially in England, was deeply implicated in the Spiritualism of the public and the psychical research tradition of the educated Victorian elites.[22] Physics and psychics, it turns out, have been reflecting one another in a double mirror for a long time, pretty much as long as we have had modern physics.

This may well be much more than a historical fluke or some rational embarrassment of the "real" history of science. It is a most surprising conclusion of some of the quantum theorists, but it has been concluded before by the (idealist) philosophers: it very much looks like *to be* is *to be perceived*. Or, if you prefer, we live in a participatory universe, one that appears to rely on our forms of awareness and observation for its very existence or coming to be, even, if John Wheeler is right, on a cosmological scale that goes all the way back to the big bang. Wheeler seriously suggested that the big bang was mathematically structured "just right" (like the temperature of Goldilocks's third bowl of porridge), because we are here in the present and had a retrocausal influence on that initial explosion. In some sense, we "caused" the big bang.

I will struggle throughout these pages with that strangest of truths—the potential participatory nature of nature, including and especially its apparent backward influences suggested by quantum theory but also well known to writers on precognition. I probably will never quite accept such strange truths in what follows, since they are *so* counterintuitive (and Wheeler himself would have adamantly opposed much that follows). Still, I really do understand that, as much as I want to fall back on notions of reality *as it is* or the world *as such* (and you will read me saying things like that), in actuality, there appears to be no such thing until we choose to make it so.

Which means, of course, that *we* might inhabit or participate in or be a part of the ultimate thing. But isn't that role in observing and so actualizing the world yet another claim on reality *as it really is*? It seems so. Talk about fantastic.

Clearly, sometimes *impossible thinking is impossible for the impossible thinker too.* I will try my best to inhabit and express this kind of awed humility on the page, or, to speak more colloquially, this honest WTF feeling. In many ways, this entire book can be captured by those three letters and the stunned vulgarism they humorously encode.[23]

Any book on the further reaches of the imagination will also be about the further reaches of myth—that is, the stories that cultures tell themselves (and do not generally recognize as story). I will struggle with the fundamental nature of myth because I *both* accept the truths of the reductive social scientific or historical readings of the term (myths are cultural representations, means by which human beings make sense of their lives, and political forms of camouflage and disciplinary power that privilege particular classes of people) *and* insist that these models are relative to a particular form of consciousness—in this case, the atomized rational academic ego.

I include and transcend these reductive social readings in the furthest reaches of the imagination that finally interest me the most here. Indeed, at least since *Authors of the Impossible* (2010), I have insisted that myth or story sometimes results in very real and otherwise completely anomalous physical effects—*that reality itself has everything to do with how we tell it.* Basically, I have read myth in realist terms as the paranormal—that is, as the very dramatic and spontaneous experience in the physical historical world of us writing us, almost always, please note, in traumatic or marginalizing contexts that have everything to do with the social readings of myth and religion of my academic training.[24] Can that double message be heard? Can the paranormal be heard as *both* a prophetic existentialist cry that involves class, gender, race, and physical death *and* the temporary transformation of matter and space-time around the gravity of these very sufferings?

Heard or not heard, that doubled vision, that physics of mystics, is what I will be about here.

I will pull no punches. This is a book about why the humanities must stop ceding reality to the sciences. We have to stop assuming that our most fantastic stuff is purely legendary, is only subjective, is mere metaphor, or can be fully explained in social, political, or moral terms. It is more than time to make a claim on material reality itself. People all over the world, for example, sometimes dream in perfect detail about what is going to happen the next day, or three years down their own space-time world-line, or, in the case of Zora Neale Hurston, decades before it happens. The evidence is *overwhelming* in these cases. If you are not convinced by the evidence, the chances are extremely high that you have not truly struggled with it. I am long experienced in intellectuals confidently expressing utter certainties concerning matters they know nothing about. Apparently, complete ignorance is acceptable in this context.

But *how* can such people cognize distant or even future physical immanent events if human consciousness and its relationship to space, time, and matter are what we generally imagine them to be—that is, stuck in the head and perceiving a strictly local physical environment through the senses?

The answer is simple: they cannot.

How (Not) to Be a Ghost

When I go on like this—thinking impossibly—something often happens.

Other intellectuals will respond to such thoughts by invoking their own expert trainings. Basically, they put me in their own boxes. Zora Neale Hurston was a Black woman who studied in New York City with a famous anthropologist, Franz Boas. Okay. So? What does that have to do with the physics involved, with the historical truth of a seven-year-old girl seeing in perfect detail twelve scenes from her future life? Or the fact that we can find similar stories in a thousand other human beings who are not Black or identify as women?

I am often not entirely certain that such colleagues accept that these impossible things happen at all. But, if they do accept them as something more than neurological fluff or cognitive mistake, it is relatively obvious that they do not think that such phenomena should fundamentally change how we understand religion, history, or the human.

I do.

And it gets worse (or better). Most intellectuals certainly do not want to admit that any of these exceptionally common experiences of the impossible have something essential to do with the physical world of space, time, and matter—that such experiences and events belong in the natural sciences as much as they belong in the humanities, in the departments of physics, biology, and mathematics as much as the departments of history, religious studies, and English. That would mean changing what we do in the humanities and the sciences in institutional ways. That would mean restructuring the very order of knowledge and our universities with it.

Instead, humanists revert to the same old tired tropes that were dominant during my own graduate training in the late 1980s and early '90s—postmodern, postcolonial, feminist, and sociohistorical ones, mostly. *Anything* to not think impossibly.

It bears repeating that it is not that I disagree, at all, with these models. Such critical perspectives are perfectly true on their own levels and within their own frameworks. But I still become bored and impatient in such moments, and, I admit, a little disgusted. No wonder we are ignored by the public. I would ignore us too.

I confess I also feel very alone. Suddenly, *I feel like a ghost*, a spectral presence haunting the academic certainties but not capable of speaking to them directly or effectively. It is as if I am dead. All I can do is haunt. I cannot quite appear. I cannot quite matter, much less materialize in full form. I suppose that is somehow weirdly appropriate in this context: I feel fuzzy. I *am* fuzzy.

But here's the thing. That is never the whole story. There are *always* intellectuals, many of them, actually, who do not try to put me in their expert boxes. They see that I am trying to say something new, something that cannot be put in the usual theoretical boxes of human, posthuman, or transhuman. They ask me, then, what I really think, or rather, *how* I think. They want me to materialize.

So I tell them. I appear. The ghost is now a person. I explain my theorizing of the superhuman and, consequently, the superhumanities—that is, the already existent humanities with their prophetic social spirit and a heavy accent on the altered states of consciousness and knowledge that produced so many of the central ideas of the humanities and their historical figures and continue to do so to this very day. Through such a model, I explain how we can think impossibly together—namely, *with* our critical theories and texts but also beyond or *above* them.

Thinking impossibly is what we are going to do here. The present book is a fairly detailed record of my own practice of the superhumanities. The book sits just to the side, but also at the side, of the humanities as they exist today within a kind of doubleness: at once a tired boredom and a genuine awe. I wrote it not to complain or to feel superior. I wish or feel neither. I wrote the book to show how we can really do this thing, in deep support *and* genuine challenge.

In the end, I am publishing this book because I think the connected dots, the impossible phenomena themselves, matter, and because it is more than time to materialize impossible phenomena in the form of new thought, even someday, I hope, in a new order of knowledge—a higher, *higher* education.

I do not know if that future form of education will include the teaching of skills like Eliade imagined in his fantastic fiction (and, I suspect, experienced in real life) like *anamnesis* (the ability to read without reading), but I am definitely imagining here a psycho-mental life "some tens of thousands of years hence," as Eliade put it. Of course, I will not live to see that day. Nor will you. I personally feel very much like the German philosopher Edmund Husserl did with respect to his "new science" of a pure phenomenology of transcendent or absolute consciousness. Speaking in the third person, he claimed to see, like Moses, "the infinite open country of the true philosophy, the 'promised land' on which he himself will never set foot."[25]

So, too, here. The School of the Superhumanities that I am imagining in these pages will not divide the cosmic human into divisions and disciplines whose constructed boundaries do not in fact exist and whose denials of the impossible have never been convincing. It is more than time to think impossibly—even if this is an order of knowledge before its time.

PART ONE

WHEN THE IMPOSSIBLE HAPPENS

ONE

Words Are Experiences

Evolutionary Origins and
the World of the Dead

Truth does not come into the word plainly, but in images and symbols, for that
is the only way that the world can perceive it.

THE GOSPEL OF PHILIP

To think impossibly is not only to think-with special human beings and their
fantastic experiences or to acknowledge the historical truth of phenomena
that emerge before or beyond our relative cognitive and sense-based divisions of the world into the mental and material temporal domains.[1] It is
also to think-with impossible words—that is, linguistic signs that crystallize
the most extraordinary forms of human experience in the codes of culture,
otherwise known as language.

To think impossibly is also to think historically. It is to know how those
words originated impossibly—how they emerged from the traumas and
most intimate details of altered states, and what specific meanings and
nuances they carried in their original historical contexts. It is also, by con-
sequence, to understand how such words dropped out of favor in the recent
past, were distorted and mangled in both the academies and the publics of
the twentieth century, and sometimes came to mean the very opposite of
their original intentions. To think impossibly is a kind of return to origins,
then, but to return smarter now, perhaps a bit more jaded, certainly a good
deal wiser. To think impossibly is to begin anew.

On the surface, this first chapter orbits around a twentieth-century
French philosopher and historian of Iranian mystical literature (Shi'i, Ismaili, and Zoroastrian) by the name of Henry Corbin (1903–1978). Still on the
surface, it traces a particular lineage of impossible thinking through Corbin
back to the Swiss psychiatrist C. G. Jung, to the Swiss professor of psychology Théodore Flournoy, and, ultimately, to the American psychologist

and philosopher William James and his good friend, the British classicist Frederic Myers. This is a lineage that thinks and writes in French, German, and English. It is very European but also more than a little American.

In truth, however, the impossible thought that I am tracing here is not restricted to these intellectuals, nor are its concerns particularly recent or, for that matter, particularly European. In the end, I wish to show that what we are ideally talking about when we use three historically and conceptually related coinages—*supernormal, imaginal, paranormal*—are some extensively theorized encounters with disembodied, intelligent, and mostly invisible presences who claim to be human, nonhuman, or superhuman. Maybe they are. Maybe they are not. But that is how these words came to be.

What I most wish to show is that *these three words have experiential origins.* They are not empty or devoid of meaning, much less are they the careless creations of the tabloids, naive enthusiasts, or conspiracy mongers. Quite the opposite. They encode the long work and careful thought of professional intellectuals, scientists, and independent researchers who had fantastically rich, if usually highly unorthodox, spiritual lives and were personally engaged with thousands of actual human experiences recorded and recovered from all walks of life and classes of European and American society. Many of these experiences, moreover, were dramatic communications from the dead, who may not be dead at all.

I suppose if you want to be traditional about it, this is a chapter about the ancestors.

Aligning and Arguing with the Ancestors

I am not sure I want to be traditional about it, though.

The context is personal enough. A little over two decades ago now, I published a study of one such intellectual ancestor, Corbin's mentor, the anticolonial activist, Islamicist, and homoerotic mystic Louis Massignon (1883-1962).[2] So I began the present project on the imaginal, in some sense, there.

Corbin, a student of Massignon, was deeply influenced by the philosophical work of the German philosopher Martin Heidegger, whom he knew personally and translated into French. Here is the thing, though. Heidegger in turn also helped form the German theologian and scholar of comparative religion Frederic Spiegelberg (1897-1994), who fled for his and his family's lives from National Socialist Germany, where Heidegger remained quite comfortable and protected as a card-carrying Nazi and an anti-Semitic academic and writer.[3]

Spiegelberg eventually landed at Stanford University in Northern California, where he would mentor the two founders of a future think tank and retreat center called the Esalen Institute—in particular, Michael Murphy (who is my own closest mentor to this day).[4] In 2007, I wrote a history of the institute, *Esalen: America and the Religion of No Religion*, and today help lead the intellectual-spiritual direction of its private symposia after Murphy's vision (and Spiegelberg's paradoxical mystical theology of the "religion of no religion"), which I have described as an evolutionary esotericism or, in Murphy's words, an evolutionary panentheism.[5] I mention such autobiographical facts because I want to make it clear that I do not write here out of some academic neutrality but out of a deeply felt and long practiced intellectual-spiritual activism.

For the same book on the history of Esalen, I wrote about Henry Corbin's category of the imaginal as a collapse of fiction and reality, especially as this collapse or fusion was known as *hurqalya*—that "heaven," interworld, alien Earth, or hyperdimensional vision of history that was so central to Corbin's *Spiritual Body and Celestial Earth* (1977) and, soon, to Michael Murphy's vision of the "future of the body."[6] Which is to say: I do not understand Henry Corbin to be writing only about Iranian patterns. I consider him to be writing about ontological truths that can only speak to us in image and narrative of the most astonishing sort and that are still very much with us today in other historical and contemporary experiential forms, including, the UFO phenomenon, to which we will turn soon enough.

Accordingly, I have recently identified Corbin as one of many partial exemplars of the superhumanities.[7] Corbin displays at least one feature of such thought almost perfectly in his erudite conviction that the human being is a kind of "bi-unity" or "dualitude," a paradoxical complex of the conscious person and the largely unconscious or superconscious angelic twin, of "consciousness and *superconsciousness*," as he put it (the italics are his).[8]

Still, I am not Henry Corbin. Nor are the superhumanities in any way restricted to the ways that the French Islamicist embodied and expressed so many features of them. Most importantly, by the superhumanities, I mean something that can take on much more social, historical, and, frankly, ethically edgy forms, something much closer to Corbin's mentor, Louis Massignon, whose queer mystical activism with respect to French colonialism in northern Africa displays both sides of the superhumanities: their socially active *and* mystically transcendent dimensions. Massignon, after all, was an ardent anticolonialist and a mystically inclined intellectual who openly used parapsychological categories to explain how he was converted by a long "dead" Islamic saint.

I certainly acknowledge and accept that the activist and the mystical do not always appear together in ways that we today can appreciate. They work in different dimensions or on different scales. Ideally, though, the super-humanities are the humanities as they are already known and practiced today in their full critical and historical scope, but they are the humanities that *also* fully acknowledge and robustly theorize a vertical or transcendent dimension of the human being. I personally see these two dimensions as ideally working together, as horizontal and vertical dimensions that need one another to make the fuller three-dimensional picture "pop." I know that is rare, maybe exceptionally rare, but that is precisely the goal and intention.

It seems more than obvious that today most professional intellectuals are working with half the picture, the secular horizontal half (which was hard won and is *a lot*). The human being variously conceived and theorized in that academy is mostly an embodied social, racial, sexual, and gender identity, a constructed and historically conditioned psychological ego, or a dying social animal dethroned from any spiritual exception and entirely embedded in an ecological, physical, and political environment. That half human being is entirely horizontal. This is all exceptionally important, of course (and this is where I would part with Corbin). But it is also half true. The other vertical dimension of the human comes crashing in, usually in some fantastic or unbelievable form (and this is where I would agree with Corbin). At that point, none of our present academic certainties with respect to the purely social human being work so well as complete or exclusive truths.

This chapter is about this not-working-so-well and what we can do about it. As I will explain again below, through my own reservations about Corbin, my impossible hypothesis is not about returning to an exclusively vertical perspective. I have no wish to return to the religions. I did not announce my book on Esalen with the subtitle (drawn directly from Spiegelberg) "America and the Religion of No Religion" for nothing. But this does not mean that I am somehow against religion as coded expression of the fantastic either. Rather, what I most desire to do is affirm *both* the horizontal *and* the vertical, the ethical *and* the mystical, and in countless formations or relationships, not just one.

The Near-Death Landscape as an Imaginal World

Toward the end of his oft-read essay "Towards a Chart of the Imaginal" (1977), and shortly before he died the following year, Henry Corbin invoked the modern near-death experience in the just-published work of Raymond Moody.[9] This brief mention is deeply significant and, to my knowledge, usually overlooked in discussions of Corbin and what he understood by his famous category of the imaginal and also in discussions of how plural, eclectic, transcendent, and, frankly, gnostic (as in "a direct state of knowledge that enacts a reversal of orthodoxy") the religious imagination can be. It certainly was in Corbin's person.

Raymond Moody is a medical doctor with a PhD in philosophy from the University of Virginia who encountered numerous striking visionary accounts in his medical school training that were difficult to dismiss (no doubt because of his earlier philosophical training). Instead of ignoring them, which is what everyone else in the medical world was doing, he chose to write about them in a little book entitled *Life after Life* (1975).

The book was a bomb detonating in the American spiritual landscape. It sold millions of copies and was translated into numerous languages as it set off a firestorm of endless enthusiasms, religious debates, and future revelations, which continue to this day. Among this firestorm were a number of theological controversies, including an outcry from fundamentalist Christians who were not at all happy that there was no hell or creedal requirement for salvation in the book, and who read the beings of light commonly encountered in the near-death experiences as the Devil himself: "These fundamentalists had little or no sense of humor." For them, Moody was "working for Satan."[10]

Despite Moody's insistence that the book did not prove the afterlife and that it was not scientific, and despite his wish to remain neutral on religious matters, the book carried a most powerful theological punch. The humorless fundamentalists were correct about that. Reading story after story, spiritual journey after journey, the message was crystal clear: belief, much less denominational adherence, was pretty much irrelevant. Everyone goes to heaven.[11] This was a modern gospel, "good news" to millions, but certainly not to the fundamentalist Christians who, apparently, wanted those who believed in more inclusive and global terms in a very bad place, forever. And we are not talking abstractions here. Such theological controversies entered Moody's personal life in some very tragic and deeply personal ways, eventually resulting in a divorce.[12]

Religious offense and conjugal suffering aside, *Life after Life* was the book to bring the phrase "near-death experience" into American English and into the popular global religious imagination. There were previous accounts of similar "near-death" experiences floating around, of course (beginnings are always relative). Some of the first in the modern world were penned by mountain climbers who had survived near-deadly falls, mostly in the Swiss Alps. There were also a few ancient and medieval accounts, like the "myth of Er" (why is it a myth?) told by Plato in *The Republic* (10:614–21): a story about a soldier left for dead who awakens on the funeral pyre twelve days later to tell a fantastic tell of the afterlife, of reincarnation, and of the nature of the soul and its moral choices. Still, that was a long time ago. Moody's book was about such visionary journeys into and out of "the Light" *today*, and in perfectly common people. The book collapsed any and all divisions between the popular and the elite—spiritual wisdom was for everyone and anyone.

The message was shocking, at once deeply confirming and yet often denying basic conservative or exclusive Christian beliefs about the soul, God, and the afterlife. The phrase "near-death experience" was also, I should add, a comparative phrase through and through. It connected instead of disconnected. It sought patterns not identities, although it also suggested an underlying sameness or unity. It definitively put the accent on what is shared, on how we are all connected to a common if also mediated afterlife—to those comforting beings of light the fundamentalists could only see as Satan himself.

And that was good news.

The Experience-Source Hypothesis

Well outside the postmodern academy and its ideological rejection of shared patterns, metanarratives, cross-cultural commonality, and a shared moral community, the literature on the near-death experience continued apace and never let up on its intensity, or its not-fitting-in. I saw this up close and personal in the near-death experience about which I know the most, that of my friend and coauthor Elizabeth Krohn.[13] In 1988, Krohn was struck by lightning in the parking lot of her synagogue on the very day she was about to memorialize the first anniversary of the death of her beloved grandfather.

Krohn's extensive experiences in the other world (which, in true imaginal fashion, she insists is *both* perfectly real *and* experienced according to the cultural and religious predilections of the seer) could finally find no comfortable place in the progressive *or* conservative Jewish traditions of her social surround. In her own words and contemporary understanding, the experience of getting struck by lightning and going to heaven made her much more "spiritual" but much less "religious." Historians and scholars can debate that distinction all they want, but it works perfectly well for Krohn and her near-death experience. It names something crucial and important about what happened, about what changed Elizabeth in a flash.

Significantly, Krohn's near-death experience also resulted in numerous paranormal capacities, including seeing auras around people, precognizing death, experiencing various electromagnetic effects involving things like lightbulbs (they would blow out in her presence), receiving a phone call from the dead (from her deceased grandfather), and the precognitive dreaming of major historical events—things like major plane crashes and earthquakes—that would not happen in space and time (that is, in "history") until the next day, when they showed up on the front pages of the international media.

My point is already a familiar one: if we can learn to think-with such a radical set of experiences (and not reduce these to Krohn's social surround, psychological makeup, or what she was reading), we can begin to theorize religious experience in ways that are very different from what we are currently doing—which is to say, we can think impossibly.

How? For one, we can seriously entertain the *experience-source hypothesis*—that is, the hypothesis that basic religious convictions around the world are cultural reworkings of a recurring set of direct experiences and not (only) reworkings of indirect historical contexts or psychosocial constructions. In David Hufford's terms, this is the idea that, "some significant portion of traditional supernatural belief is associated with accurate observations interpreted rationally. This does not suggest that *all* such belief has this association. Nor is this association taken as proof that the beliefs are true."[14] In short, by adopting the experience-source hypothesis, we can move away from a strict social production thesis and toward a global phenomenology of religious experience. We can begin to notice *both* real differences *and* genuine similarities. We can begin to compare again.

I have long taken Elizabeth Krohn's visionary landscape as a particularly acute, but implicit, lesson. Her own imaginal landscape was basically divided into two parts: the garden, where her own long conversation with a gentle male presence took place, and the high mountains on the horizon, behind which a light, or Light, shone. Krohn had the sense that if she moved into and then beyond those distant mountains, she would not return, but that, if she stayed in the garden, she could continue to interact with its doubled presence and return to this life.

Her culturally specific postmortem landscape thus encodes both the human and the superhuman aspects of the soul—the imaginally constructed

and the apophatic or mystical Light. I should add that this is my own reading, but I think it is perfectly resonant with Krohn's experiences and what
they portend for us all, some of which, please note, she formed in conversation with a scholar of religion, in this case me. This, in truth, is what I think
many such experiencers should do with willing intellectuals, not so that the
experiencers can be told the truth by the intellectuals (we don't have it) but
so that we can all come to a more adequate, more three-dimensional theory
of religion together. This, again, is precisely what I mean by thinking-with.

The near-death experience account is now an established genre. It has its
own peer-reviewed professional journal and its own collective of scholars,
medical professionals, and public activists, including a man who is probably
the academic doyen of the field, the psychiatrist Bruce Greyson, professor
emeritus of the University of Virginia.

Greyson's *After* (2021) constitutes a kind of magnum opus and an excellent description of the historical, medical, and philosophical landscapes.[15]
It begins with a striking personal story involving a spaghetti stain on Greyson's shirt that was seen and described by a patient who was unconscious
and in a distant room at the time. It was this original, perfectly empirical,
perfectly impossible event (impossible because it involved both an altered
state and a perfectly empirical spaghetti stain on a shirt) that would set
the psychiatrist off on his own lifelong intellectual, medical, and spiritual
journey. Bruce has not stopped since.

Such stunning historical empirical scenes, of which there are *endless*
examples, collapse the imagined and the material within what Corbin would
have certainly called the imaginal—the middle place where spirits take
on bodies, where matter becomes mind and mind becomes matter. Such
events are simply impossible to explain in the terms of our present order of
knowledge (with internal "subjects" perceiving external "objects" in three-
dimensional space and the imagination as nothing more than an internally
trapped spinner of sense-based fantasy).

I am honored to know Bruce Greyson.[16] I have also participated inti-
mately in the near-death experience genre and its public adjudications.[17]
Still, I do not want to address further the near-death experience. Instead,
I want to employ the genre, explicitly invoked by Corbin in his "Towards a

Chart of the Imaginal," as a kind of swirling pivot into the immediate past, the present, and the near future.

The Imaginal World of Henry Corbin

One of the very few tools that historians of religions have in their theoretical toolbox that is even remotely capable of handling the extensive mental-material paradoxes of the near-death experience is the category of the imaginal. The word and idea are associated with Henry Corbin.[18] And there is very good reason for this, as long as we stick closely to the meanings that Corbin himself intended. Technically and historically speaking, however, the claim or assumption is simply false. Henry Corbin did not coin or invent the category of the imaginal. It had been in circulation in the study of unusual or extreme religious states for some *eight* decades before Corbin began to use the term in his own specific senses in the 1960s.

Corbin himself explains that he derived the term from the French, as in his own neologistic expression *le monde imaginal*, or "the imaginal world," a phrase that is not common in French and that Tom Cheetham dates from 1964 on.[19] Corbin also tells us that he struggled with this word choice, as he did not want to signal the "imaginary," as the French *imaginaire* certainly would.[20] He was painfully aware that a kind of spiritual abyss had opened up in the metaphysical possibility of Western cultures between, on the one hand, the "intellectual" or "intelligible" ground of Being (he used such terms in the Neoplatonic sense of a cosmic intellectual structure, a *nous* or, to be more Christian about it, a *logos*) and, on the other, the material-based senses tuned to the physical world and the cognitive rational habits of the Western mind shaped and limited by those same senses.

By the twentieth century, there simply was no longer any culturally available model of the imagination that could relate or connect these two dimensions of reality—the ontological and the sense-based material world. To speak in Corbin's beloved Latin (he had been trained by the great medievalist Étienne Gilson), there was no longer any *imaginatio vera*, or "true imagination," available.

It is worth pointing out that such a concern was by no means unique to Corbin or, for that matter, medieval theology. It can be found in the German idealism of the late eighteenth and early nineteenth centuries, for example. It echoes particularly strongly with Arthur Schopenhauer's German *Wahrträumen*, or "dreaming the truth" (he was writing of his own precognitive dreaming that contained perfectly precise and banal details about the next day's events, much like the very specific flowers and physical posture of the woman that Zora Neale Hurston saw in her last life-event that would fulfill and end the twelve precog visions of her childhood).[21] It also resonates with the more recent English expression "veridical hallucinations," which was common in the Victorian psychical research of the late nineteenth and early twentieth centuries (they were writing of apparitions of the dead, or the presumably disembodied, that carried accurate empirical information—like Bruce's spaghetti stain seen by the woman in a coma in another room).[22]

It is perfectly true that this was not the intellectual situation of the mid-twentieth century in which Corbin was writing. Such things could only be construed in imaginary or purely subjective terms now. They were "all in the head." The mediating spiritual symbol had become the socially constructed metaphor. All such a secular subject could be in touch with now is the physical world through the senses and a rule-bound cognition. Since there was no longer any way to connect the human to the larger metaphysical world, it was sincerely believed and vociferously declared that there is no metaphysical reality. The material world is all there is, and objectifying science is the *only* way of knowing what is real. Agnosticism and nihilism were the results. A particular reductive reading of Aristotle had, in effect, erased the spiritual reception of Plato.[23] Or so it seemed.

Corbin wasn't having it. In his own specific (Platonic) terms, he understood the imaginal to name a noetic organ or intellectual-spiritual capacity that was mediating an actual dimension of reality, whose appearances were nevertheless shaped by what he called the *imagination créatrice*, or "creative imagination," an astonishing imaging or mythmaking dimension of consciousness that acts well outside any conscious control or apparent human agency—hence the phenomenology of revelation as something given or shown. As Cheetham has argued, Corbin used this key category in his own very specific ways, but he was likely inspired and enthused by Carl Jung's very similar idea of the "active imagination," at least from 1949 on, which is when Corbin appears to have first used the expression in a lecture at Eranos.[24]

Eranos was a kind of intellectual retreat center at the villa of Olga Fröbe-Kapteyn in Ascona, Switzerland, that hosted a set of annual meetings and lectures from 1933 on, some of which were essentially seed essays for later projects and books that would come to define the comparative study of religion, and especially the comparative study of mystical literature, in the twentieth century. Luminaries included Carl Jung, Mircea Eliade, Gershom Scholem, Henry Corbin, and Joseph Campbell.[25]

Jung's creative or active imagination, which was so influential on these meetings, was understood to be a kind of unconscious faculty that works effortlessly and expertly with the stuff of its social surround, combining and connecting things "just lying around" into new syntheses, collages, or artforms that show little respect for cultural, religious, ethnic, or political boundaries. This is all done quite autonomously and independently from the social ego, "from above," as it were.

Here was the intermediate world of the *mundus imaginalis*, the "imaginal world," which for Corbin is the "place" where myth and symbol happen, where they are true (but not to be taken literally or exclusively), where visions occur and the spiritual body is resurrected. Put in metaphysical or theological terms that were central to Corbin, the human cannot experience with the senses or its ordinary cognitive capacities the apophatic (unspeakable, unthinkable, impossible) depths of the Godhead, which is known, or unknown, as a kind of brilliant darkness or black light—"black" because it can never become an object of perception. There is a kind of transcendent blackness here.[26] Blackness is ontological.

The structure of the human, as already noted, is thus a kind of paradoxical "dualitude" or "bi-unity," with the human person accompanied by an angelic twin or companion who can mediate between this *Human as Two*.[27] This is uncannily similar, I must add, to what Elizabeth Krohn saw in the afterworld of her own visions (where every human soul was accompanied by a partner or double). It also relates deeply to the ancient Mediterranean world that Charles Stang has studied around the theme of the divine double.[28] In truth, this doubleness can be seen almost anywhere, probably because it is true anywhere.

I have five nuances or corrections to make before I move on. Each of these interventions is very important for the impossible thought that I am developing.

Romantic Origins of the Imaginal

First, it is important to realize that, although Corbin's notion of the imaginal may have been invoked to interpret Iranian mystical literature and possesses its own distinct features, it is very much in line with earlier Romantic philosophy, poetry, art, and music, which sought to vatically—that is to say, out of its own inspiring mystical states—express what M. H. Abrams paradoxically sums up as a "natural supernaturalism." This Romantic paradox resulted in a series of artforms and philosophies that preserved *but also transformed* the earlier spiritual modes of humanity, now understood and so expressed in a secular or naturalistic key. Nature is "the perfect image of a mighty Mind."[29] This was to be a synthesis of idealism and the natural sciences.[30] Nature is a poem and history an epic of God.[31] This was about begetting a *new* world and so a new mythology. Always, "the best is yet to be."[32]

I cannot stress enough these mystical states toward some kind of futurity, what Abrams elliptically calls "moments," when eternity breaks into time.[33] Abrams, indeed, begins his classic study with the English poet William Wordsworth, who was granted "an internal brightness" that was "shared by none," and that inspired him to teach "of what in man is human or divine."[34] Hence Abrams drew on earlier scholarship to provide a two-word definition of Romanticism: "spilt religion."[35]

Out of these same altered states, at once human and divine, the Romantic movement defined the "symbol" as much more than an arbitrary linguistic sign or artificial metaphor. The symbol *participates in that which it expresses*. Words are experiences. The Romantic writers also fully recognized that symbols grow cold and stale and must be replaced by other newer symbolic real experiences, which is what they themselves were all about. Hence their poetic, visual, and musical arts. Abrams argues that this, in the end, is a "total revolution of consciousness" that took place, not at all accidentally, during a historical period in Europe and America of political and moral revolution.[36] The "High Romantic Argument" is at once a scientific, a spiritual, an artistic, a philosophical, and a political argument. I mean them *all* when I insist that "words are experiences."

One of the greatest poet-thinkers in this cultural stream, who tried to integrate and express all that came before him in terms of his famous distinction between "fancy" and "imagination," was none other than Samuel Taylor Coleridge (1772–1834), the writer with whose dream-flower we began this book as a potent exemplar of the fantastic.[37]

It is no accident, at all, that Coleridge turned to religion to find the ultimate examples of what a symbol is and, more specifically, to Ezekiel's scriptural vision of the wheels or disks in the sky. Symbols are not God, who cannot be imagined, but neither are they constructed products of reason. They are forms of real energy that transmit what they themselves are pointing us toward. They are "consubstantial with the truths of which they are the conductors. They are the *wheels* which Ezekiel beheld, when the hand of the Lord was upon him, and he saw the visions of God as he sate among the captives by the river of Chebar. *Withersoever the Spirit was to go, the* wheels *went, and whither was their spirit to go:—for the spirit of the living creature was in the* wheels *also.*"[38] Note that wheels, disks, and a living creature are the Romantic symbols par excellence in this passage. These will all return in our second and third chapters.

This was precisely how another key Romantic figure, even more radical than Coleridge, saw the imagination. William Blake saw the creative imagination in the grandest of terms—that is, as the "Divine-Humanity," as the "Eternal Body of Man." And he meant all of it. For "All things Exist in the Human Imagination," and "Mental Things are alone Real." The rest is "this world of Dross," far below Blake's notice and not worth his time.[39] I can hear Henry Corbin saying something like that, in his own later philosophical terms.

In short, there is a *vast* prehistory of the imaginal in the eighteenth and nineteenth centuries, and it lies squarely in English Romanticism and, though I have not treated it here, German idealism.[40] Indeed, as Engell has argued extensively, it is the creative imagination that is the "hinge" between the two great eras of European thought that initiate and announce modernity: the Enlightenment and the Romantic movements. Such an idea, after all, embraces and includes Enlightenment reason, even as it attempts to push it further beyond mechanism and reduction into individuality, intellectual freedom, artistic expression, and the most sublime reaches of the natural cosmos.

Such ideas carry, again, *vast* theological, political, and moral implications. The Romantic movement is what Harold Bloom would call "Prometheus Rising," the man who would defy the gods in Greek mythology, which is also to say, defy the social community. These were artists and intellectuals whose work was inspired by the political dawns of the American and French revolutions, those "terrors" that aligned themselves with the satanic idolatry and blasphemy of self-exaltation and what William Blake would call "vision"—that is, the experienced truth that poetry is often given or heard, and, as such, is prior to and more basic than religion, moral philosophy, or anything else that claims authority over the human.[41] For Blake, the imagination is, in truth, the Imagination, as he had it in *Milton*. It is not a psychological function or tangential byproduct of neural firings. It is "Human Existence itself." It is what we might now call "consciousness."

This same creative imagination, or Imagination, was a most effective means to overcome the dualisms of Cartesian doubt and Newtonian mechanism, in the profound sense that its endless historical formations relied on a linking or relating of the subjective and the objective, time and eternity, matter and spirit, the natural and the supernatural. Such a vision of the imagination and the symbol had "one foot in the empirical and one foot in the ideal or transcendental."[42] It was a third space, a middle world that linked all human experience and all human beings.

It was also, I should point out, the "boldest reply" to the problems of human diversity and unity, since such a model could explain why the religions were so different and yet so very similar: they were all expressions of the same creative imagination working its wonders in different contexts and cultures. Little wonder that Blake would pen an early little book called *All Religions Are One* (1788). They were, and still are, one, and precisely because the religions are of a poetic symbolic nature, as figures like Blake and Coleridge meant these terms—living creatures, wheels in the sky seen within altered states around the world and throughout time, including and especially today.

Docetism and the Monotheistic Symbol

Second, Corbin's imaginal was deeply informed by his own docetic Christology—that is, his convictions that the doctrine of the Incarnation should be interpreted symbolically but not literally: divinity cannot become literal flesh for Corbin, *ever*. It only "appears" to have done so. Here he is following Islam and an important Koranic passage about how they did not really kill or crucify Jesus but another person or an appearance (4.157). In most forms of Islam, the doctrine of Incarnation is a classical heresy, association (*shirk*), or untruth.[43]

The situation is complicated, though, since what Corbin was also arguing is that *the symbol is true*, that *the image is an actual manifestation of revelation*. This is terribly important when interpreting Corbin's use of the imaginal (within the Romantic tradition), since it appears that Corbin used the expression paradoxically, *both* to keep apart *and* to relate the material world of the flesh and the transcendent world of the spirit. Such a monotheistic paradox could be somewhat mediated through a philosophy wherein the symbol or the icon *is* the veiled reality becoming unveiled or revealed, but the same monotheistic mystical model also carries a significant tension that it is difficult to see transcended or completely overcome, at least in Corbin's thought.

I confess some discomfort or uncertainty here, since my own theorization of the fantastic insists on both the mental and material dimensions of that which appears in particularly extreme moments of materialization or physical expression: the poltergeist presence literally throws *pots and pans* around the kitchen to communicate or signal its *emotional distress*. This is not a simple physical event, but neither is it a simple emotional feeling.

It is both. But is the icon of the ghost real? Well, yes. And no. It's fuzzy. Something is indeed expressing itself, but in terms that cannot, or should not, be taken literally. Here is where I probably agree with Corbin.

I also have no doubt that some people in fact experience divinity or "God" as utterly transcendent to the physical world, but I personally hesitate to sign my name to such a conviction, as I consider these phenomenological reports to be partial or culturally determined ones, not universal truths. If one can speak in Christian theological terms (that is, with reference to the human and divine natures of Christ), I am a thoroughly heretical thinker who unapologetically affirms the ultimate unity of the two natures of *everyone* and *everything*. No exceptions. It is all One. If one wants to speak in more Muslim theological terms, the One God is *everyone* and *everything*. No exceptions.

Because of these transtraditional monistic convictions, I see no reason why we need to follow Corbin's Docetism in order to employ the imaginal for our own purposes toward our own ends. On the contrary, it seems to me that his own Christological conceptions in fact hamper Corbin's category of the imaginal from the beginning, since Docetism is hardly the only option Christian culture possesses to relate the human and superhuman natures, nor—much more importantly—is it reflective of the larger history of religions, including some of the Islamic traditions. In the end, such an incarnational theology is simply one of countless systems in the history of religions to express and delimit our superhumanity. It is relative and temporary, not absolute and permanent. That is my concern or hesitation.

Parapsychology, Eclecticism, and Politics

Third, it is very much worth underlining that Corbin's understanding of the imaginal assumed the presence of parapsychological phenomena. This is evident in his repeated and rather matter-of-fact treatment of clairvoyance, mind reading, telepathy, synchronicity, photisms, and even materialization and a kind of apport or manifested object in the life and teachings of the great Islamic mystical philosopher Ibn 'Arabī.[44] Such paranormal phenomena display the same paradoxical kinds of spiritual bodies or embodied spirits that the intermediate world does. They are "imaginal" in structure and intention. They are also tricky.

This, of course, is not exactly a criticism of Corbin. I think he was moving in the right direction with his embrace of the parapsychological within the imaginal and the mystical. I suppose I just wish he had moved further and said more.

Fourth, I have a problem with the ways that Corbin seems to imply—by the very content of the books he published—that the imaginal is somehow reserved for traditional symbols and historical mystics, that the imaginal cannot appear, as it were, in popular or nontraditional sources and "ordinary" people. I think this preference for the historical and the established is simply mistaken, both historically and morally. Indeed, I have insisted on this elite-popular collapse. I would insist that the very same imaginal or unconscious processes that Corbin treated in his Shi'i, Ismaili, and Zoroastrian sources also appear in contemporary science-fiction writers or comic-book artists.

Harold Bloom, the late Yale literary critic, makes a related case in his admiring preface to an edition of Corbin's *Creative Imagination*, pointing out that both the Iranian Sufis and Corbin himself were much more eclectic than Corbin wants to admit.[45] Indeed, Bloom wants to push Corbin's imaginal history and hermeneutic into the professional study of Shakespeare and modern English literature, both *well* outside any Iranian or even religious register. I think Bloom wants to do this, partly or mostly, because of his own gnostic revelations, with which he tells us, point blank, he was gifted in "three timeless moments" in the same preface.[46]

For Bloom, then, and certainly for me, Corbin underestimates the eclectic habits of his Iranian sources. The creative imagination is, by nature, wildly appropriating. It does not care *a whit* for the orthodoxies of the religions, the languages of cultures, the borders of nation-states, or, frankly, the moral concerns of the academy. It endlessly and effortlessly combines anything and everything to express itself.

Fifth, and finally, this insistence on the eclecticism and transcendence of the creative imagination has real and important implications for the relationship between scholarship and politics. More specifically, Corbin has been criticized for any number of perceived sins: that he was too aligned with the shah and therefore somehow participated in the European colonial project in Iran or, from the opposite direction now, that his work on Iranian mysticism ended up supporting the Iranian revolution of 1978, which deposed the shah in order to establish a theocratic system of government. The latter charge involves the observation that some of the revolutionaries read Corbin's work as supportive of their own theocratic goals to reenchant and sacralize the secular world.

I understand that Henry Corbin would have strongly resisted these political criticisms from either direction on the grounds that his work was about the spiritual individual and so fundamentally apolitical.[47] I do not want to issue any quick and simplistic judgments. I must make two observations in this context, however.

First, it is perfectly possible to be a gnostic intellectual (by which I mean someone whose thought emerges from extraordinary experience or direct *gnosis*) and to care deeply about politics, ethics, and this world. Scholarship on ancient, medieval, early modern, and contemporary gnostic forms of spirituality has advanced significantly since the 1970s, and we now know that not every form of Gnosticism is world-denying or anti-body. We also know that there is no single political or ethical expression of the gnostic orientation: such expressions can range from radical politics, social justice activism, and ethical humanism on the progressive left to dangerous conspiracy-thinking, antimodernism, and authoritarianism on the far right.

The study of religion is especially instructive here. I made the argument almost two decades ago that the ancient Gnostics were some of our first and still some of our most radical theorists of religion, particularly in their rejection of the moral monstrosities of the biblical God and their dramatic reversal of key religious assumptions: God did not make us—we made God, the serpent was a figure of gnosis not of temptation, and so on.[48] I still think that this is the case. If you want to take religion critically but seriously, to take it apart (and have something deeply meaningful left over), there is no better place to begin than with the gnostic writers, be they ancient, medieval, or modern.

Second, as I have already repeatedly observed, the religious collages or artforms of gnostically inclined individuals and communities are famously eclectic and wildly combinative with respect to the local social and cultural surround. This matters a great deal. If, after all, one does not see and say that religious traditions, and *especially* gnostic traditions, are appropriative and eclectic all the way down (and then some), then fundamentalist or right-wing reception histories of one's work on the same are always a possibility. These kinds of orientations, after all, need "essences" and "identities," not collages and combinations.

This fundamentalist linkage of religious essentialism and nation-state happened in Iran in a Muslim context in 1978. It happened in India in the 1990s and continues to this day within a Hindutva nationalist context (a context in which my own scholarship, by the way, has been condemned, censored, and rejected by specific publics and conservative politicians). A similar project is being attempted in the United States within a white, racist Christian nationalism, where it continues to do endless damage and threaten the very structures of democracy and the intellectual freedom of intellectuals, particularly those analyzing this very white form of racism.

I observe all of this to reflect on the political and moral implications of the impossible thinking I am trying to trace, theorize, and model in these pages. Such thought, precisely in its emphasis on the theoretical importance of cross-culturally available anomalous experiences, does not and cannot confirm the exclusivity of *any* religious tradition (including Islam, Hinduism, or Christianity). Hence it stands fundamentally against these fundamentalist projects, be they in Iran, India, the United States, or anywhere else. I cannot repeat this enough: impossible thinking is an expression of a deep and most profound humanism, a paradoxical superhumanism that recognizes that what we think of today as the "human" cannot be identified with any culture, religion, or society and that understands the further reaches of this human have indeed been most fully imagined and practiced—if, yes, usually in unconsciously projected and specifically culturally refracted ways—in the symbols, rituals, and myths of the history of religions.

"It Matters Not by What Name It Be Called"

I have stated in print that the category of the imaginal was used earlier by Carl Jung's mentor, the Swiss professor of psychology and interpreter of Spiritualism Théodore Flournoy.[49] This is wrong. Flournoy never used the term to my present knowledge. *However*, the three related authors (Corbin knew Jung well, who knew Flournoy well) all meant something very similar by the active or creative imagination, even if they interpreted its products in different ways.

Such an autonomous or unconscious imagination certainly laid at the center of Flournoy's work. Indeed, he himself called it the *imagination créatrice*, or "creative imagination," more or less exactly like Corbin and Jung. And he argued for hundreds of pages that this same creative imagination could and does act as a mediating, spontaneous, or independent organ of spiritual romance (what we would today call "science fiction"). He even wrote of likely empirical or "supernormal" cognition in especially gifted individuals. Flournoy was thinking of mediums. This was essentially his argument in his classic study of just such a spiritual prodigy, *From India to the Planet Mars* (1900).

From India to the Planet Mars is Flournoy's study of a Swiss medium named Élise Müller (called Mlle Hélène Smith in the book), who claimed to be both the reincarnation of an Indian Hindu princess and a regular visitor to Mars (hence Flournoy's telling title, moving a kind of orientalism into outer space). Flournoy's novelistic retellings of the medium's various personalities and channeled adventures made his book an instant hit. In the process, Flournoy also helped to revolutionize the study of mediumship and trance formations through a relatively new tool: hypnosis. This is quite significant for the history of extraterrestrial esotericisms (the subject of our next chapter), since the same technique, hypnosis, would later play a major role in the production of the alien abduction literature in the 1960s, '70s, and '80s.

I must emphasize that Flournoy's use of the supernormal is very much related—historically and conceptually—to Corbin's imaginal, this time through yet another earlier intellectual who in fact brought both terms into wide use well before Flournoy and Corbin, the Victorian classicist and psychical researcher Frederic Myers. Significantly, Myers's original English expressions *supernormal* and *imaginal* collapsed any final distinction between the mental and the material dimensions of reality and carried strong evolutionary impulses, as we shall soon see.

Like Myers before him and Corbin after him, Flournoy made it very clear that his subject matter, in his case mediumistic phenomena, was of a deeply personal nature. He was not entirely divorced from the claims of the phenomena that he studied and wrote about. Here are some lines from the very first pages of his book. Note the academic context of the first encounter, the way it invokes another unspoken academic and superhumanist context (Mrs. Piper of Boston was studied by William James of Harvard), and, above all, the prominence of the supernormal:

> In the month of December, 1894, I was invited by M. Aug. Lemaitre, Professor of the College of Geneva, to attend some seances of a non-professional medium, receiving no compensation for her services, and of whose extraordinary gifts and apparently supernormal faculties I had frequently heard. . . . we seated ourselves in a circle, with our hands resting upon the traditional round table of spiritistic circles. Mll. Smith—who possesses a triple mediumship: visual, auditive, and typtological [rapping]—began in the most natural manner, to describe the various apparitions which passed before her eyes in the partially darkened room. . . . I was greatly surprised to recognize in scenes which passed before my eyes events which had transpired in my own family prior to my birth. Whence could the medium, whom I had never met before, have derived the knowledge of events belonging to a remote past, of a private nature, and utterly unknown to any living person? The astounding power of Mrs. Piper, the famous Boston medium, whose wonderful intuition reads the latent memories of her visitors like an open book, recurred to my mind, and I went out from that seance with renewed hope of finding myself some day face to face with the "supernormal"—a true and

genuine supernormal—telepathy, clairvoyance, spiritistic manifestations, it
matters not by what name it be called, provided only that it be wholly out of
the ordinary, and that it succeed in utterly demolishing the entire framework
of present-day science.[50]

Flournoy, it should be noted, did not interpret these apparitional forms
the way Corbin did with his Iranian materials. Flournoy was both devastatingly skeptical and deeply sympathetic when it came to the contents of Müller's visions and channelings. Hence, he could demonstrate how Müller's
claims to speak a Martian language, to remember a past life in India, and
to communicate with dead spirits were no such things. They were creative
imaginative products of the medium's present-life memories that she had
forgotten and unconsciously wove back together in her various sittings with
the hypnotist. Or rather, the creative imagination did this through her. She
herself was innocent.

Flournoy marveled at the astonishing unconscious creativity of this
process: it was like being present for the inspiration and production of some
serially published Jules Verne novel—in short, science fiction (although the
genre would not be so named until 1929, by the American magazine editor-
publisher Hugo Gernsback). This nearly unbelievable productive ability of
the human mind was what Flournoy meant by the *imagination créatrice*.
Significantly, both Flournoy and Müller would become respected patrons
of surrealism, a twentieth-century art movement particularly resonant with
my present theorization of the fantastic.[51]

Although he was extremely suspicious of the mythical content of Müller's mediumship, Flournoy took seriously what he called the "supernormal
appearances" of her life. By "supernormal," Flournoy meant things like
telepathy and telekinesis, which he concluded were likely genuine phenomena. His final conclusion? That the common assumption that one must
choose between the "brutal alternatives" of Spiritualism (the medium was
really talking to dead people) and materialism (there is nothing to any of
it) "is surely puerile."[52] That is to say, childish and stupid.

Note the same epistemological dilemma that Corbin would observe and
write about to such effect. At the turn of the twentieth century, supernormal phenomena were caught between literal belief and rational debunking,
neither of which were convincing to those who worked closely with the
experiencers. There simply was no way to mediate between the two dimensions of human experience. Corbin would certainly offer a way out half a
century later, but he would be mostly ignored.

We are still at a loss, an impasse.

Flournoy would have none of this either-or thinking. Only both-and thinking, or what I am calling impossible thinking, was remotely adequate. Nor, though, would he jump to conclusions about what it all meant and where it was all going. He refused to take a position about whether these visions and supernormal abilities were "forerunners of a future evolution" (which is what Frederic Myers thought), or evolutionary survivals from some previous condition (which is what Sigmund Freud would suggest), or "whether they are purely accidental" and so meaningless.[53]

He did not know, and so he left it at that.

The Supernormal and the Imaginal

Flournoy clearly and confessedly borrowed his central category of the supernormal from the Cambridge-trained classicist Frederic Myers (1843–1901), a superhumanist if ever there was one. He could have just as easily borrowed the imaginal from the same man, as it was Myers who first defined and used the term in a very extensive way in the early 1880s. Myers in fact understood the imaginal and the supernormal in very similar, if not identical, ways.[54] The imaginal implied the supernormal. The supernormal implied the imaginal.

Moreover, and perhaps most importantly, both terms—the imaginal and the supernormal—carried definite *evolutionary* dimensions for Myers. Basically, he used both words to describe what we would think of today as paranormal phenomena as evolutionary "buds" or early immature developments of superabilities, like, say, telepathy, that would someday become integrated into consciousness and controlled in an intentional way.

Myers actually coined the word "telepathy" ("pathos-at-a-distance") in December 1882 and theorized it extensively through literally *thousands* of case studies, linking it in the process to strong emotional connection between loved ones or a strong erotic attraction toward a beloved—"erotic" here understood in the sense of Plato's *Symposium*—that is, as a metaphysically oriented *eros* that can be "held in," sublimated (Freud was thinking of the *Symposium* too), and directed ecstatically toward the divine world of Forms. Indeed, telepathy was probably the signature idea of Myers, along with his notion of the *subliminal*, which, much like Corbin's imaginal, could mediate under specific conditions between metaphysical (Platonic) realities and the psychological experience of a historical individual. The subliminal region of the psyche was that realm "below the threshold" (*sub-limen*) where telepathic communications took place before they emerged up into a dream or vision in symbolic form. Today, we might say that psychical phenomena were mediated by and through the unconscious, but only if we understand that unconscious in terms literally transcendent to space-time.

Myers appears to have coined his super-word, the supernormal, around 1885. He was trying to get away from the dualistic or theistic connotations of another earlier super-word, the supernatural, with its centuries-long insistence (it was coined in the thirteenth century) that extraordinary events must possess an agent outside the natural order to be considered genuine miracles. Put simply, they must be from God—"God," of course, defined by the teachings and authority of the Roman Catholic Church.

Myers the Protestant pushed back on this and then moved in a very modern direction. He turned to early evolutionary biology and abnormal psychology and saw the abnormal and supernormal as existing on a shared developmental spectrum. Here is Myers turning:

> When we speak of abnormal phenomenon we do not mean one which *contravenes* natural laws, but one which exhibits them in an unusual or inexplicable form. Similarly by a supernormal phenomenon I mean, not one which *overrides* natural laws, for I believe no such phenomenon to exist, but one which exhibits the action of laws higher, in a psychical aspect, than are discerned in action in everyday life. By *higher* (either in a psychical or a physiological sense) I mean "apparently belonging to a more advanced stage of evolution."[55]

In short, the supernormal was entirely "normal," but it remained nevertheless "super," super, at least, to a species that had not yet fully manifested and stabilized these future evolutionary traits. As I have put it in a kind of meme, "the supernormal was super natural but not supernatural." Here is Myers again defining his key super-word:

> *Supernormal.*—Of a faculty or phenomenon which goes beyond the level of ordinary experience, in the direction of evolution, or as pertaining to a transcendental world. The word *supernatural* is open to grave objections; it assumes that there is something outside nature, and it has become associated with arbitrary interference with law. Now there is no reason to suppose that the psychical phenomena with which we deal are less a part of nature, or less subject to fixed and definite law, than any other phenomena. Some of them appear to indicate a higher evolutionary level than the mass of men

have yet attained, and some of them appear to be governed by laws of such a kind that they may hold good in a transcendental world as fully as in the world of sense. In either case they are above the norm of man rather than outside his nature.[56]

It was in this same super natural context, still in the front matter of his masterwork, that Myers defined the *"Imaginal"* as "a word used of characteristics belonging to the perfect insect or *imago*;—and thus opposed to *larval*;—metaphorically applied to transcendental faculties shown in rudiment in ordinary life."[57]

Myers was thinking of two distinct bodies of knowledge when he wrote such lines: (1) the extraordinary data on telepathic communications from the dead or from loved ones in danger that he and his colleagues had been collecting for years, and (2) the young science of entomology.

On the telepathic side, Myers was well aware that the percipient of such telepathic communications received and often transformed the empirical information through the imagination—a dream or vision, for example, that would often get specific details wrong but would clearly carry the "punch" of the message in fantastically precise, even empirical ways. The imagination in these specific situations—which the Victorian researchers clumsily called "veridical hallucinations"—was somehow operating as a supernormal organ of cognition. It was translating and projecting, for sure, but it was translating and projecting real and accurate information about the external environment.

Myers was also thinking of the science of entomology when he wrote of the imaginal. An imago is the final adult form of an insect's metamorphosis, during which it, for example, develops wings and becomes sexually mature. This final stage is called the imaginal stage. The insect's immature or adolescent feeding form is called the larval stage. Just as the larval stage of an insect looks nothing like the imago of its adult form (which indeed appears "bizarre" or alien-like in comparison to the larval slug), so too the functioning of the human imagination can metamorphize into extremely strange but astonishingly effective forms, which Myers called "imaginal," after his beloved bugs.

Joseph Maxwell and the Paranormal

It was in this same turn-of-the-century period that the lighting flash of a new word, "paranormal," first appeared in the cultural sky. The word's inventor was a highly educated French lawyer and medical doctor named Joseph Maxwell (1858–1938). Maxwell coined the French term in 1903, in his book *Les phénomènes psychiques*.

Maxwell was almost certainly looking for a French word that could translate the English supernormal, since he writes about Myers and the Cambridge group often and in fact uses French versions of the English term, as in "the supernormal fact" ("*le fait supranormal*"), "supernormal phenomena" ("*les phénomènes supranormaux*"), and "supernormally" ("*supranormalemente*"). These all appear to be synonyms of his own adjective *paranormal(e)*.[58] As we have seen just above, Myers's category of the supernormal was suffused with esoteric evolutionary meanings. These are not quite as apparent in Maxwell, but they are definitely there.

The French *paranormal* was a philosophically careful word. It was, in fact, an impossible one in the precise sense that I have defined the term; that is, it inhabited a third epistemic space beyond belief but also beyond reason, at least as the latter was understood and practiced in the contemporary sciences of the time.[59] As much as he revered science, and he really did, Maxwell thought that the contemporary scientific view of nature was clearly "erroneous or incomplete." After all, as it stood then (and still does today), such a science could not accommodate the patently obvious paranormal facts that he had witnessed with his own eyes many times.[60] The same word, *paranormal*, would, of course, be endlessly misunderstood and misused later in the twentieth, and now the twenty-first, century until it became a virtual synonym for *supernatural*, which is *precisely* what Maxwell was firmly rejecting and moving beyond with his new coinage, as was Myers before him with his own *supernormal*.

In Maxwell's mind, something paranormal is very much a part of the natural world but is also currently beyond our scientific understanding. He was being very careful. He meant "to the side of" (*para-*) the normal or the natural, *not* something outside that natural order, which is what the Latin *supernaturalis*, or supernatural, had come to mean for specific theological and hagiographical reasons—the miracle of the saint had to come from outside nature, from God, to be a true miracle and so give witness to the supernatural authority of Roman Catholicism.

This hard distinction between the paranormal and the supernatural is fundamental to Maxwell's project. Indeed, he would often write blunt things like, "I do not believe in the supernatural (*surnaturel*). I do not believe in miracles."[61] Or he would write of "so-called miracles" ("*soi-disant miracles*") that religious people mistake as supernatural but that are actually natural forces that likely come *from them*. More specifically, these real forces come from their own unconscious intensities and convictions, which religion nevertheless shapes and encourages on the social plane in misdirecting and deceptive ways.[62]

Following Myers, Maxwell also writes of a "subliminal" level of consciousness, and he speculates that many of the psychical phenomena are products of an impersonal consciousness ("*conscience impersonelle*") or the collective consciousness of the séance circle, not of a particular person or spirit, as is falsely believed.[63] In this same sociology of the impossible, Maxwell would point out that five or six people are better than one or two to produce the phenomena (since the phenomena are social).[64] Moreover, it is best to alternate the genders around the circle and have them in equal presence. So there is a kind of gendered or subtly sexualized nature to the séance circle (although he never really says this the way we might today).

Maxwell would write of the endless séances in which he participated and in which ghostly personalities and even materializations "pretended" to be the spirits of the dead, God, the Devil, or faeries, or whose supernormal movements involved religious objects, like statues of the Blessed Virgin or a crucifix. For Maxwell, of course, these religious meanings were all ruses, pretenses. The phenomena were also "capricious" in their very nature.[65] He would even write of a "pseudo-entity."[66]

Such religious objects, pretending spirits, and pseudo-entities focused the energies for the group on a social plane, *but they were not the energies.* It is much more likely, he suggested, that the paranormal movements of objects, apparitions, and voices were the projected or "exteriorized" manifestations of the emotions, sensitivities, motor capacities, and intentions of the living human beings—or what he sometimes called "organisms"[67]—seated around the table working together, often unconsciously, with the medium as catalyst and focus. The people were, in effect, haunting themselves without knowing it. Religion was show, but it was a showing of something real in pretended guise.

It is really quite obvious—by the paranormal, Maxwell meant easily observable physical effects, which could be framed with other newly coined words, like *parakinesis* and *telekinesis* (the word "paranormal" first appears in a section on these very two topics). Parakinesis involved movements of objects with contact that was insufficient to produce the observable effects—think floating, levitating, or spinning tables with hands resting *on top* of the table. Telekinesis involved movement of objects without any physical contact at all.

Probably the closest Maxwell comes to an actual definition of the paranormal appears at the very end of his book. It goes like this: "Paranormal, that is to say implying the existence of modes of perception of which the normal personality is not aware (*étrangère*), clairvoyance, clairaudience, telesthesia, telepathy (Myers, Gurney, Podmore), exteriorization of the sensibility (de Rochas)."[68] The clear reference to the three major figures of what was then known as the London Society for Psychical Research is significant—these are the authors about whom he was thinking when he coined and defined his own "paranormal." Hence he writes, for example, of "paranormal facts" ("*des faites paranormaux*") and then immediately point to the same British Society and their star subject across the ocean, Mrs. Piper of Boston.[69]

Maxwell is very insistent that he does not know the ultimate agency of these paranormal facts. For all he knows, they may in fact be the workings of "some mischievous Kobolds."[70] But he rather clearly does not think this. What he thinks is that the agency of such paranormal phenomena is impersonal (which we might better understand here as superpersonal, since this force is collective and independently intelligent). He goes on to speculate, much in line with Myers, that such an impersonal consciousness will normally communicate with a person through dreamlike narratives and symbols—that is, through the imagination.[71] He also describes these paranormal facts as a process of "exteriorization" of the motor functions or nervous energy, a process which can result in actual materializations. He finally states flatly that these phenomena are true and are not fraudulent.[72]

It is worth repeating: Maxwell did not claim to understand *how* this all worked—that is, how the transformation of the "will" ("*volonté*") of the collective group manifested in seemingly independent and intelligent paranormal movements ("*mouvements paranormaux*") of things like raps, levitating tables, blowing curtains, cold breezes, and other exteriorized or even materialized forms.[73] But he knew that it did. He *saw* it happen, many, many times, and in full light, and with no reasonable chance of fraud (his book is filled with details about how to prevent and catch the fraudulent, and he distinguishes between voluntary and involuntary fraud).

Maxwell also goes into great physiological detail about the phenomena, like how the energies seem to be emitted from the palm of the hand instead of from the hand's backside.[74] He writes astonishingly reflexive things, like how the ghostly personification of the séance is a kind of co-experimenter and often responds to the intentions and thoughts of the sitters and medium.[75]

In the end, Maxwell certainly did not turn back to the religious past for his answers. He *turned around*—to the future and to better science. He called for a "future Newton," who would explain such obviously physical but completely inexplicable forces in a formula more complete than our own, forces that Maxwell recognized were often likened to electricity or a subtle vibration (*"une sorte de vibration très faible"*) but were clearly not the same thing that the physicists were measuring.[76] Again, he wanted a fuller science, not more religion or belief.

Renaud Evrard has probably written more than anyone about the doctor-lawyer. Working with Maxwell's great-granddaughter and acknowledging that much of Maxwell's life is still not known (partly because he had three children out of wedlock, none of whom he could recognize officially), Evrard has filled in the basics of the writer's career, ideas, and publications.[77] Evrard explains that Maxwell had a long and productive career. He worked in law, including progressive attempts to push the legal landscape into reproductive and abortion rights and sex education. He worked in medicine, particularly around psychopathology. He worked in the psychical research of the place and time. He also wrote a book on modern forms of mysticism within Theosophy, *Le mysticisme contemporain* (1893), which he discusses again in *Les phénomènes psychiques*, confessing that he came upon these texts by accident and that he felt some surprise that "a mystical movement" could find clients at the end of the nineteenth century.[78]

This latter point is important, as Maxwell clearly believed (and I think he was correct), that "mysticism" was already waning as a viable category around 1900. The implication was that we needed new terms to discuss these and related phenomena. Maxwell would give us one.

Toward the same end, Maxwell even wrote paranormal novels. Evrard explains: "In these adventures of his alter ego, Dr. Heurlault, Maxwell described the situations of a man of science engaged with the paranormal in a romantic atmosphere."[79] I take it, then, that these paranormal novels were what we might call true fictions, ways that the scientific researcher could express his deeper convictions and struggle openly with his thoughts and materials.

Maxwell certainly knew of what he was writing. He worked with numerous mediums and psychics, including and especially Eusapia Palladino, whom he called a friend. He thought Palladino was the real deal but also acknowledged that she sometimes cheated. He did not hesitate to expose fraud when and where he found it, but he was not so naive as to equate a failed trick, as in Palladino's case, with an entire panoply of real effects, again as in Palladino's case. Nor did he equate the psychical phenomena around mediums and psychics with their oft-noted psychopathology;

indeed, he was of the strong opinion that one could not study supernormal appearances without a knowledge of mental pathology, and that a certain nervous instability was a condition favorable to the phenomena.[80]

Moreover, much like Arthur Schopenhauer and a long line of intellectuals before and around him, Maxwell concluded that the anthropology of magic, divination, and witchcraft needed to be rethought, since all these demeaned and dismissed phenomena actually happen. He certainly did not believe the beliefs of these cultures, but he had no doubt that the phenomena happened because of such beliefs. Such beliefs are necessary tricks, but they *are* necessary for the phenomena to appear.

Evrard speculates about why Maxwell chose the new word *paranormal*. He thinks that Maxwell did not want to use the obvious French term for "anomalous" since it is "*anomal*," which is indistinguishable to the ear from the French "*anormal*," or "abnormal" (a word that Maxwell did occasionally use as a synonym for "paranormal" or "psychical"[81]). This discomfort with the "abnormal" is probably one reason Maxwell came up with a new and more neutral French word: *paranormal*.[82]

But there are certainly other reasons, including Maxwell's obvious desire to affirm that these movements and energies are entirely natural, as well as his equally obvious admiration for Myers, the London Society for Psychical Research, and the specific evolutionary undercurrents of Myers's supernormal. Again, Maxwell shows every sign of looking for a French expression that carries similar meanings.

Perhaps, in the end, it really does hinge on what is "normal" or, better, how one day what is now experienced as paranormal will be experienced as normal. Consider this key passage:

> It is more reasonable to think that our nervous sensibility will become more and more refined. It would be rash to believe that the current type of human is the final outcome of evolution. Our species is not only a ring in a chain of beings: the causes that have brought the perfecting of the human species are still active, and it is logical to think that there is in nature above us the same means as there is below us. The latter represent the ancestral types, which recall past forms; the former are perhaps the precursors. They present to us what are considered abnormal faculties today but that will become normal one day.[83]

The paranormal, then, would signal something like "perfectly natural forces that we presently do not understand but that we will someday in the future evolution of the species." It is not an embarrassment here, something ridiculous, a word to remove from a book or essay title. It is part of the future order of knowledge.

Still, even after we put as much as we can back on our comparative table, we need to be very careful with words like *supernormal*, *imaginal*, and *paranormal*, especially the last one. The twentieth- and now twenty-first-century transformations of the paranormal back into what is basically the older dualistic notion of the supernatural are clear pop-regressions of the term. They should *not* be taken as educated uses of the term. These transformations should be noted and studied as such, for sure, but such linguistic regressions should not be taken as absolute or even necessary by readers and writers who know their history of science.

Much less should they be understood as reflective of the English, French, Swiss, and German intellectuals, Frederic Myers and Joseph Maxwell in particular, who first so carefully coined and fashioned these three related impossible words—*supernormal*, *imaginal*, and *paranormal*. The third space that they carved out beyond all good reason and all good faith is *precisely* what makes them so amenable to impossible thinking.

The origins of these three key terms lie not in abstract thought or "primitive" cognitive error but in the endless traumatic and cognitively shocking experiences of countless and often unnamed individuals who, in their own earnest understandings, had made contact with what they believed were other human, nonhuman, or superhuman intelligences. To put an accent on it, these words were not made up out of thin air or fantasy. Individuals *suffered* them, *heard* them, *saw* them, *loved* them, and even sometimes had sex with them.

Words encode such experiences. Words are experiences.

We can interpret these experiences in all sorts of ways (with more words), of course, and we should, but we must begin with this fundamental lived context: the presence, urgency, and agency of what we have come to call the dead.

So speak the ancestors, fantastically.

TWO

Why They Don't Land

Mantis, Mystical Theology, and Social Criticism

Protest all you want. We are at the end of being us. . . . The first and final alien disclosure is not that aliens don't exist, it's that you don't. It's that no matter how loudly you protest, you're not ready for true disclosure, nor will you ever be.

JEREMY VAENI, *Aliens: The First and Final Disclosure*

The main thing to understand is that we are imprisoned in some kind of work of art.

TERRENCE MCKENNA IN ANDREW R. GALLIMORE, *Alien Information Theory*

Eerily, the insectoid, telepathic, and evolutionary themes that fascinated Frederic Myers so and were secretly embedded in the French coinage of the paranormal from his own supernormal would later show up in spectral form in the twentieth- and now twenty-first-century alien abduction literature, where gigantic telepathic insectoids are extremely common, particularly in their mantis forms.[1] In other words, the modern abduction visions replicate, in precise detail, the imaginal of Myers. What was an entomological metaphor in Myers has become a literal visionary form or physical encounter in the alien abduction literature. The larval stage of the human being has morphed into the future stage of the imaginal, which appears to the present human mind as a superinsect with the exact supernormal power that Myers himself first named in 1882: telepathy. The paranormal is insectoid.

Put a bit more responsibly, the category of the imaginal has morphed considerably over the last century and a half, since the 1880s, when it was first introduced. It starts out within the Victorian psychical research tradition in an explicitly evolutionary framework deeply indebted to Charles Darwin but not technically Darwinian (since it possesses a spiritual telos, or purpose). It then appears, perhaps most famously, in the middle of the last century within Henry Corbin's well-known work on his Shi'i, Ismaili, and Zoroastrian materials. It then morphs again over the last five decades up to the present, this time in the ufological and abduction materials, where the imaginal displays *both* of these earlier meanings—that is, both evolutionary patterns suggestive of a future superhuman state and dialectical or apophatic sensibilities often explicitly indebted to the mystical traditions of the past.

I want to say something more specific in this chapter about this last imaginal morphing.

From the World of the Dead to the Halls of Congress

As counterintuitive as it might sound, it is well known among seasoned researchers that the UFO is very much connected to the near-death experience (NDE)—the heavenly saucer and the transcendent soul.[2] Hence, for example, the recent two-volume study of Joshua Cutchin, *Ecology of Souls*, which is our fullest and most far-reaching comparison of the flying saucer and the transcendent soul in the histories of global folklore and mythology.[3] This massive work was inspired by two individuals who are very relevant to the present project: the psychedelic author Terence McKenna and the abduction researcher Anne Strieber (the wife of Whitley Strieber, who collected the "*Communion* Letters" that now sit in our Archives of the Impossible at Rice University). As Whitley often comments, Anne once emerged from her research into these letters (one can still see her colored penciled notes on the letters themselves) stating, "This has something to do with what we call death."

The comparative pattern of the soul as a UFO or the UFO as a soul is similar, if seldom described as such, on the technological or scientific side. For example, the parapsychologist and skeptic George Hansen has spoken of how numerous writers, major funders, and public activists have linked parapsychology and ufology.[4]

Here is what has happened most recently (as of the time of this writing).[5] On June 25, 2021, the US Office of the Director of National Intelligence published a much-anticipated report on what many people customarily call UFOs (unidentified flying objects) but that it insisted on calling UAPs, or unidentified aerial phenomena (a clear political dodge, not so different from calling psychedelic plants "drugs," although the argument was that some of them come out of the water and so are not technically flying—okay). Its "Preliminary Assessment: Unidentified Aerial Phenomena" was a slim, nine-page document referring inadequately to 144 events involving these UAPs, collected from military sources between 2004 to 2021.

The news was not good for the deniers. Of the 144 events noted, 80 involved multiple sensors and were therefore likely caused by encounters with "physical objects." Many of the UAP events interrupted pilot training or military activity. Some resulted in near collisions in the sky. Most importantly, the new report made clear that *only one* of the 144 events and encounters reported in the past seventeen years could be explained. Just one.

Many of the explained cases, moreover, involved multiple human visual contacts or dramatic radar videos, a few of which had earlier been reported in the *New York Times*, where two bombshell articles in December 2017 revealed that the Pentagon had, in fact, been taking UFOs seriously enough to spend $22 million on a program to investigate their existence.[6] Notably, the tone of media coverage toward UFOs has shifted since the initial *New York Times* articles from one of derision and eye-rolling to one of puzzlement, occasional seriousness, and clever dodges.

Congress engaged in official hearings on the UFO (or UAP) subject in May 2022. Alas, this was mostly a repeat of the report of the previous summer and its dodge. Politicians and military officials (and the media) continued to focus almost exclusively on the physical aspects of the phenomenon and continued to ignore their paranormal or impossible effects. Much of the hearing was about a well-known 1967 incident at Maelstrom Air Force Base in Montana where a UFO shut down a nuclear missile base (yes, it gets that serious, fast).

In the media coverage that followed, I could find only one politician who openly disagreed with the general tone and direction of the conversation. Republican representative Mike Gallagher from Wisconsin explicitly raised the time-traveler hypothesis—that is, the idea, which is quite common in the literature, that UFOs might be "us from the future." That did not go anywhere, of course. As a result, Gallagher was profoundly unhappy with the hearings, which he described as "fundamentally unacceptable." He correctly observed that the "genie is out of the bottle," and that if we really want to get to the bottom of this mystery, we need to release all the data into the public sector, where serious people can study everything and come to their own conclusions.[7]

Representative Gallagher might want to question whether his colleagues are ready for the conversation that would ensue, were his excellent question taken truly seriously. That conversation, after all, would have to include the history of religions as well as the neuroscience of temporality and the physics and cosmology of space-time, not to mention the utter disregard the objects and presences show for our national borders and very recent notions of capitalism, moral agency, and personal possession.

Things continued to heat up, considerably, in the summer of 2023. In early June, a UFO whistleblower came forward: United States Air Force officer and former intelligence official David Grusch. Amid serious retaliation and harassment, Grusch spoke openly to the American journalists Leslie Kean and Ralph Blumenthal, and to the Australian journalist Ross Coulthart, of "retrieved craft" and "non-human intelligences," a phrase which, by this time, had its own acronym (NHI).[8] On July 26, 2023, Grusch, along with two other retired military professionals, retired US Navy commander David Fravor and former US Navy pilot Ryan Graves, all spoke to a special session of the House Oversight Committee in Congress about the need for more openness and democratic oversight over the UFO or UAP issue.

In a major online appearance in September 2023, with Jesse Michels of *American Alchemy*, Grusch went on to mention several things, all relevant here, including: the deep historical linkages between atomic and UFO secrecy, possibly going as far back as the Manhattan Project; the relationship between the UFO phenomenon and the history of religions (Grusch has come "full circle" on his religious beliefs, which he now takes much more seriously after his work in intelligence and the UFO phenomenon); and the usefulness of hyperdimensional spatial and temporal models in coming to understand what looks very much like the manipulation of the space-time continuum by the craft and their presumed controllers.[9]

As I watched the proceedings of the congressional hearing live on my laptop, I saw a few of my own colleagues in the chamber audience and realized that impossible thinking had entered the center of American politics (though not very effectively). I realized that we need much more thinking with *way* more philosophical nuance, historical awareness, scientific humility, comparative reach, and, frankly, impossible paradox.

The simple truth is that I am very much a part of these developing conversations, as is this very book (I am finishing it as the UFO, or UAP, is zigzagging its way into the center of both the mainstream media and American politics). I have been seriously thinking and writing about the UFO phenomenon since about 2004. I first struggled with it because I had to do so: it was simply everywhere in the historical and ethnographic sources with which I was working at the time—in this case, the history of esoteric spiritual currents in the California counterculture and, more specifically, the history of the Esalen Institute in Big Sur. I have also cohosted four different private Esalen symposia on the UFO topic, "beyond the spinning," as we sometimes put it, and with many of the central scientists, intelligence professionals, writers, and scholars.[10]

Before December 2017 (that is, before the *New York Times* articles), it was not uncommon for people to assume that the UFO phenomenon was not serious, that it was "New Age," or that it was somehow a "California thing." This is not entirely wrong, since the New Age movement in fact took the UFO phenomenon seriously (it was correct about so many things) and spun out a number of UFO religions. Some scholars of the New Age have even argued, quite convincingly, that particular currents or strands of these esoteric movements *began* among UFO contactees of a particular Theosophical persuasion in the 1950s (this fits the case of one of the earliest witnesses, the businessman Ken Arnold, whose family story sounds *very* Theosophical to my ear). There are also contemporary scholars who link Theosophy and the UFO, especially the Swedish scholar and archivist Håkan Blomqvist.[11] There are even humorous postcards in the Archives of the Impossible at Rice University to this effect—ones that comedically identify flying saucers as "Californians."

It *is* funny. But it is not true, *and it has never been true.* Among the firm takeaways from spending days, then weeks, then months in the Archives of the Impossible, are just how many contact and encounter cases actually exist in the historical repository (thousands upon thousands, it turns out); how extensively documented many of them in fact are (in typed or handwritten letters, in carefully drawn maps and diagrams, in endless newspaper reports from around the world, in periodicals and self-published zines galore); how these same phenomena go back to the nineteenth century, and further back, *way* further back, into the medieval and Renaissance European worlds, and into the African, Asian, Australian, and South American worlds of the present; and how they repeat, over and over again, the same comparative themes: the large almond eyes, the diminutive and lanky bodily form, the idea of some kind of hyperdimensionality or time travel, the presence of "humanoids," the religious ambiguity of the messages revealed, the presence of death and the ancestors, the apocalyptic vision, and the radiation burn, swelling, or vomiting. And on and on we could go.

In historical truth, the phenomenon is *everywhere* and *everywhen*, and its comparative patterns are astonishingly stable and consistent. Indeed, it is remarkable how often the act of comparison leads to a growing realization, which is not always welcome, that these things (which may not be things at all) are very, very real.

Of course, the way the phenomenon is interpreted, and to what degree it is even noticed, is always a function of the local culture and politics and, more recently, the advancing technology that can detect more and more of what is present in our skies (in this, the UFO phenomenon is not unlike the near-death experience, which relies on advanced biomedical technology to bring individuals back from the brink of death from which they seldom would have emerged in any earlier period). The coming of the saucers in twentieth-century America was definitely tied to the Cold War and its various fears and anxieties. It was the Soviets and, before that in the late nineteenth and early twentieth century, the Germans.[12] (Now it must be the Chinese.)

To begin in the United States, which is where the UFO phenomenon has had the most play in media, popular culture, and entertainment, some of the earliest and most dramatic documented modern encounters have been near nuclear military sites around the country, as the Congress hearings of May 2022 gave clear witness.[13] Roswell, New Mexico, for example, was just such a nuclear site, where one of the earliest alleged crashes took place in early July of 1947, just after the sighting of private pilot Ken Arnold, which took place on June 24, 1947, near Mount Rainier, Washington. It was the latter event that in fact gave us the expression "flying saucer" through a journalist reporting on the incident the next day. New England has also played a special role. The region boasts the first iconic American case, of the night of September 19–20, 1961, involving a mixed-race couple who were civil rights activists, Betty and Barney Hill.

But the phenomenon extends far beyond US borders. The Indigenous communities of the Americas are filled with UFO lore, as are Canada, Mexico, Argentina, and Brazil, to name some of the most reported. Other signature cases have occurred in places like Australia, the Czech Republic, England, France, Japan, Romania, Russia, and Zimbabwe.[14] The UFO, moreover, is everywhere in Islamic science fiction and its various astrobiological visions in languages as distinct as Arabic, Bengali, Malay, Persian, Turkish, and Urdu.[15] To put an accent on things, the UFO phenomenon is in fact global, although, of course, its reporting and interpretation are always contingent on surrounding cultural influences.

This is why to study the UFO phenomenon adequately is to study pretty much everything. It is also to come up against, hard, the realization that the institutional or university order of knowledge within which we work and think today, an order that effectively splits the sciences off from the humanities, is simply not helpful, and certainly not reflective of the material-mental reality we are trying to understand. The difficult truth is that the

UFO phenomenon has both an objective "hard" aspect (think fighter-jet videos, photographs, alleged metamaterials, apparent advanced propulsion methods, missile silo shutdown, and landing marks) and a subjective "soft" or "human" aspect (think close encounters, multiple and coordinated visual sightings, altered states of consciousness, subsequent paranormal powers, visionary displays, and experienced traumatic or transcendent abductions). And *both* sides—both the objective material and the subjective mental or spiritual dimensions—are needed to get a sense of the fuller picture.

This essential doubleness of the UFO phenomenon became particularly obvious in the secret US government studies contracted by Robert Bigelow, the aerospace entrepreneur in Las Vegas, Nevada. Much in line with the argument of the present chapter around the linked subjects of the UFO and NDE is the fact that Bigelow has been a major supporter and funder of research into both the alien presence and the experienced reality of the afterlife. It is simply not possible to read about the stunning insiders' accounts of the still-classified research program that Bigelow contracted with the US government and not confront this doubleness head-on.

Indeed, three of these insiders go to great lengths in their coauthored *Skinwalkers at the Pentagon* (the very title fuses the two aspects, since a "skinwalker" is a shape-shifting superwitch in Indigenous American lore) to describe and discuss things like the "hitch-hiker effect." The latter phenomenon involves military and intelligence professionals apparently being "infected" with occult presences on Skinwalker Ranch in Utah and bringing the phenomenon back home, literally, to their East Coast families. One of the authors, Colm Kelleher (with a PhD in biochemistry and an extensive background in cancer research and immunology), theorizes the human witness as a kind of biochemical readout instrument, whose immunological system is particularly sensitive to these hidden presences and forces.[16]

Kelleher, with whom I have worked on an advisory board for Mr. Bigelow, is explicit about what *Skinwalkers at the Pentagon* calls "blue orbs." These are basically intelligent balls of plasma that interact with humans, often at close range, and occasionally penetrate their bodies, sometimes with cancerous effect. Kelleher once advised me over dinner, "If you see one, turn and run like hell. That's my advice."[17]

Of course, one can slice up the UFO phenomenon into the scientific and the humanistic, or the technological and the paranormal (and then ignore or dismiss the second dimension). But one will never understand the full phenomenon by doing so. One will just continue the nonsense, the game, the distraction.

And that is precisely why I think the UFO subject is so incredibly important: it bears a particular power to challenge, even abolish, our present order of knowledge and its arbitrary divisions into the objective and the subjective. *Whatever* the UFO is, after all, it simply does not behave according to such rules and assumptions. It certainly couldn't care less about the concerns and categories of our militaries, politicians, and academics.

But it is even stranger still. As Hussein Ali Agrama has argued, the UFO also calls into serious question the presumed divisions between science and religion, the two reigning epistemologies of the premodern and modern worlds. The UFO basically abolishes what Agrama calls the "secularity of science"—that is, the manner in which science is believed to be a purely material or physical pursuit without spiritual or, in this case, esoteric or mystical dimensions. The anthropologist thus writes of a powerful "uncanny science" that is very much active around the UFO but also inevitably maligned by conventional scientists. Accordingly, this uncanny science ends up secreted in intelligence, corporate, and military technologies, thus creating a separate esoteric world, a set of hidden truths or rejected forms of knowledge that are sharply distinct from those presumed in the public sphere.[18]

If I may speak for a moment in the language of some of my own intellectual ancestors, particularly the quantum physicist Wolfgang Pauli and the depth psychologist C. G. Jung, the UFO, or flying saucer, is the ultimate modern "symbol" that participates in *both* the mental or social world of the subjects who encounter it *and* in the material world of their own epistemic or sensory experience. Ezekiel's wheel or disk is *the* Romantic symbol that participates in what it communicates for a reason. The UFO phenomenon appears to have emerged from a third realm and so manifests obvious mental, social, and historical elements *and* physical, material, or sense-based features. In Jung's mature terms, the UFO is *psychoid*, at once psychic and physical.

There is also humor here. As I already noted, we have postcards in our Archives of the Impossible. One of them says something like, "Blurry photos of flying saucers available in the bookstore." This reminds me of the late American comedian Mitch Hedberg on the omnipresent figure of Bigfoot: "I think Bigfoot is blurry. That's the problem. It's not the photographer's fault. Bigfoot is blurry, and that's extra scary to me. Because there's a large, out-of-focus monster roaming the countryside."[19] The point is the same: *the fuzzy nature of the sightings is part of their point*. The entire realm *is* "out of focus." Maybe it must be.

Allow me to speak from the humanities—thought not in the way my colleague Timothy Morton jokes of the humanities as being "candy sprinkles on the cake of science." I am not here to drop candy sprinkles on your cake. I am here to say something about that cake, or maybe its eating.

Take science fiction. The genre has not always been entirely helpful. Indeed, it has too often distorted and exaggerated the phenomenon beyond all recognition. Hence, what I call the "Cold War invasion mythology" has been dominant until very recently. In this mythology, the UFO is seen as the sign of an "alien" invasion that must be fought back or resisted by patriotic, God-fearing citizens. This is basically what we still hear too often today, including from the American military and intelligence communities. It goes at least as far back as H. G. Wells's 1897 novel *The War of the Worlds*, an invasion story (inspired by British colonialism and the overwhelming technological force of the British military in places like New Guinea). Or, on the American side, think of the movie *Independence Day* from 1996. Only recently have writers and filmmakers begun to move away from this Cold War invasion mythology and toward what we actually see in the encounter cases, if we really study them (as opposed to just assuming things about them). Hence the movie *Arrival*, by Canadian director Denis Villeneuve, in 2016.

The latter movie borders on the profound, since it orbits lovingly and contemplatively around the paranormal transformation of its central character, a gifted humanist or social scientific linguist, Dr. Louise Banks. And, indeed, in historical fact, human witnesses are often radically transformed by their encounters. They experience, either within the event itself or later develop, new astonishing abilities—think telepathy and, as in *Arrival*, precognition (the same, again, is true of near-death experiences). Such encounters can also be of a deeply spiritual nature, by which I do not mean "good" or "nice." People experience awe, fear, uncanniness, and absolute terror. Their worlds are turned inside out and upside down. What they thought was real is no longer so. What they considered to be imagined is actual. Sometimes they adopt new, much more cosmic worldviews.

Hence, in *Arrival*, Dr. Louise Banks develops the ability to precognize the future as she gradually learns that time, like the grammar of the alien language she is deciphering, is circular not linear. Such a circular vision of temporality is an idea about space-time that is well known to humanists, from ancient Greek philosophy to Friedrich Nietzsche.

My own position? It is always developing and confessedly tentative, and I would be most happy to change it tomorrow, but my general sense is that there is definitely something being covered up or lied about (the misinformation is grotesque and obvious to serious scientific and humanistic researchers), but—and here is the thing—*no one knows what that something is or what it means.* Put a bit differently, there is a *there* there, and it may well possess a very physical dimension, but no human being knows what that *there* really is, including those institutions (government or corporate) that naively think they control it toward some kind of national security interest or profit motive.

The present situation is so confused partly because our culture has grown incredibly stupid around the nature of the soul, or what we would today call "consciousness," and its fantastic relationship to the cosmos, which is to say to space and time. We simply no longer understand the shocking multinaturalism (nature behaves differently in different bodies and places) and very physical efficacy of belief, symbol, and myth—how they, in effect, really *in effect*, create different realities. Too many of our political and media elites think it is all nonsense, except, of course, for *their* beliefs, symbols, and myths—that is, their own realities.

It is in this way that we are forever limiting ourselves to this or that identity, culture, or religion, exclusively believing our belief systems and mythologies (including our scientific ones) and not looking "behind" them to know the presence that is creating or projecting them all. It is rather like a series of movies in which the characters can never come out of the screen to know the projector at the back of the theater. They naively think that the movie they are in is the only movie the projector is projecting or can project. That is simply not true—*whatever* the movie is.

In such convictions, my own position is very close to that of the astronomer, computer scientist, and medical research investor Jacques Vallee and his "forbidden science," an artful and lifelong combination of astronomy, computer science, information theory, esoteric practice, and *profound* skepticism around all UFO belief systems.[20]

I want to stress that skepticism, as does Vallee. I am *extremely* dubious of any and all talk of the "extraterrestrial." I don't believe in gods, deities, demons, or aliens. It's my job. If you gave me enough beer and asked me to be blunt about the situation, asked me what I *really* thought, I would say that we need to flip from a spatial register (it is not *where* they are from) to a temporal register (it is *when* they are from). I would say that the UFO often appears to be a time machine and that the so-called aliens look more than a little like human beings from the future—or, in some contact cases, a sophisticated form of future artificial intelligence. If I were from the future studying the violent hairless monkeys of the past, I certainly would not go into their troop either. I would send a very smart but very expendable machine.

My point? That we should be approaching the total UFO phenomenon much more like *Arrival* and much less like *The War of the Worlds*. But we cannot seem to do that. Instead, we go on and on about potential "threats," enact endless security and secrecy measures, hold congressional hearings about machines in the sky caught on fighter-jet radar systems, and seek knowledge of things that are likely hidden now in confidential corporate interests. And then we wonder why no one understands this? Let me be blunt. I strongly suspect that what we finally have in the UFO phenomenon is a physical-spiritual phenomenon that is being grossly misinterpreted as a conventional military threat or potential corporate technology but that no one really understands, much less can reverse engineer with our present strictly physicalist assumptions. To speak paradoxically and, I hope, shockingly, we are trying to shoot down souls.

Good luck with that.

Man Meets Mantis

Allow me to say something more specific now about the recent transformation of the imaginal, as it morphs (back) into some entomological forms in the contemporary UFO and abduction literatures.

Stuart Davis is a spiritual seeker exceptionally well versed in the American spiritual and Buddhist scenes and a full-time artist who has worked in music, television, film, and language (he invented his own language). Davis has recounted and shared his encounter and UFO experiences at some length in a number of places, foremost among them an audio salvo, "Man Meets Mantis," which is the first episode of his remarkable podcast series on the relationship between nonordinary experiences and artistic creativity, *Aliens & Artists*.[21]

Much of what I describe below is based closely on that podcast episode, but it is also shaped and guided by conversations I have had with Davis over the years, including on *Aliens & Artists*, and with a most remarkable group of individuals of the Experiencer Group. The latter, led by cofounders Kirsten Blackburn, Jay Christopher King, and Davis himself, is a sophisticated collective of individuals from all walks of life who create community and smart conversation around superhuman experiences, often of the high strange type.

I happen to think that the basic thesis of the *Aliens & Artists* podcast series—namely, that creativity and paranormal experiences are deeply related—is perfectly correct. I was especially surprised by the story of "Bobby." Stuart Davis tells us that Bobby was staying at friend's house in Aspen in the late 1990s. Sometime between three thirty and four o'clock in the morning, his "mind woke up," but his body "was completely paralyzed." Beside the bed was a six- or seven-foot mantis, "ghostlike," not quite physically there but *there*—another life-form. Bobby was in what he called "a fear state." The altered state lasted for about five minutes before Bobby gained control over his limbs again and the terror subsided. Bobby had this sense that the mantis was taking samples of his emotional, subtle, and causal bodies.[22] Bobby later jokes that, whereas Davis was issued onto a long artistic life path by his encounter, all he got from his was a blood-pressure reading.

Joking aside, I am personally stunned by Bobby's experience because it replicates, in near-perfect phenomenological detail, what happened to me in one early morning in Calcutta in November 1989. Unlike Bobby, though, the results of what I call "that Night" were more than a blood-pressure reading. The result was a lifetime of scholarship (including this book). I saw no such insectoid entity in that literally shocking event (I initially thought I was being electrocuted), even if Kali, literally the "Black One," the Hindu Tantric goddess I was trying to understand at that time, is sometimes compared to a spider with eight arms.

Still, there was no visual content. Okay, it *felt* like I was having sex with a many-armed superspider (maybe I was), but I saw no such thing. Vision or no, the experience convinced me, quite completely, that nonordinary encounters can result in new forms of insight, wild intuition, and powerful creativity that lie at the very core of some types of cultural production, including, believe it or not, nerdy scholarship. It happened to me. I am who I am, I think how I think, because of that Night. Bobby's story is more than "just a story" for me. Something like "sleep paralysis" does not do it either. No doubt, that physiological condition is what "let it in," but it says little, or nothing at all, about what came in.

Here is Stuart Davis's story. Very much unlike me, he saw the thing.[23] At midnight, on the eve of the New Year 2010, Davis had what he called "the strangest experience of [his] life." He was meditating in a high fever (a physiological pattern he had as a small child and grew to appreciate as "glimpses into another reality" before the pattern receded in his teen years). The fever was back. He decided to do what he had learned as a practicing Buddhist— meditate through it. He also decided to set an intention: he asked his fever to meet his spiritual guides. That seemed reasonable enough. Davis was expecting dead Buddhist monks, or perhaps some lesser Buddhist deities, maybe a bodhisattva—you know, something traditional and respectable.

Instead, Davis found himself face to face with an eight-foot-tall praying mantis entity standing at the foot of his bed—basically a gigantic insectoid, who was wearing a purple robe with a high collar, no less. He was *shocked*. Literally. The mantis shot a "powerful signal" into him. It was "like being hit by an invisible firehose." The signal was as sonic as it was shocking, since it was filled with inexplicable data: "clicks, pops, information, something akin to music, and it is all nonverbal." Then, as it ended, a thought lands in Davis's head, telepathic style (as is often the case in these alien abduction experiences). He hears a single sentence, in English now: "Remember who you work for."[24] Then it was over. The whole thing lasted, maybe, one minute.

That was enough. But it did not end exactly there. It just went on and on, as the insectoid or imaginal presence entered the daily lives of Davis, his family, and, soon enough, his cinematic colleagues, resulting in what he calls "super synchronicities" mostly with—what else?—praying mantis insects. These super synchronicities in the social and biological worlds are what made it so difficult, impossible really, to take the easy way out and read the mantis as his higher spiritual self, as "imaginal" (he will use the word), as some kind of bizarre psychospiritual projection, the debunker's rhetorical ploy of "hallucination."

What to do with all the *physical* events, with "history," as academics like to say? (As if they know how *that* works.) These were not just dreams or internal altered states. These were very physical and often literally entomological events. The praying mantis insects, for example, began showing up at key moments to and around very specific individuals, including a number of artists (a producer, a director, a cinematographer, and a line editor, who had terminal cancer and would soon die) working on a trilogy of films—a trilogy, by the way, that Davis firmly believes that he wrote out of the signal shot into his body that night (again, I feel the same about my books, including this one).

The opening scene of the first film in the trilogy features a teenage victim of human trafficking, a young Russian woman who is about to murder her captor when she finds a praying mantis insect that triggers what Davis calls a "paranormal awakening." He saw the whole movie in his head, and he knew who would play its lead when he met her, the singer-songwriter Jasmine Karimova. Davis wrote the abstract for the filmscript and then set it aside, laughing that "no one will want to make a paranormal sex trafficking movie in Russian." Well, that wasn't true. The scripts were purchased, the mantids kept showing up, and the movies are half funded now.

The details, at once spiritual and empirical, go on and on. Indeed, Davis recounts in this opening episode many other historical events that happened around these extraordinary inner experiences. It all seemed linked: an intelligent horizontal-shaped UFO appeared in the day sky, moving at ninety-degree angles, and was witnessed by Davis's two daughters and wife as it moved and mind-melded with him (it is not just an object for Davis but also has a subjective feel to it); a locked door swung open and lights flickered in a house for an executive producer reading Davis's script before she encountered a praying mantis eating a dragonfly at her door the next morning (a dozen of the twenty people working on the movie at the time had related mantis experiences when they began working on the film); the sound of a massive tuning fork was heard, in reverse, and then a hearing-test sound vibrated over the family house (such sounds, including the reversibility of the first, had deeply personal meanings for Davis), while coyotes wailed in the distance to announce the appearance and departure of the sounds; a family dog refused to investigate a second occurrence of the reversed tuning fork while his ears perked up to the eeriness (one can imagine what the dog, a large Doberman pinscher, was thinking: "No way, human—I ain't going up there"); and, as if that were not enough, Davis tells us of an old scoop mark in his left leg and a BB-like lump in his right shoulder (mysteriously, dreamed by a movie producer, despite the fact Davis had never told her, or anyone else, about his uneasy suspicions from childhood)—possible physical implants as quintessential imaginal or mental-material technologies.

To quote Davis's exact response, "What the actual fuck?"

And this was all before Davis met Jacqueline Smith, who revealed at a 2016 UFO conference that she had been visited by mantis beings at noon on—get this—New Year's Eve. She also claims to have mantis DNA. She suggested that Davis himself was a mantis in a past life in another dimension—that is, between the physical and the spiritual realms. They are interdimensional, not strictly physical. Davis is stunned, goes to the gym to process everything, only to encounter a praying mantis on the keypad, staring at him. It flies down onto his left foot, stares at him intently, and then flies away. He remembers Jacqueline's teaching that the immense mantids can temporarily look through the eyes of the little mantis insects.

You know, typical stuff.

Yes, Davis knows, and he swears again to get the point across. It's all a "labyrinthine mindfuck." He does not dismiss the craziness. He embraces it and then tries to think-with it: part of this thinking-with is deep suspicion around the behavior of all such abducting entities (we would be arrested for such behavior), and part of it is deep creativity and a visionary film in process that is now the first of a trilogy. Part of this thinking-with is his awareness of the deep historical etymologies that seem to be at work in such moments. He knows, for example, that the word *mantis* is related to *mantic*, itself derived from the Greek *mantikos*, as "relating to divination," as well as to *mania*, which comes from a Greek verb for "to be made mad."[25]

We are back to the deepest sources of human civilization in inspired madness and altered states of consciousness. This is what Stuart Davis calls the "primordial" lineage that produced religion, philosophy, and civilization itself. It is superhuman to the core.

The Mantis in the Chinese Martial Arts and European Surrealist Art

It might seem to the reader that the mantis encounters of Stuart Davis are somehow without precedent or comparison—that they are inexplicable precisely to the extent that they are incomparable.

This would be wrong. Very wrong.

The mantis, it turns out, appears often in the history of religions, the history of art, and human culture more broadly. Indigenous folklore in the Americas and Africa is a particularly fruitful place to look. For example, many of the testimonies collected by the folklorist Ardy Sixkiller Clarke are about what Indigenous Peoples call the "star people," which are described as "more like bugs than men" and as "big insects" with faces that are "half human, half insect," and even as "mechanical bugs" (this latter expression will soon take on a new force in the present chapter).[26] The praying mantis itself is a significant figure in southern African folklore, particularly within Namibian culture, where it functions as an omen or serves an oracular function and possesses a particular trickster nature.[27] It also takes on the role of a benign god or deity associated with nourishment, resurrection, and creation among the Indigenous Peoples of this same part of the world.[28]

The praying mantis, known as the king of insects, is also central to the Chinese martial arts, or kung fu, which are derived from various imperial, military-monastic, Daoist, and Buddhist influences. Here, in this martial arts reference to the praying mantis, is what one insider ethnographer, Douglas Farrer, has called "the unleashing of hidden human potentials" and "the continual revelation of esoteric embodied knowledge through practice"— that is, anomalous forms of embodiment that continue to morph and evolve, that are never exhausted.[29] Such an ongoing revelation takes place within a Chinese cosmology in which humans and animals move between levels of becoming and interchange their respective powers through the processes of reincarnation, and in which the vital force of the body is increased through cycles of extreme asceticism and mysticism, in particular, breathing exercises, restrictions on sexual intercourse and masturbation, a particular orbital mediation or circulation of energy in the meditating body, and the sheer pain of physical training.

The results are, quite simply, superhuman. Here, after all, we encounter the manifestation of "dark or hidden powers (*um ging*) of the body," like the "death touch" (*dim mak*) of mantis practitioners, the telepathic ability to read the opponent, external force transmission, and the inexplicable ability to become very heavy.[30] These are *techniques du corps*, technologies of the body, "inherent in our molecular animality but hidden by our molar [or molecular] human form."[31] Farrar asks a series of rhetorical questions of these powers hidden in the human: "Might practitioners acquire the strange capabilities of becoming-telepathic, becoming-invulnerable, or becoming-immovable through some bizarre animal or insect agency?"[32] The answer, we are led to conclude, is yes.

But it is probably the twentieth-century surrealist art movement that is the most immediate, and most significant, precedent for the mantis theme in the modern abduction experience. Ostensibly, this was because of the extraordinary interest the surrealist artists took in Freud and psychoanalysis and the unconscious relationships between eating, sexuality, and violence. In particular, the female mantis eating the head of the male after copulation was taken as a prime example of the *vagina dentata*, or the teethed vagina, that was said to be a common castration fantasy, or fear, of the human male within psychoanalytic circles at this time. This entomological observation (female decapitation of the male after coitus), it turns out, was not unique to the mantis in the insect world, and it was also exaggerated, since the gruesome postcoital act is largely a function of captivity. But the associations between the insect, eroticism, and beheading nevertheless stuck. The anthropomorphic, or humanlike, form of the mantis also played a major role. The mantis just looks, well, *human*. Not to mention pious. Hence the "praying" projection part.

The French writer Roger Caillois was one of the first commentators on the surrealist mantis in an important 1934 article in *Minotaure*, a key surrealist publication. His argument and, perhaps more importantly, the larger artistic practice around the insect in major surrealist figures are directly relevant to our present discussion:

Caillois claimed for the mantis a prestigious ancestry, surmising that it was probably the first insect to have appeared on earth. He described the prolific meanings associated with this creature in the folklore and myths of diverse

cultures in which it plays both a divine and diabolical role. . . . He also mentioned the fascination of his artistic contemporaries with the mantis. André
Breton and Paul Eluard were described as raising the insect. . . . Caillois's
discussion of contemporaries who were inspired as well as intrigued by the
mantis climaxed with Salvador Dali.[33]

Other major surrealist artists that used the praying mantis in their painted
visions include Pablo Picasso, André Masson, and Max Ernst.

These related themes of death and eroticism were in turn taken up by
the surrealist writer Georges Bataille (Masson's brother-in-law) in his highly
influential *Erotism: Death & Sensuality*.[34] Bataille's argument is that the human is a "discontinuous" being and seeks a kind of mystical "continuity"
with the cosmos in two ecstatic acts: death and orgasm. Such a mystical
continuity was in turn reflected in the surrealist mediation on the mantis.
The insect's mimetic ability to merge with the natural environment, to
mimic a flower, for example, in order to hunt its prey, was emphasized to
transform the deadly insect into a kind of surrealist symbol of total cosmic
communion, of the magical pantheistic fusion of the animate and inanimate
worlds.[35]

One could posit some kind of mantis experience, much like those recounted above with Stuart Davis, among the surrealist artists (I certainly
would), but such thoughts must remain speculative. The extraordinary
importance and presence of the mantis in surrealist art is nevertheless
striking, obviously relevant, and, I should add, mostly unknown to the contemporary abduction literature.

The Insectoid in Rock Music, Comedy, and Painting

Consider the fact that insectoid or mantis entities and rock music commonly coincide or play together. John Lennon of the Beatles famously saw
a small UFO just outside the penthouse apartment that he shared with
his personal secretary and then girlfriend, May Pang, on August 23, 1974,
at nine o'clock in the evening. John was naked, in a kind of poetic reverie,
and called Pang from the shower, "like Adam and Eve!" as he put it. The
object appeared to be trying to communicate with the naked couple.[36] Lennon makes a point of insisting that he was not on anything at the time. He
tried to photograph the craft, but the pictures came back "like they had
been through the radar at customs." Still, he was obviously stunned by the
shared event.

Lennon also reportedly described how "an extraterrestrial entity" gave him a particular egg-shaped object, which he then gave to the Israeli psychic and magician Uri Geller (to muddle the waters further, Geller describes obtaining his powers from what is essentially a UFO when he was a child in Tel Aviv, and he once appeared as a real-life superhero in Marvel Comics' *Daredevil* #133). The story, as most such stories are, is complicated. Lennon was asleep when a "blazing light" shone around his door. There were four people out in the hallway. Sort of. They weren't people. And they could use their minds in extraordinary ways.

> "Well, they didn't want my f—in' autograph," said John. "They were like, little. Bug-like. Big bug eyes and little bug mouths. . . . I was straight that night. I wasn't dreaming and I wasn't tripping. There were these creatures, like people but not like people. . . . They did something. But I don't know what it was. I tried to throw them out, but when I took a step towards them, they kind of pushed me back. I mean, they didn't touch me. It was like they just willed me. Pushed me with willpower and telepathy.[37]

When the musician woke up (the narrative is not at all clear, since Lennon states clearly that he was not dreaming), he had "this (egg-shaped) thing in my hands. They gave it to me." Here is the fantastic in physical form: Coleridge's dream-flower; Eliade's materializing rose. Whatever it was, Lennon gave the object to Geller: "Keep it, it's too weird for me. If it's my ticket to another planet, I don't want to go there." Geller had the sense holding it that Lennon knew more than he was saying.[38] Lennon was assassinated a few weeks later.

Stories like this just go on and on. In one related vein, Jerry Garcia of the Grateful Dead claimed to have had a two-day abduction experience on a futuristic spaceship involving insectoid beings, as well as an out-of-body experience and some kind of creative contact with a higher intelligence, which he acknowledges might be "another part of my mind."[39]

It would be easy, I suppose, to dismiss such accounts as exaggerated bits of promotion or, if you prefer, "drug-induced hallucinations" (that's always an easy one), except that, often enough, no drugs are involved, and we have endless comparative cases elsewhere. Stuart Davis, for example, reports that he has heard similar stories about "eggs" being presented to experiencers, often in childhood and without the slightest knowledge of the Lennon story.

Indeed, after sharing his own story online, Davis "heard from Mantis entity experiencers who live all over the planet." Some encountered such beings in entheogenic or psychedelic states (for example, after drinking the thick syrupy "tea" that is ayahuasca). Davis continues: "Almost no one who contacts me wants any kind of public exposure. They just want to share and be heard. I have even found in the literature another experiencer who encountered an eight-foot Mantid being in a purple robe, with a high collar. I mean, come on." He lists numerous people on his *Aliens & Artists* podcast who describe mantis encounters or suggest they may well have mantid DNA. Genetic hybridization is a relatively common theme.[40] For the historian of religions, such things are not entirely without precedent, as they echo and update in a new form the much older theme of divine-human hybridity or human deification.

Consider the life and work of the late Houston stand-up comic Bill Hicks. Hicks was a deeply influential American comedian who could get very metaphysical on his otherwise raunchy stage and just blurt out things, like how we are all the temporary imaginings of a single immortal consciousness. Hicks also, it turns out, had some extremely impactful UFO experiences in the Hill Country of Texas, outside of Austin, after taking five grams of dried magic mushrooms.[41]

Of course, they involved alien bug-like beings. As one biographer, Cynthia True, put it with respect to Hicks's experience with a friend, "It wasn't easy to describe what they had witnessed (neither reported physical contact) but Bill described it . . . as having seen insect beings inside balls of light." According to True, Hicks "found it hugely frustrating when people assumed his encounter was a joke." For him, "the encounter was absolute confirmation that we are not alone."[42]

Hicks is a glowing presence where I write, both in Houston, Texas, and at Rice University, whose live-oak-lined walkways he used to tread, high on mushrooms, dreaming about getting into the elite school. His grades would have prevented that admission, alas, but I nevertheless wish he would have been admitted. I also wish everyone would read Kevin Booth's chapter in *Bill Hicks: Agent of Evolution* on the comedian's psychedelic opening, which was actually a mushroomed ufological experience on a ranch near Fredericksburg, Texas (up in the Hill Country, not far from where I write these lines), on the Harmonic Convergence of 1987 no less. The planets had lined up, literally. So had the plants and the humans.

Booth begins the chapter with these two sentences:[43] "It was the most important event in Bill's life. It was the moment all the possibilities he believed in and had searched for became a reality." Booth and Hicks took five grams of dried mushrooms (enough, Booth comments, "to punch a hole through the fabric of space-time") and sat in the lotus posture near a pond to wait—for what, they were not quite sure. They also wandered around the pond as Hicks asked Booth to explain Einstein's theory of relativity.

The two then sat down again: "The next thing we knew, we opened our eyes and we shared this UFO experience." Booth walks the tightrope of the impossible thinker, acknowledging, of course, the strong hallucinatory potential of the mushrooms (including the "bad science-fiction movie" nature of the vision that they shared) but also wondering if these molecules might open some door in the mind to interact with actual entities and presences *through* the hallucinations—that is, whether the drug is a "key opening a door" that is always there, always waiting to be opened into "the next room."

The UFO ship itself was shaped like a conch shell, with glowing beings inside. Booth came out with "a firm belief that the barriers to time travel and communication were all inside your mind. Basically, anything was possible." Afterward, the two friends were both strongly telepathic (a most common result of ufological contact and psychedelic experience).[44] They had "a perfectly normal conversation without either one of us opening our mouth."

Booth continues: "But the spaceship, that was the most important thing that ever happened to Bill. He saw the 'source of light that exists in all of us.'" And then the evolutionary punchline, the reason Booth calls his friend an "Agent of Evolution" in the subtitle of his biography: "So when Bill talked about mushrooms and evolving and UFOs he wasn't joking. Okay, he found a way to get a laugh out of it—"That's why people in rural areas always see UFOs, they are always tripping on mushrooms that the cows just shit out of their butt"—but Bill wasn't kidding when he talked about it from the stage. . . . Bill believed that he had direct communication with another intelligent life form that was trying to show him what the future could be like."[45] Note the theme of an evolutionary esotericism.

But there is another point to be made here, and it is this: impossible phenomena can come with some very strong social, moral, and political consequences. Was this not why the comedy of Bill Hicks was so fiercely critical of everything from American patriotism and nuclear armament to the drug war? Here is the result of the Fredericksburg event, of what Hicks humorously describes as getting their third eye "squeegeed quite cleanly": "I realized our true nature is spirit, not body, that we are eternal beings, and God's love is unconditional. . . . In fact, the reality is we are one with God, and he loves us. Now if that isn't a hazard to this country. . . . What's gonna happen to the arms industry when we realize we're all one?"[46]

As many such experiences do, this one came with a very strong critique of religion. A friend of Booth and Hicks spoke of reading from the Bible and the Koran and brought along New Age crystals: "Oh my God. None of that matters. Let it go. Let it all go," Hicks exclaimed.[47]

For yet another clear instance of the alien insectoid, we might look at David Huggins as portrayed in Brad Abrahams's documentary film, *Love & Saucers* (2017).[48] What Huggins calls "Insect Being" is a large praying mantis alien who plays a fairly major role in his abduction memories and paintings, including those memories in which he has sex with an alien entity named Crescent, the large female lover of his contact experiences who, according to Huggins, was the mother of his many hybrid children.

As I explained to Abrahams for the film, it is a serious (and inhumane) mistake to dismiss and demean Huggins's memories and experiences. It is also a serious mistake to take them literally. We somehow need to move out of both reason and belief to orient ourselves here. For now, only a kind of real paranormal science fiction can begin to do these events some justice. Significantly, Huggins owns literally hundreds of sci-fi movies. This, after all, is the only genre that can do some modicum of justice to his experiences, and no doubt shape them at the same time. *Something* has to be as wild and bold to capture something, anything, of these events.

The Mantis in the Communion Letters

We might also recall what is probably the most well-known and most sophisticated account of an abduction experience, that of the science-fiction writer Whitley Strieber in his nonfiction *Communion* (1987). Much of the book is a series of transcripts of hypnosis sessions he underwent with Dr. Donald Klein to make sense of his abduction and bring it into consciousness (recall Théodore Flournoy from our last chapter). During the hypnosis session dated March 14, 1986, Dr. Klein asks, "The vision you had of this praying mantis thing—is that the same as the others you've seen?" An entranced Strieber answers: "They all look like that. Yeah. I thought at first it was like a skeleton on a motorcycle or something. It was flying—no, it wasn't flying. I could just see it, and I see it almost does, really does, look like a praying mantis, only bigger. It's got great big eyes that just scare the hell out of you."[49]

Mantis beings also occasionally appear in the letters of our Archives of the Impossible here at Rice University, particularly those in the Anne and Whitley Strieber Collection, an archived store of letters that were culled from about a quarter of a million that Whitley Strieber received after *Communion* was published. Whitley's wife, Anne, took it upon herself to read through all of these, notate a number of them, and organize the keepers. She also hired a secretary, Lori Barnes, to type up about eight hundred of them. Anne eventually crystallized the massive store of letters down to the five thousand or so that we now have in the archive after Whitley gifted them to us a few years ago.

Anne and Whitley also coedited a book about the letters Whitley received, *The Communion Letters*, with sections entitled "Knocking Down the Walls of the Soul," "Sexual Experiences," and "Visitors and the Dead." The last section opens with an obvious question: "Why would aliens appear in the context of ghosts and apparitions?" The couple offers an immediate and perfectly plausible answer: "Probably, because they are not aliens in the simple, conventional manner that is usually assumed. . . . Indeed, what if the visitors *are* the dead . . . disguised, perhaps, to avoid revealing their true identity to the living?"[50]

I cannot also help but remember in this context one of my favorite lines from Whitley's many books, this one about (or coming from) Anne, now deceased. Whitley describes Anne saying to him, "I'm not Anne anymore, I'm me. But I'll always be Anne for you, Whitley." In more metaphysical terms, she also told Whitley, "I am the part of me that's part of you."[51] Whitley still wears their wedding ring.

To call the *Communion* letters "rich" would be an understatement. Nicholas Collins points to one letter writer who describes the female mantis figure she saw as "wonderful," leaving her feeling "so loved and cherished . . . that everything was harmonious and good." This woman then relates the encounter to a more traditional Roman Catholic world, to "the Blessed Mother herself."[52] Believe it or not, this fusion of the extraterrestrial in and as the Marian apparition is not at all unusual. Marian apparitions have often been read as coded UFO events, at least since Paul Misraki's chapter on the Fátima apparitions in *Des signes dans le ciel: Les extraterrestres* in 1968.[53] This same comparative connection—Mary and the UFO—is indeed a staple in our Archives of the Impossible collection, present in dozens, if not hundreds, of magazines, newsletters, and commentaries.

The Mantis and the Modern Space Opera of DMT

Related to some Indigenous groups (this time in Latin America, mostly), the mantis and insectoid alien are perfectly obvious again in the contemporary psychedelic research and religious scenes, particularly around those altered states involving the intravenous injection or oral consumption of DMT (dimethyltryptamine), present in what is commonly known as ayahuasca in the Indigenous cultures of the Amazonian basin but also related to the psilocybin molecule, another tryptamine that occurs in magic mushrooms.[54]

The contemporary Brazilian religion of Santo Daime, which traces its roots back to the middle of the last century and a seven-foot-tall Black prophet by the name of Mestre Irineu, born Raimundo Irineu Serra (1890–1971), is positively filled with insectoid creatures and "clicking" or buzzing revelations, as well as UFOs. The "Daime," filled with the entheogenic DMT molecule, is its central and most sacred sacrament.

Within this same Latin American context, also consider the collection of essays that relate these inner altered states and the outer realities of science fiction, *Inner Paths to Outer Space*. The book opens thus:

> Perhaps, instead of searching for aliens in the sky, we should look much closer—inside ourselves, in those places from which we have been receiving messages of alien intelligence for millennia, though few of us have ever paid attention. . . . One notable exception: science-fiction writers who have traveled through these gateways in search of inspiration. There is more science-fiction literature, art, and film that may have been inspired by mind-expanding experiences than most of us suspect, from Lewis Carroll's *Alice in Wonderland* to Philip K. Dick's brilliant novels such as *Valis*, *Ubik*, and *The Divine Invasion*.[55]

The thesis, which is deeply resonant with my own writing (and that of Stuart Davis's *Aliens & Artists* podcast series), is a dramatic one. In the blunt terms of the four authors (three of them medical doctors): "DMT often induces experiences that belong to hard science fiction or space opera."[56] And, indeed, many such reports of DMT parallel strongly those of "alien abduction," leading the researchers to suspect an endogenous release of the molecule in the brains and nervous systems of those in whom such encounters take place. Rick Strassman goes as far as to call DMT "the brain's own psychedelic."[57] The molecule has been found in every mammal in which we have looked and in innumerable plant species. It is pretty much everywhere in the natural world and appears to play some central role not only in these altered states but probably in the construction of the consensus state we call "reality" as well.

Perhaps it should not surprise us that the subject reports of people taking DMT are filled with insectoid presences and mantis-like entities. One

subject describes how an "insectlike thing got right into [his] face," suck-
ing him out of his head "into outer space . . . a black sky with millions of
stars."[58] Another subject, this time a carpenter and singer-songwriter named
Rex, described "insect creatures all around [him]." These were "insectoids"
who fed on his heart and were most interested in the complex energies of
his emotions. In a pattern that is familiar to the historian of comparative
mystical literature, Rex made love to the insectoids as they ate him, with the
thought that "they were manipulating [his] DNA, changing its structure."
Rex then found himself in "an infinite hive" where "there were insectlike
intelligences everywhere, in a hypertechnological space. . . . It wasn't at all
humanoid. It wasn't a bee, but it seemed like one. It was showing me around
the hive. It was extremely friendly and I felt a warm, sensual energy radiating
throughout the hive. It said to me that [that] this is where our future lay."[59]
Note the clear evolutionary and entomological themes that are so clearly
encoded into the experience. Frederic Myers and Joseph Maxwell all over.

Perhaps the takeaway is what Sara, a mother and freelance science
writer, learned—namely, that "humans exist at many levels."[60] Rick Strassman certainly struggled with the reality of these experiences, finally concluding with his test subject Jeremiah that "this is not a metaphor." In other
words, the encounters, abductions, and even anal rapes were quite real and
could not be reduced to brain chemistry; rather, they were happening on
another level or in another dimension of reality that was not this material
world known to the sciences and their sense-based knowledge. In the end,
Strassman adopts what I would call a filter or reduction thesis of the brain-
mind relationship, where the "wetware of the brain is modified, leading to
the opening of other realms."[61] In short, specific molecules can turn the
brain into a kind of door or receiver station.

A forty-five-year-old American mechanical engineer, for example, reported an abduction that did not take place on the physical level but on
some other. There he encountered beings, some of which he described as
reptilian, some as octopus-like, and some as "tall, skinny, kind of lanky,
very thin-featured but with a praying-mantis head, triangular and very bug-
like." He could also hear the "hum of a ship, structurally strong yet hollow-
sounding, vibrating with energy."[62] Another subject reported "A True Hallucination by Mantid" that culminated in "a metallic buzzing sound" like
"the sound of a cicada, but with many other elements added." This subject
felt "as a bug making the sound" and "had an intuitive understanding of
metamorphosis."[63]

Gnostic bugs and humanoid mantids are pretty much everywhere in
these DMT states, then, including in the art that such states produce or

inspire. The science-fiction artist Stephen Hickman illustrated the cover of *Inner Paths to Outer Space*, which also includes a discussion of futuristic and flying saucer themes in the ayahuasca art of Pablo Amaringo Shuña and some digital paintings by coauthor Slawek Wojtowicz on the insectoid and mantis themes, including one called "Mantids Rule!"[64] According to the artist Amaringo Shuña, "They are not machines. . . . They are spirits, and come from other dimensions."[65] Hickman himself stresses the "psychic component of the evolutionary process," as distinct from the physical or biological aspect, "which has gone quite far enough."[66]

A related linkage of the extraterrestrial and the insectoid-psychedelic was certainly behind my own participation in a four-day symposium on the DMT molecule and entity encounters at Tyringham Hall in England in 2017, which was led by the entrepreneur and English visionary Anton Bilton.[67] The organizers contacted me with the specific hypothesis stated above—namely, that abduction experiences are correlated with endogenous DMT release in the human body and brain. More specifically, they wanted me to ask Whitley Strieber to attend the symposium (they knew I knew him because of a shared book Strieber and I had just published). I knew that Strieber had never taken such a psychedelic (and he won't—his life is crazy enough, as he jokes), so I was a bit skeptical of the request, or at least surprised by it. I was in no way dubious of the thesis, however. They explained again the obvious parallels between the abduction experience and the DMT state. I did not need to be convinced.

I should finally add that "buzzing" and entomology have been associated with the flying saucer phenomenon at least since 1951 (there was no "UFO" yet). It was that year that Gerald Heard (an author who also wrote extensively of his own evolutionary esotericism and helped inspire the founding of the Esalen Institute) speculated that the crew of the flying saucers were not only highly evolved but also bee- or insect-like. Heard's discussion of this insectoid theme is quite extensive, as is his use of the adjective "super" to capture something of what he is trying to say.[68]

Why the Aliens Don't Land

I want to go elsewhere now, where I might have something more specialized to say. Specifically, I want to go back to my own training and ask what all of this has to do with comparative mystical theology. I want to ask what the dead, spinning disks, and big- or bug-eyed insectoid aliens have to do with Meister Eckhart, Plotinus, Shankara, Dōgen, Nāgārjuna, and Ibn 'Arabī. On another level, much of the modern paranormal material, and *especially* the ufological material, looks a good deal to me like what a traditional theologian might call angelology or demonology (the Christian Evangelicals are not wrong to see the latter, although their too common paranoia and quick moral judgment with respect to these other spiritual intelligences are not helpful). In these instances, it is not about the Godhead or the true nature of reality. It is about invisible entities and spectral beings, some of them noble and beneficent, some of them misbehaving *very* badly. Stuart Davis is correct: we would get arrested for this kind of behavior. So how might we contextualize all of this within a bigger theology of the Godhead or a philosophy of ultimate reality? Can we?

The experiencers themselves certainly care about this question. A figure like Stuart Davis certainly contextualizes the most fantastic aspects of his encounters in a much larger Buddhist cosmology and philosophy. Dōgen and Nāgārjuna are in there. It is impossible to listen to Davis for long and not come away with the distinct impression that he is making a claim on the nature of reality itself, that his entomologically imaginal experiences possess a kind of spiritual intentionality—the insectoids want to change the very texture and reach of our experience of cosmic reality, much as they have changed Stuart Davis.

But other contemporary experiencers have also been explicit in their mystical theologies, philosophies, and intentions. Consider, as a second chapter exemplar, Jeremy Vaeni, my friend and teacher.[69] I read him in a long line of nondual and idealist visionaries, none of whom, I must add, sound *anything* like Jeremy Vaeni. He is both an alien abductee and a kundalini experiencer, two points along the anomalous spectrum that we will encounter again together in our next chapter. Among his many artistic endeavors (from comedic writer to conference organizer to podcaster), Vaeni has written four books on the abduction phenomenon and a fifth book, a novel, about the same.[70] He has made a documentary about his own abduction experience.[71] He has even written a metaphysical children's book about a big toe coming to realize that it is part of a larger loving body.[72] He has also gotten kicked off a major paranormal radio show (that somehow does not surprise me).

It is not possible to summarize all that Jeremy Vaeni has written or spoken (or joked) about in so many genres. There are five motifs I most associate with him: (1) human deification (Vaeni is God, or better, God has manifested through him); (2) bizarre energy effects in the spine, brain, and physical environment, which are somehow connected to human sexuality; (3) apparent alien abduction, which is somehow connected to human sexuality; (4) humor, which is definitely connected to human sexuality (Vaeni thinks I am obsessed with sex—this is what we call projection);

and, easy to miss but obvious when one looks carefully, (5) the final death or absorption of literary characters in his raucous books.

In terms of the latter theme, Vaeni enacts as the page what he is saying on the page (by having some of his central characters disappear at the end): that the ego does not survive, that it is an illusion, and that all personal forms, including and especially us, are basically actors in some gigantic cultural story, which we would do well not to believe for both moral and philosophical reasons.

Not to believe. The thing about Jeremy Vaeni is that he can and will laugh at just about anything, even things you definitely do not want him to laugh about. Which might mean that he is truly trustworthy. But it might also mean that his humor is functioning sometimes as a kind of defense mechanism (to keep him from taking seriously what happened to him— hey, it is not so easy being God, especially when you are writing an email to your mom about it, which he has also done, or tried to do). But Vaeni is perfectly aware of the defense mechanism thingy, and he often says the same. No one seems to take him seriously. Jeremy Vaeni does not even take Jeremy Vaeni seriously.

But I do. Because he makes me laugh *and then think again about that laughter*. Laughter is not just a defense mechanism here. It is also what he is trying to get us to see, or, rather, to be. I have sincerely called Jeremy Vaeni my teacher because he makes me think that maybe all those previ- ous deified human beings in the history of religions took themselves *way* too seriously, did not have to die like that, and certainly didn't have to be tortured by their unfunny, much too serious contemporaries and then get gory theologies wrapped around their basic unfunniness.

The history of religions is really sort of morbid if you think about it.

Jeremy Vaeni is also my teacher because he knows that what he is say- ing ("All is God") is so unbelievably serious that he has to hide behind the laughter to endure it and then to say it in a way that also unsays it. This "it," of course, will largely escape most of Vaeni's readers ("the five of us left who read," anyway, as he puts it in one of his gazillion one-liners). It, after all, is not an it. It is more of a No-Thing, a Oneness, a Nothing, a Mys- tery that includes everything in order to transcend everything, an uncanny Other that is also oneself, or Oneself, a oneness of multidimensional being, as what Vaeni calls the "Mystery Voice" puts it toward the end of *I Am to Tell You This and I Am to Tell You It Is Fiction*. The laughter subsides.

Still, the mystical punchline is real, and it is very consistent in his books. Indeed, Vaeni's first book encodes the punchline in the very title: *I Know*

Why the Aliens Don't Land. Why don't they land? Because there are no aliens, not at least any that are ultimately different from us. "ALL is ALL," as he puts it in so many places. As for the ego that wants to separate everything and delights in difference, it is little more than insecurity and fear, really a kind of existential terror wrapped into a psychological ball: "Here's what you need to know right now. The ego is an unstable, insecure illusion. It sees itself as an individual and will do anything to protect that individuality. It takes a look around the room and sees what appears to be individuality. Other bodies with other thoughts and different skin colors and sexualities. It mistakes this for fragmentation. . . . Fear sets in, which naturally leads to aggression. . . . Fear is you. You are fear—you are that, every bit of it."[73]

And this includes the religious ego. Vaeni has a fairly distinct theory of religion. And it is not pretty. All the roads (of religion) do not lead to God. They cannot. All religions are human constructions and social functions, "thousands of years of crapola, friend." Religion is basically us protecting ourselves from us. And so Vaeni speaks out:

> Do you get how when the little self sees the big Self, it flees? It erects gods and churches and rules. It worships. What is worship? Worship is a great wall created to divide you from YOU. Words, rituals, rites. . . . That's all trash, all means of denying our Whole Self. All is All. That necessarily includes the little self—the ego—and its traps, its illusions. This is why you can live ignorantly-ever-after in the shallow end and be every bit as worthwhile as those yucking it up in the deep: it's all one ocean. . . . Don't look to Mommy, Daddy, Priest, Jeremy, or Carl Sagan's corpse. The kingdom is found within, not without. This is why the aliens don't land.[74]

There you go.

There is more, though. Popular culture commonly considers such aliens to be "superior beings." They communicate telepathically. Okay. They zip around in ways that defy what we know about gravity, space, and time. So? Still, it is all one ocean:

> **Universal** Mind. There is no intellect in this universe or the next that can tell me anything about spirit—about my Original Face—that I do not already know NOW, should I awaken to it. Moreover, because this insight is by definition inner sight, there is nothing they could flat out tell me that would wake me up. . . . they don't land because landing won't make a bit of difference![75]

A difference in what? The coded reference to Zen Buddhism answers the question: our awakening into the true nature of reality and our own cosmic condition.

So how do we know the "aliens" are worthy of our spiritual and intellectual respect, that, in the pop-speak of our present-day secular culture, they are "enlightened"? "Precisely because they don't land. Look to the literature. When asked where they are from, the answer is sometimes a specific star system or a planet. Often, they say they are from 'Everywhere' or 'Nowhere' or 'Around.' None of this conflicts when you are awake to One Ocean."[76] Such "aliens" are "midwives birthing us out of our egos."[77] Or better, "They want us to wake up out of our egos so we can join the sentience of the cosmos (or Kosmos, if you like). It's not that they refuse to give us this, it's that they can't. All they can do is point it out in a way that bypasses ego. They can bypass it, but we have to transcend it."[78]

Bypassing the ego. That is a pretty good description of what is at work in the unbelievable beliefs of the abduction experience, the sci-fi visions born out of our own fears, projections, and warlike natures. There is no alien invasion. There is an attempt, largely a missed opportunity, to tell us that we are not, and that our violent nation-states and horrible histories are not really worth preserving—that those histories are there for *mourning*, not celebrating, that we need to go back to *remember* and *redeem* the past, not to deny, preserve, or conserve it.

Vaeni is fierce here. He writes of the "sheeple" who dominate the UFO subculture, the bigoted racist who is Donald Trump, and the simple truth that if you voted for this man, you are "a selfish racist, sexist, homophobic destroyer of asylum-seeking families, American society, and Mamma Earth."[79]

None of this is irrelevant. It speaks directly to the mythological features of the contact and abduction phenomena. The early contactee movement of the 1950s was often about what Vaeni calls "Nordic supermodels": blond, blue-eyed, buxom women—basically "Hitler's wet dream."[80] In the second round, now more or less equated with the abduction phenomenon that was now "baked in hypnosis" (Flournoy again), we wove conspiracy theories as the aliens "treated us like lab rats, forcibly taking our sperm and ova to create alien-human hybrids." This was the story in which "the intergalactic praying mantises were space relatives after all."[81] In the third round of storytelling, we no longer believe the conscious tale *or* the unconscious one. Now we believe they are somehow us, perhaps an us visiting from the future.

All three stories are still expressions of "our racist colonial conquests" for Vaeni. Hence the term "alien," which is what "we also use to describe brown people who want to move to America from other countries we hate because FOX News told us to because they're selling our racism to us."[82] The "Gray" (a displaced mixture of "black" and "white") for Vaeni is another obvious racist trope. Here are some typical lines: "In the end, Grays were here to emasculate White men, impregnate White women, and steal from us. We abductees were like slaves taken by foreign colonizers. This is what 'higher beings' look like when filtered through the inherently racist, colonialist lens of Western mind. They look like how we victimize. They look like enemies."[83]

Vaeni also struggles with the obvious violence and the utter lack of consent in the phenomenon. I mean, it is called an *abduction* for a reason. But there is always this "out," this loophole the abductee and the culture as a whole can jump through to avoid being some kind of prey. That out, or cop-out, is this: the abduction experience is madness, not real . . . it never happened.

Well, it *does* sound crazy, but it *did* happen. Vaeni goes back and forth here within a paradoxical intention, denying that any of it is real but also writing for pages about the phenomenology of his own contact and abduction experiences, which included a shared sighting with his mother of a massive green, port-holed, double-spinning UFO hovering over the family car somewhere between Massachusetts and Vermont;[84] a spectral, red-colored and clawed lizard man or "Gargoyle demon dude" possessing him for a time within what is basically a superimposition over his own body;[85] the splitting open of his spine or back via an "invisible slit" that resulted in utter bliss pouring into his backside and a later bodily practice of flowing movements that he attributes to an "invisible ecology" and the original intentions of those who first invented the movements;[86] and a full analysis whose conclusion is that we interpret these as violent terrifying abductions mostly because they are taking away our egoic sense of reality and showing us that the laws of physics are not really laws "but actually mere suggestions."[87] In truth, or in Truth, "you are a thought construct maneuvering in the world from a basic fear of annihilation."[88] The whole point is annihilation or abolishment, in the liberating sense of those expressions. We've got it all wrong, including, by the way, what the future already is.[89]

Vaeni will thus write lines like these: "I'm talking about aliens here. The real deal. The ones who land and the ones who steal us in the night."[90] But he will also point out that the perceptions of the abductees are almost certainly being manipulated and controlled: "By all accounts these beings are telepathic and can make you see what they want you to see."[91] Such sentences, of course, imply that the telepathy and mind manipulation—the psychic powers—are quite real but are also being employed to trick us.[92] This is a pretense, a ruse, but a pretense and a ruse of something very real.

In another current, Vaeni will also deny that there are any such aliens and write eloquently of these visionary forms coming to us to teach the Truth with a capital *T*, the Truth of Timelessness appearing in time.[93] This is why their thoughts are not our thoughts, why their lives are not our lives, why they play by entirely different rules. How could they not? They are coming into space and time from outside space and time. The result is a kind of overwhelming paradox. Here he is on the doubled truth, what I have called the Human as Two, or what Vaeni prefers to call a "multidimensional being":

> What about our timeless, nondual aspect? Not even *aspect*—that doesn't describe it. How about the very sense of being that transcends and includes time, transcends and includes us? Can we *be* that? Not visit it as if it's another state of consciousness we slip into for a bit and then come back to normal and bore people with our excited story. Not have flashes of "my" truth. Can we puny humans forever be the 1st-person self-consciousness of timeless, nondual nowness?
>
> If we can, what becomes of the human?
>
> Anyone here find it the least bit peculiar that we have an easier time proving aliens are here over any other hypothesis than we do proving what a human is?
>
> Perhaps aliens are here. But humans haven't arrived yet.[94]

Vaeni calls on words like "transrational," "translogical," and "transpsychological" to name this experiential paradox—how people like himself make sense, and do not make sense, of what happened to them; how Truth is not outside us, not an object or thing or anything that can be languaged or conceptualized but might nevertheless communicate with us as egos through unbelievable truths, images, and nutty stories; and why the mental illness explanations ("you're crazy" or "this is a hallucination") simply do not work to explain their experiences and are often just empty rhetoric.[95]

In the end, Vaeni's thought is profoundly hermeneutical, by which I mean that he understands that meaning is actually there in life but that we normally block it out with our crap, including our intellectual crap.[96] Part of this crap is that we are living in stories we have told—that we create—and these are some pretty nasty stories. As he has long suspected (originally inspired by Whitley Strieber, who has argued much the same), Vaeni thinks that we speak and imagine the aliens into existence. We let them in, as it were.

That is about as hermeneutical as it gets: they are real, and they are here because we believe they are here. And then it gets more so. We speak them into existence, *so that they can speak us out of existence*. We are not living in a horror movie with the "aliens" as the monsters. These so-called aliens are showing up in the horror movie that we ourselves have made to get us out of our own stupid movie.[97] We believe them so as not to believe them, or ourselves.

Here is Vaeni again, in a more personal voice still, but also with the fear-projection model well in place: "It's no overstatement to say that I *am* science fiction: one part tentacle monster, three parts futurist, all too human. . . . The only fact we may discern about abductions is that something is happening that we cannot experience directly because we filter it through our own fiction, composed of stories, assumptions, and visualizations conjured in fear."[98]

Accordingly, Vaeni invokes the theme song of *2001: A Space Odyssey*, with the implication that the movie basically had it right: we are not destined for "organhood" but for the "multiverse."[99] Vaeni is joking, again, but perhaps it is significant that the theme song for the film is thoroughly Nietzschean. Indeed, it is Richard Strauss's tone poem "Also Sprach Zarathustra," or "Thus Spoke Zarathustra."

Well, so did Jeremy Vaeni.

In his most revealing moments, Vaeni will speak even more directly and admit that he is coming from a place that most cannot—a place of doubleness or inclusion. "My understanding is different and is inclusive of yours, meaning that I can understand exactly what you're saying and why you're saying it. You on the other hand, are circumscribed to thinking I'm nuts or lying."[100]

Can we come to terms *with that*? That paradox? I am to tell you this, and I am to tell you it is fiction. I suppose that makes many people feel better. Still, what Jeremy Vaeni is really saying constitutes the true end of the (Western) world as many have come to know it, the true overcoming of the (Western) human as so many have constructed it over the centuries in Europe and its former slave-owning colonies, including and especially the United States. This is "the human," a term that was not granted to the enslaved peoples and the Indigenous "savages" so that they could be enslaved and savaged, killed as nonhuman or subhuman. This is Western civilization as a colonizing, enslaving, polluting, bombing, nature-destroying mental illness and narcissistic ego, a "sick, separative mind."[101] This is the terrible truth so many cannot face but some of us are fated to be abducted and terrified by.

Yes, there is also a superhuman side, a mystical unity beyond the contact, the abduction, and the coming to terms with Western civilization. Vaeni will also joke about "that one time I saw and was the universe exploding into existence from nothingness and Spirit rushing through the entirety of creation in an unstoppable burst of joy. I try not to brag."[102] This was his I AM experience, his becoming God or "ultimate consciousness," with invisible presences in the room, no less.[103]

That is why the aliens don't land. That, after all, would simply confirm our illusory separateness. It would tell us what is not so.

The Mechanical Insectoid Encounter of Karin Austin

It is not just the men. Women have these insectoid encounters too. And these sometimes function as a most powerful challenge to the spiritual or religious reading of the experience, which I have engaged in above. Indeed, I worry about this. I worry that the spiritual dimensions might overtake or even erase the physical dimensions in the mind of the reader. I do not wish this to be so. Indeed, my dual-aspect monism—which denies any final distinction between the "mental" and the "material"—implies that these two domains cannot be separated, since they share a fundamental ground or superreality that is both and neither.

The encounters of Karin Austin are most instructive here. Austin was the personal assistant of the Harvard psychiatrist and abduction researcher John E. Mack in the last years of his life. It was Austin who drove Mack to the airport on his way to London, a trip from which he would not return alive (he was struck and killed by a drunk driver on September 27, 2004). Austin was also one of the experiencers about whom Mack wrote with such conviction in his books. Indeed, Mack includes in *Passport to the Cosmos* Austin's drawing of what Austin calls "the fucking bug" (she likes to swear). I used the drawing, not knowing whose it was (Mack only cites "Karin"), in *Mutants and Mystics* to make a case that the iconography of Spider-Man displays an eerie familiarity with the contact experience.[104]

Karin Austin comes from what she calls a "very straightforward materialist background." She was raised in Florida in a Southern Baptist household (in the early 1990s she would listen to Rush Limbaugh). There was no place in her life for what she calls "California woo" or "New Age babble."

Still, before she even had language, friendly gray beings with big black eyes floated through a window into her childhood bedroom one night. She called out to her sleeping mother to alert her to the "clowns" in her room. The beings were gone by the time her mother arrived. And she, like most mothers, assured the little one that she was just dreaming. Later, Austin would feel a surprising connection to the childlike alien beings in the final landing scene in *Close Encounters of the Third Kind* (1977).

The childhood encounters portended more, as they often do. Around age 25, in 1993, Austin became increasingly anxious, "uncomfortable in my own skin," as she puts it. Something was building. The future was reaching back. Then the future happened:

> Suddenly, in the middle of the night, I came into full consciousness—as if from anesthesia—to find myself standing in the middle of some trees just inside the curved edge of a forest clearing. As I looked around, I could see about a hundred other people spread out in the woods along with me. About half of them appeared to be dazed human beings in various stages of dress and undress, some wearing pajamas. . . . It was a full-moon night, so the clearing was very well lit. In it, I could see activity and flashes of moving figures. As I struggled to process what was happening, a 5'2"-ish woman who seemed like a nurse walked toward me with a shorter individual beside her. When she stopped in front of me, I glanced down and noticed the smaller person was a child about ten years old. She silently and gently pushed him toward me. Looking into her face to understand what she was asking of me, I realized she was not entirely human. Which was terrifying. Without acknowledging my fear, she nudged the child in my direction again to suggest I had some relationship to it. I was deeply confused. I was not a mother then (or now), and I didn't have any nieces or nephews. As I looked down at the child to consider our connection, I flushed with panic and distress as I realized he wasn't entirely human either. I recognized familiar human features in his face (and hers), but the eyes were much bigger than normal, and his hair was white-blond and wispy like a cancer patient's. He was listless and seemed emotionless, as if autistic.[105] My stomach churned. And I'm embarrassed to say that I was disgusted. And angry. . . . With my thoughts scrambling, the non-human "nurse" tried one more time to help me understand. This time, there was an intrusion into my consciousness. She was saying, "Yours."[106]

Austin notes that in the moonlit night she had seen in the distance an old wooden rollercoaster that seemed to be nestled in an abandoned amusement park. Shortly after her interaction with the nurse and child, she was led to a curved, highly polished, sliver-metallic vehicle that reminded her of an aluminum Airstream travel trailer. As she stepped into it, she blacked out. She came to in an instant to find herself back in her bedroom, as if there had been some kind of film splicing or edit (the film metaphor seems significant). Suddenly, she was awake. Austin rushed into the bathroom and checked her body all over, aware her behavior was odd. What was she looking for? She also drank about a gallon of milk. She was *so* dehydrated.

Austin and her fiancé at the time decided to move back to Florida. While temporarily staying at her mother's house, she slept on a blow-up air mattress in the living room for a few weeks. As she tucked herself in on her first night there,

the room suddenly flooded with blue light that streamed through vertical blinds hanging in front of a sliding glass door. My body started to vibrate, as if it were being electrocuted (though it didn't hurt). I immediately thought I should get up to investigate the light as the apartment was facing an interior, common garden. There was no road there. No way the blue light of a police car could be its source. But I instantly passed out again. I awoke a second time to my body vibrating, unable to move. This time I felt myself lifted out of bed as I was floated out of the room and into a very exotic space where I encountered a particular gray being with big liquid-black wraparound eyes who telepathically connected with me (it felt as if he merged his consciousness with mine). Impossibly, I recognized him with some part of myself I did not know. It was as if we had known each other for millennia. He was more real to me, and more my family, than my biological family had ever been. He was a soulmate beyond any conception of the term that I had ever imagined. Within a moment, an almost unbearable, overwhelming wave of love traveled through my body in the recognition of our reconnection. And just as I was taking in the fullness of this transcendent encounter, in the very next instant—again, as if a film had been spliced—I was awake. It was morning as if there had been no night. Yet, images of the gray being lingered with me. I considered whether or not I should tell my family about what happened. I decided it best if I keep my mouth shut.

It was a secret.

The young couple soon moved into an apartment. Then, all holy hell broke loose for about six months. "All manner of psi phenomena started happening to me: telepathic communications with people and the beings, out-of-body experiences, the sudden ability to 'hear' dead people, channeling and automatic writing about a coming earth emergency, an almost religious concern about the wellbeing of all life on the planet, art that I created compulsively." As Austin emphasized: "None of these things were familiar to me or supported by my worldview. It was like the mashed potatoes scene in *Close Encounters*. My fiancé thought I had lost my fucking mind."

Austin had become an insomniac. She could not sleep since she was worried that she was going to be awakened "in that vulnerable position again—with the electrical current running through [her], a helpless specimen of the beings." The events were "smacking into [her] worldview." Physically, emotionally, and psychologically worn down, she reached the end of her tolerance for the ambiguousness of her encounters. So, she spoke: "If this is really real in the way I understand the meaning of that term, you guys need to show yourself in a way that makes sense to me." They did a few nights later:

> I felt a tickle on the back of my head, twice. It felt like a bony finger. I was instantly awake and adrenalized. But *again*, nonsensically, I immediately fell back asleep.[107] What seemed about two minutes later, I awoke to the electrical current coursing through my body again. I was not able to open my eyes. Lying there hyperventilating, I felt this very real and careful depression of weight on the end of our bed, the sensation was similar to that first moment when a cat lands on a bed. . . . I felt another very deliberate step, this time on the left side of my feet, which were straight out in front of me while I was on my back. Then, after a moment's pause, there was another step forward on my other side, this time it stopped at my calf. I could now feel this thing was about twenty-five pounds, as its weight depressed into the duvet. It remained still. It was clearly waiting to make sure I wasn't going to move. It was about to commit to a course of action that would put it directly in danger had I been able to regain control of my body and throw it against the wall. When it started forward again, it continued on its slow, deliberate path in the direction of my face. One step. Then another, and another, and another. Finally, it stopped at my chest and rested one knee on top of ribs. It placed its hands on either side of my head, about chin level. At this point, I am losing my shit. I'm hyperventilating so hard I think I might die. But bizarrely, I hear myself think, "Holy shit, one of these things is actually in my room, right here on top of me . . . and *I have morning breath!*" No sooner did this embarrassment flush through me when I felt its consciousness swoosh into my own with the thought, "This is of no concern." That did little to abate my hyperventilation or the very raw, animal fear that I was in at that moment.

Just when I thought I might pass out from the overwhelm of the encounter, another thought whooshed into my mind, something to the effect of, "I mean you no harm." The vibration of its words was instantly soothing and much to my surprise, I felt a warm, calming wave of love pass through my body. Finally quiet, I received another telepathic thought, "Good. Now you can open your eyes." Expecting to see the same grey being I had previously encountered, I was utterly shocked when I did open my eyes. The very first thought that burst into my mind was, "Geezus, you look like a fucking bug!" It was about eight inches from my face, this insect-looking thing with a too-skinny neck and spindly, segmented arms straddling my head. Before I could think about how to react, "the bug" initiated a rushing "data packet" download of colors, shapes, images, and information into my mind. They came in so fast I thought I was going to vomit. Precisely at the moment when I couldn't take it any longer, he stopped and then activated the same process in the opposite direction, as it vacuumed up the contents of my memories and life experiences.

What Austin calls the "bottom line" is that she woke up the next day and knew, beyond a shadow of any doubt, that "whatever is happening, has a very physical component." More to the hyperdimensional point: "It has these other ways of showing up in our 3D reality. Somehow, probably technologically, the beings can shift into what feels like a higher vibration (for lack of a better way to say it) to the place where they generally exist. At the time, what mattered most to me was that I understood I wasn't crazy, that the phenomenon does, at least occasionally, manifest *physically* in our world."

Austin has her hunches. She thinks it is entirely possible that the gray beings are some iterations of "human" that reached an evolutionary, biological dead end in our very distant future (thus the interest in, and compatibility with, human DNA), and that the bug was biomechanical, a kind of safety device for those future humans who were intervening in this world of ours. Don't mess with the violent monkeys, not at least in person. They will throw you against the wall you just came through.

And there is more. Austin ended her story with a very physical "hook." Five years after these events, around 1998, she was walking down a hallway toward John Mack's office. As she stepped in, she saw a book manuscript written by another experiencer open to a page with drawings on it. As she looked down to view the pictures, she was stunned to see a rendering of the exact moonlit rollercoaster venue of her original event. Each of the main details she remembered from her middle-of-the-night excursion into the woods were in the illustration. Another person, a perfect stranger, had been there and clearly had the same or a similar experience. In 2020, it occurred to her that Google Earth could help her find the actual, physical place if it still existed. It turns out that the theme park with the campground, the opening in the trees, and the old rollercoaster is in Connecticut. It was temporarily closed down between 1992 and early 1994, so during her experience in 1993. She concludes: "So, if we want to say this is just happening in our minds, I am going to say, 'Fuck no it isn't.' If anything, these encounters have taught me that we do not understand the true nature of reality. Most likely, and perhaps most profoundly, so many things about how the phenomenon operates suggests that consciousness is fundamental and gives rise to the material world, not the other way around."

The drawing of the rollercoaster and campground in the trees is a contemporary equivalent of Coleridge's dream-flower that becomes physical after the dream, which was no simple dream. Crucially, Austin points out that the details related above are from explicit memories that did not require hypnosis to recall. The imagined is the real. The contact experience is physical: indeed, it is superphysical in hyperdimensional ways that we cannot imagine with the three dimensions in which our brains and cognitive capacities have evolved and to which we have culturally adapted.

Maybe it is time to get out of that.

Alien Gnosis

I want to tie together these first two chapters and conclude this one (it won't conclude, really) by asking a single, not so simple question: What are we to make of this apparent phenomenological link between Frederic Myers's entomological imaginal and his coinage of telepathy and today's entomological abduction, contact, and psychoactive events involving large telepathic insectoids and mantis-like creatures?

I see four basic options.

Option 1 is fascinating but, as far as I can tell, without any smoking gun. This option comes down to the speculation—and it really is just that—that Frederic Myers experienced (maybe through an available psychoactive substance, like opium, maybe through an endogenous release of DMT[108]) some kind of telepathic insectoid-shaped spectral event and so came to coin the imaginal after what he knew of entomology and his own unspoken experience. In this suggestion, when he wrote of the wormy "larval" and strange "imaginal" maturity of the insect, or told his well-known parable about the caterpillar munching on the leaves and only vaguely recognizing

itself in a butterfly that lands nearby (a parable for him about our supernormal capacities and how we barely recognize that these are signs of our
future evolution), he knew personally of what he was writing. Myers, in
this model at least, linked his three new words so tightly—supernormal,
telepathy, imaginal—because this is what he knew in his own direct but
private experience.

Maybe. Maybe not.

A weaker form of the same option, let us call it option 2, would be to
suggest that Myers saw this pattern in his ethnographic cases and textual
materials in some unspecified but nevertheless consistent way. Myers basically intuited the link between the imaginal and the insectoid in his thought
and writing, largely unconsciously or, as he would have said, subliminally,
below the threshold of personal awareness.

Option 3 argues that the connection is simply accidental, a kind of historical fluke that is simply a function of my own overreaching mind. In this
model, Myers just happened to pick an entomological term that just happens to be connected, visually and conceptually, to the abduction, contact,
Indigenous, and psychedelic cases, and I am now inappropriately relating them because these just happen to be literatures about which I know
something.

Option 4 counters with a strong physicalist challenge. It is most powerfully represented in this chapter by Karin Austin. It says (or swears)
simply: "These things are physically real, and they are entering our three-
dimensional reality in ways that we do not understand, and maybe cannot
understand at our present evolutionary bottleneck." But they are very real,
more real than real in fact. Bug-like beings keep appearing in our folklore,
mythology, and scholarship because there are bug-like beings. "Get the fuck
over it," Austin would say (and I would laugh).

I honestly do not know what to think for certain, but the only option
I think is simply mistaken is option 3. As much as someone might want it
to be so, these comparative reflections are not accidents, are not flukes. I
also want to affirm Jeremy Vaeni's strong postcolonial and racial readings
of much of the abduction and UFO material. And I end where Vaeni ends—
with a nonduality in which we share, or are, the Other, in which the aliens
do not land because they are finally us. This is why I have long insisted
that Vaeni is my teacher. He finds that remarkable, but I mean it. It's the
nonduality and the humor, *together*.

I do think these apparitional forms are often very deep, unconscious
forms of cultural and historical memory of the Atlantic slave trade and

Euro-American imperialism. That is why Black people and Indigenous Peoples have *very* different responses to those disks in the sky. But I also recognize that much of the general history of religions has been defined by a depressingly similar "servant spirituality," to use the phrase of my colleague, the historian of early Christianity April DeConick.[109] To be blunt about it, this servant spirituality, which is oh so biblical (with God as a king or lord in the heavens), has also been a *slave mentality*, as someone like Friedrich Nietzsche saw so keenly. I happen to think DeConick and Nietzsche are simply correct.

In my own developing thought, I probably lean further to a model wherein the Victorian typology emerges from the thousands of case studies that Myers and his colleagues studied so carefully, and where there was something there in those sources—maybe not this specific or conscious—that led Myers to coin, define, and then narrativize the imaginal the way he did, much as we have it in option 2. Perhaps, for example, a psychoactive release was entirely endogenous—that is, internal to the traumatized or dying bodies of Myers's subjects. This nineteenth-century ocean of the human experience of the dead or dying then flows into the ocean of the near-death, extraterrestrial, and psychedelic literatures of the twentieth- and now twenty-first centuries, and these waters mix. In the end, it is all one thing, one swirling ocean. This explains the patterns we see and can easily identify.

In short, I am suggesting a shared but largely unconscious experience-source hypothesis, what DeConick might call a "gnostic spirituality." The different sets of cultural materials resonate so strongly because they are all based on the same kinds of human experiences, which themselves correlate very strongly with a shared neurobiology or human mind, which I understand in transmissive or filter terms, not in productive or epiphenomenal ones. In short, what we have here is a type of alien gnosis.

Still, I am challenged by Stuart Davis, Jeremy Vaeni, and Karin Austin. I worry that the above hypothesis is too mushy, not physical enough, too "spiritual," to use a much-abused word. To cite Davis, there is not enough WTF in my thought. I also worry, with Austin now, that it is much too "male" and not surrendering or receiving enough. I have not let go. I suppose that is what I am trying to say with Vaeni's nonduality material—that we need to be abolished in some profound way, because "we" are the problem itself. I stand challenged, then, and I am most open to be put into question, to be overcome.

THREE

"That They Are Not Human"

Thinking on the Autistic Spectrum

> There is a beauty in being partially unplugged from the human condition. . . . This is how many autistics feel; that they are not human. Now, of course, genetically I'm human. I just sit on the outer edges of the bell-shaped curve as to what a human is, at least psychologically. It seems certain to me that shamans and mystics throughout the ages probably had Asperger's syndrome or something of that nature. If you are "born spiritual" like I was, you can make a claim that you are some old soul incarnating again. Or you can just admit to being on the autistic spectrum. It's hard to say which is or is not true on a case-by-case basis. In any case, spirituality, genius wild creativity, and high-functioning autism seem joined at the hip.
>
> KEVIN, "Kundalini as Self-Induced, High-Functioning Autism"

> When you let your "personal soul die," then the only soul that remains is "the Universe"—and then you know ME.
>
> KEVIN, May 20, 2022

One basic feature, really the first, of impossible thinking is its desire to go big, to make the grand comparison or see the big picture. Impossibility itself, one realizes with something of a shock, is mostly a matter of scale and place. It is also a function of exclusiveness and inclusiveness—how much one is willing to take in and try to understand *together*.

This third chapter begins with such a grand comparison, but it quickly leads into some even grander thoughts—about the roles of sexuality and hormones in the spiritual life, the nature of awareness and reality itself, the imagination and its summoning abilities, the irreality of the soul and the nonexistence of death, and the inspiring importance of neurodiversity. It is a rather direct development of what has come before in the previous chapter, but I now think-with not a few people, as I did in my last chapter, but with just one.

World Mythology and Modern Science Fiction

First, the grand comparison.

I personally have this lingering conviction that I have never been able to shake. It is this: whatever produced the astonishing and multiple Sanskrit and vernacular literatures of the medieval and early modern South Asian Tantric traditions are now producing the equally astonishing and multiplex global literatures of the modern UFO encounter and abduction experience.

I am not quite alone in this lingering guess. A few historians of South Asian art and religions have engaged in playful comparisons between modern ufology and medieval Indian temple, ritual, and religious experience. The late art historian Michael Rabe, for example, described the sacred *vimana*, the interior "chariot" or "vehicle" of the Hindu temple complex, as "UFO-like."[1]

Such a reference can be found in David Gordon White's erudite chapter, "The Flight of the Yogini: Fueling the Flight of Tantric Witches" in *The Kiss of the Yogini*, in which he demonstrates the relationships between flight and cremation grounds (as we have already seen, the alien and the dead, or the UFO and the near-death experience, are commonly related in the modern materials) and the manner in which the extraction of blood and semen (seen often in the modern abduction literatures) is said to produce the *siddhi*, or superhuman, power of flight.[2] White's chapter even ends with a description of how "the circular Yogini temples, open to sky, were landing fields and launching pads for Yoginis and their consorts, the Virile Heroes and Perfected Beings."[3] It is difficult not to hear Rabe's "UFO-like" between the lines. It is difficult not to think of the modern abduction.

Are such resonances simply playful metaphors that a few Indologists have used in what are essentially one-offs? Or is there something deeper at work in these folklores across the centuries and cultures? My questions, of course, are mostly rhetorical ones. In the end, I think the South Asia Tantric materials and the contemporary abduction reports are connected, but I also wish to exempt Michael Rabe and David Gordon White of my sins. I think such inappropriate thoughts for a few reasons.

First, many of the mythemes of the UFO phenomenon have long or anciently been present in the world's mythologies—from very obvious ones, like floating ships, sky baskets, flying boats, sex with the spirits, and star ancestors, through ghost lights, out-of-body journeys, and "implants" (shamanic crystals), down to very specific details, like small humanoids, owl people, mantis beings, large eyes, and even paralysis by a magical object held in the hand.

As I argued above, I suspect such patterns exist not because people make up the same stories everywhere, or because of any material transmission (oral tales, migrating texts, or cultural products), but because many of these folktale features are *experience-based*—that is, rooted in universal human potentials (in embodied consciousness). If you prefer a more contemporaneous framing—these are all expressions of a shared neurobiology and sexual physiology.

With the Dutch Africanist anthropologist, initiated sangoma diviner, and intercultural theorist Wim van Binsbergen, I am of the conviction that "culture" and "identity" are themselves modern constructions, that human beings are far more potential and cross-culturally capable ("performative" in Van Binsbergen's ritual terms) than our present models allow. I am also of the conviction that the present intellectual focus on "difference" is productive of a particular kind of scholarship and moral concern but is finally not adequate to our actual histories and shared human nature. Regardless of what many would want to insist, *humanity has long been a single species*, a species moreover that has absorbed, killed, or had sex with other related hominid species. More to my present point, the human species has also always been a cosmic one that has evolved among the stars.

Precisely because of this evolved human nature and this shared cosmic environment, extraterrestrial mythemes are ancient and stable patterns, not modern, much less American, inventions. This is why such patterns have also long been systematized in the academy, as far back as the 1950s, by Stith Thompson in *Motif-Index of Folk-Literature*, his oft-cited and near-exhaustive mapping of world folklore and mythology.[4] The obvious American cultural prominence of the UFO phenomena today appears to have everything to do with who is creating which cultural productions (that is, who is writing the science fiction and making the Hollywood blockbusters), not with what has long been so around the world. Let us first get that crucial distinction right. Let us not confuse our very local and very modern cultural productions with what is actually so—and has been the global pattern for a very long time.

Historically speaking, the first widely reported "alien abduction" case (which is not to say, at all, the first abduction case) was in Brazil, not in the United States. This was the 1957 abduction of the farmer and eventual lawyer Antonio Vilas-Boas. Significantly, its sci-fi mythology already involved an implied racial mixing (Spanish and Indigenous, hence the doubled ancestry of Vilas-Boas), an electromagnetic beginning, the production and exchange of sexual fluids, and an evolutionary esotericism in the form of a divine-human offspring pointing toward some kind of "humalien" futurity.

In this case, we hear of a landed spaceship on a farm in the middle of the night, a stalled tractor, a physical abduction involving some kind of goop spread over Vilas-Boas's body (to prepare him for sexual intercourse, we assume), a growling and sexually aggressive female alien, subsequent forced intercourse (consent was irrelevant), and the female's final signaling of a future hybrid star-baby in her womb before she takes off in the ship into the sky.

It was the slightly later American case of the civil rights activists and mixed-race couple Barney and Betty Hill (another racial mixing), which took place, as already noted, on September 19–20, 1961, that entered American public culture through John G. Fuller's *The Interrupted Journey* (1966) and reinforced many of the narrative elements already present in the earlier Brazilian Vilas-Boas case, while also laying down new ones. The forced intercourse became a kind of sexualized experimentation in a laboratory setting, again without the consent of the human subjects. The eyes became more prominent.

This abduction narrative would branch out, deepen, and get stranger as the 1970s, '80s, and '90s ticked by, particularly in what is probably the best-known and certainly the best-written abduction case of the twentieth century, that of Whitley Strieber's *Communion* (1987) and a string of subsequent books, which continue from Strieber to this very day.[5] The cover alone of *Communion* is arguably the most famous and the most culturally reproduced cover in American publishing history. I certainly could not name another to match it. The Ted Jacobs painting of the eerie bald, bug-eyed alien on the cover is probably what set the icon into the popular imagination of the ensuring decades. Another insectoid.

What people do not generally know about that iconic cover image is that Jacobs was a crime artist—that is, an artist who recreated the visages of criminals for the sake of police investigation.[6] Certainly some assumed that Strieber's case may have been a potential violent crime scene. His anal region had been damaged, and he was clearly suffering from traumatic shock after the event. Strieber would later bravely describe the original abduction as a rape. Again, there was no consent. Vila-Boas and Strieber were both raped by female entities. Let us be clear and blunt about that.

Obviously, such contemporary cases are not the same as the Sanskrit medieval stories, but they appear to be phenomenologically related in some uncanny ways: the large almond eyes of the flying female abductors; the energetic effects in the lower spine that are often reported at the beginning of such abductions; the endless reports of flight, be it out-of-body or empirical; and the production of bodily fluids—blood, semen, and even feces.

Indeed, Whitley Strieber even suggested in *Communion* that the triangular form imprinted on his body during his original abduction account may be related to the yantra, or "sacred diagram" or geometrical form, of the Hindu goddess Kali and what he describes as the "Indian Tantras."[7]

Yes, things occasionally get *that* close.

There is also, of course, a most obvious interest in human sexuality and hybridity evident in both cultural sets of abductions. The modern alien abduction cases involve endless sexual experimentation (sometime of an "out of date" laboratory kind) and both bright and dark fears of hybridity and reproduction. Similarly, the Indian Sanskrit stories draw on millennia of mythical exempla and philosophical thought on the experience of human deification and the identification of seminal retention and superhuman power (*siddhi*), powers which traditionally include and indeed foreground the spiritual or physical power of flight. We will return to that superhuman power of flight, big-time, in our last chapter. There it will become actual, physical, realist, historically inescapable. Historians will still try to avoid it. They always do.[8]

Archives of the Impossible

Here is something to consider for now: there are easily as many mind-bending sexual-spiritual complexities in the modern UFO and abduction literatures as there are in the medieval or early modern South Asian materials or, for that matter, as there are in *any* culture or literature. As I explained in my introduction, I have been leading an archival collection of such modern materials at my own university for about eight years now. The Archives of the Impossible contain *thousands* of reports, case studies, and exempla of modern abduction accounts among *millions* of documents. For all practical or individual research purposes, this material is endless.

I really have no idea how any single person can possibly be aware of all these cases, much less read them all and adequately theorize the global situation. Moreover, I see no way to affirm one such cultural complex, say, the medieval South Asian Tantric complex, as worthy of our historical attention and professional expertise and another, say, the modern ufological complex, as less so. I think such moves, which, alas, are all too common, are entirely arbitrary and are forms of boundary policing and cultural essentializing (not to mention colonizing) that prevent us from understanding what is actually going on in our folklore, our cultures, and, indeed, in our own backyards and bodies.

I often joke that if an entity from the sky lands in your backyard and engages you in supersexual intercourse, this is obviously hallucination or pathology, or just a wish (there is the psychoanalytic theory of the fantastic). If, however, this takes place in an ancient manuscript, preferably one in a difficult language, then it is "religion" or "history" and can be analyzed endlessly (but never really taken truly seriously; never taken as actually having happened). When we put these two events *together* and refuse to privilege one over the other, all hell breaks loose.

Kind of my point.

Allow me to introduce the single subject of this chapter, a man I will call simply "Kevin" (since that is his name). Kevin has asked me to keep his identity anonymous (so "they don't do 'The Life of Brian' thing on me"—in other words, so the peace of his family is not disturbed for his honest, and often hilarious, views).

Kevin's case displays in crystallized forms many of the ideas that I have explored here and elsewhere (he jokes that if an idea cannot be put on a T-shirt, it is not worth having). After beginning with the context within which our friendship began and the nature of our two-person reflections, I will treat a series of Big Ideas in Kevin's writing. These are (1) the central role of autism in the global history of religions; (2) the difference between "energy" in a yogic and physiological or hormonal sense (and by "yoga," Kevin means "the more philosophical kind, not the mere stretching of the body for exercise"[9]); (3) a model of the Human as Three—that is, his understanding of consciousness, Awareness, and metaphysical Mind; (4) the importance of what is basically a dual-brained imaginal creativity that refuses to literalize its contents; (5) the central role of trauma, suffering, and death as means to realize the true nature of reality within which "humans do not have souls"; and (6) some of the sci-fi special effects that the human being can magically summon into reality, including, in Kevin's case, a gigantic black triangular UFO gliding silently over his backyard.

As big as these ideas are, they are all simply means or ways to explore and better understand Kevin's developed worldview, a set of new realities that he will variously call an emergent monism, a neutral monism, or, perhaps most accurately, a Platonic surrealism. The fantastic foundations of reality. Obviously, this is all more than a little complex and deserves a full book (you should see my printed file of Kevin's essays and emails). So let us consider this a beginning and not an end.

The Imaginals

I often laugh reading Kevin's emails and essays, mostly in a kind of recognition. But the guy is also just funny. As with the cases of Stuart Davis and Jeremy Vaeni, this laughter is philosophically important. It makes fun of and so transcends the silly social ego—my own, included. Impossible thinking involves not just a thinking-with but also a laughing-with.

Kevin often writes me that "reality is what two or more can experience together."[10] He has quantum mechanical reasons for saying that: he is of the conviction that a physical event comes into existence through participation—that is, through a certain twoness but also through a kind of occult sociality that materializes physical reality. There is no ultimate subjectivity or objectivity deep down. Both emerge together from a deeper source, as does any friendship or society. This is his neutral or dual-aspect monism. As he put it in an email, "at the quantum scale, there is no clear separation between subject and object, 'hardware' and 'software.'"[11] Philosophy of physics aside (or not) for a moment, this chapter is a very much an example of that "two or more together." I am not interpreting Kevin as some kind of silent object of analysis. You are reading both of us, even when you are reading just one of us.

Kevin originally reached out to me via email, partly because of my early writings on mystical forms of eroticism and the sexualities of Jesus in texts like "The Apocryphon of the Beloved" in my book *The Serpent's Gift*. As he explained to me in those original emails, he figured that if I could handle those stories, I could handle his stories. Such sexual-spiritual origins of our friendship and subsequent correspondence are neither random nor tangential, as we shall see in a moment. For what it is worth, I have long been of the strong opinion that my earlier erotic interests are connected to my present paranormal ones: I went from queering religion to weirding religion.[12] Same thing.

The correspondence and conversation themselves have worked on many levels. Kevin has written well over one hundred rhetorically and intellectually distinct emails. He has also written over forty brief, often aphoristic, essays on specific questions that emerged from our conversations.

I have not hesitated to incorporate this email correspondence and these essays when appropriate.[13] All of these printed essays also now exist in my files, which will eventually be deposited in our Archives of the Impossible.

For my own part, I have found Kevin's work clarifying, sometimes disarmingly simple, challenging to my own egoic assumptions, but also cogent and crystallized in surprising ways. Kevin, I think, would immediately describe each of us as "an imaginal," by which he means a person who can think and speak with both sides of the brain—critically and rationally but also nonrationally and fantastically. There are two ways to become an imaginal for Kevin: one can be autistic or one can be nonautistic but work to think and imagine like someone who is. He would put himself in the first category and me in the second.

What a great title for a book, *The Imaginals*, particularly if such imaginals say things like this: "In many ways I am the Book, and you have been reading me before we ever met."[14] There is always, of course, the slightly embarrassing possibility that one thinks that someone is onto something big because that someone happens to think like oneself. Scholarship can be ventriloquism, and agreement can be camouflaged narcissism.

Shamanism, Computer Science, and Autism

Kevin grew up dirt poor on a farm in Minnesota. His paternal grandmother was a full Lakota Sioux medicine woman. Perhaps because of this Indigenous family background or genetic line, he writes and speaks often of a shamanic worldview, including a ritual conviction that the proper intentions and rituals can manifest physical reality. Toward these same ritual ends, he does not hesitate to combine Hindu subtle physiology and Indigenous American ritual forms. These are available cultural tools for him. Nothing more. Nothing less.

As a young man, Kevin was trained as a nuclear mechanic in the navy and commanded eighty men. This scientific and technological training shows up consistently in his correspondence. He clearly knows his science, and he deeply admires it all. Not that his military service was all rosy. It in fact was forced on him, on the pain of death no less, to get him out of the house. No one yet knew that he was autistic, or why he found social interaction with human beings so terrifying, and often literally vomit-inducing. He ended up abusing alcohol and was hospitalized in the Oak Knoll Naval Hospital with a diagnosis of "schizotypal personality" and "magical thinking" (there's an easy out for the debunking reader). Kevin insisted that these facts be included, so as not to gloss over his fundamental humanity. He completely charmed the hospital staff.[15]

Today Kevin is an IT specialist who works for a major American state university system; he founded a successful dot.com venture and is happily married—not exactly your typical expression of mental illness. Kevin, however, is indeed autistic and self-identifies as being diagnosed with Asperger's syndrome, a generally high-functioning form of autism on the spectrum that results in general impairments in social functioning but also the ability to focus intensely on a single subject. I should add that I never made such an observation, much less such a diagnosis, until he offered it to me. Kevin didn't even know about it until 2001, when he was diagnosed by a man he describes as a gentle Quaker psychiatrist.[16]

The autism, it turns out, is one key, and not just to Kevin's life story but to spiritual, visionary, and trance accomplishments in the history of religions in general. Kevin firmly believes that many of the greatest saints and mystics in history were autistic as we understand that term today.

He is hardly alone with such thoughts. The contemporary American psychiatrist Diane Hennacy Powell has argued in similar directions. She has written extensively that autism and psychical gifts commonly appear together and can be psycho-physiologically related (she relates both to the mediation of a dominant right brain hemisphere). She has also argued, in a more esoteric vein that is especially evident today, that these superhuman powers have something to do with the future evolution of the species. She would know. Powell has interacted with psychiatric patients with obvious paranormal abilities who, among other things, saw the details of her own future, and she herself has worked extensively with autistic children since her own conversion to the shocking reality of such impossible phenomena.[17]

Darold A. Treffert has made a related case in his *Extraordinary People: Understanding Savant Syndrome*.[18] He points out that such a savant syndrome, which he also describes as "the genius within us," likely correlates with a dominant right hemisphere and may, in some cases, have to do with excessively high testosterone levels in the developing fetus (we will see this focus on hormones return in Kevin's thoughts).

The Fortean folklorist and comparativist Joshua Cutchin has argued something similar again, including in a section entitled "Aliens, Autism, & Shamanism" in his book *Thieves in the Night*.[19] After relating a series of fascinating case connections between aliens, autism, and childhood abuse (significantly, all three are major themes in Kevin's life), Cutchin discusses Grant Morrison's thoughts on how the "I" is a kind of lie that has gotten us into the absolute mess that is the twenty-first century, and why we should instead begin to see and embrace those "mutants living among us" who have been pathologized and dismissed. Cutchin also notes the likely relationship between autism and sainthood and speculates about the rocketing rates of autism in the population coinciding with the sharp rise of the alien abduction experience.

Kevin will also relate aliens, autism, and shamanism, even the lie that is the "I," but he will speak a somewhat different language again. He will invoke his training in computer science and write of two very different kinds of being: a constructed or conditioned consciousness that defines the social ego and a cosmic but nonhuman Mind, which are in turn constructed or mediated by the two hemispheres of the brain, one "digital" (the left hemisphere that constructs the nonexistent social ego and works on an on/off or yes/no logic) and one "analog" (the right hemisphere that mediates a very real Mind and is more holistic, pictorial, and atemporal).

Kevin explains the basic difference between nonautistic people and autistic people like himself: "Most people are digital computers, haunted by flashes of analog Mind. I live in a pool of Mind but have difficulty interfacing with digital (non-autistic) humanity." Or now, in a rather stunning few sentences that combine the history of philosophy, art, computer science, *and* evolutionary biology: "Humans are analog/digital processing units for the Cosmic Unconscious of Platonic Surrealism." The digital interface is still under active development: "it was under Darwinian evolution at first, but now with the modern world, it's becoming self-directed evolution."[20]

A bit of immediate commentary is in order here.

By the phrase "Platonic surrealism," Kevin means to theorize what I would call a fantastic idealism, a vision about the fundamental nature of Mind exteriorizing itself in and as physical reality, often—particularly in autistic states—in a surreal or sci-fi way. For Kevin, all patterns preexist in the "Cosmic Imaginal" and bubble up into new life as a result of both Darwinian and non-Darwinian interactions. That is simply what is.

The cosmic interface of which he writes that is currently evolving in a self-directing way is the much maligned but eminently necessary social ego. Kevin is hardly arguing for a one-sided or one-brained view of things here, then. He means to affirm both sides, both brains—one that constructs the social self, the other that mediates Mind. The two hemispheres of the brain are physiologically "to the side" (*para-*) of one another, even as they work together as a single brain. Hence the Human as Two (and One). The right side mediates Mind. It is *literally* "para-normal," to the left hemisphere, which constructs and projects the rational ego as a still-evolving and now self-directed cosmic interface. Here is the neuroanatomy of impossible thinking. It's all in your head, for sure. Or much better, it *is* your head but *all* of it, not just one side.

Super Sexualities

Such autistic and neuroanatomical themes play out in Kevin's life. By his own memory, Kevin was aware in the womb. As the Hindu yogic tradition had it, he was "born with an active kundalini."[21] In Kevin's mind, this early awareness was likely a result of "autistic early brain development," which can also be seen in his childhood nickname, "Pumpkin Head."[22] He possessed full technicolor sensory modalities in this autistic brain and big head as far back as he can remember. Still, he did not speak until he was five years old. He was endlessly bullied and beaten as a young boy in a cruel and stupid society that did not understand him, that could not understand him. He writes of "extreme child abuse growing up."[23] I am sure that is an understatement.

Kevin jokes that there was not a great deal of occult information in Minnesota (another understatement). So, around the age of twenty, he reached out and contacted an international organization that worked directly with kundalini experiences, which he had already known in his own body and brain. He remained with this group for eighteen years, from the ages of about twenty-two to forty, before leaving, mostly because of what he experienced as the community's dogmatism and insistence on an exclusive model of kundalini awakening. He did not fit, again.

Kevin obviously and humorously rebels from *any* religious exclusivism (some of his funniest lines I probably should not repeat). Which is another way of saying that his related sensibilities around religion, spirituality, and comparativism are acute, fierce, and transparent. This same sensibility plays directly into his use and understanding of language, including the use of technical philosophical language.

For example, he uses several Sanskrit words and categories to describe his practice and experiences (like *urdhvaretas*, *kundalini*, *pingala*, and *ida*), but he cannot always pronounce them correctly. He knows that, and he tells me. *And that is a big part of his point.* These are just words, after all, words that, for him, point to something physiological and hormonal, something *physically real*. So, as words, be they Sanskrit or English, and however they are pronounced, they are simply not important. In his own terms: "Wrapping things in Sanskrit doesn't do anything for me; I'd rather experience it for myself, not be lost in cultural concepts."[24]

Take *urdhvaretas*. This is a common Sanskrit term in Indian thought and mythology that refers to the "seed" or "semen" (*retas*) being directed "upward" or "backward" (*urdhva*), often, it is claimed, through a subtle spinal channel, which is sometimes considered to be the spine itself. In any case, this is an especially common category to explain the ascetic redeployment of semen or "seed" toward spiritual or nonsexual ends, up the spinal column, as it were. For Kevin, such an idea "is both bullshit and entirely true." Literal semen does not go up the spinal channel, which is what yogic subtle physiology strongly suggests (with all that "seed" talk), *but sexualized energy does*. More specifically, testosterone, which can easily cross the blood-to-brain barrier and does indeed travel from the genitals to the brain and back under the right conditions. The yogic tradition is entirely correct, then, even if it sometimes literalizes its sex-to-spirit claim in ways that are physiologically false (this would also, by the way, explain the maleness of some of these yogic traditions—they would be about the supereffects of testosterone, after all).

And so we come back to the sex hormones. When Kevin was twelve years old, he felt a "warmth in [his] forehead." He then felt "this warm liquid go back to [his] naughty bits." It was "a message from [his] brain to start puberty." He doesn't "give a flying ass" about the *prana*, *apana*, and so on—that is, the specific ways that the yogic traditions have mapped the movements of the "breaths" or "energies" in the body. Still, again, there is truth to such language. Here is Kevin on the actual process that he experienced: "An awakening kundalini makes you as horny as anything. . . . I learned I could redirect that whole current. I felt hot and cool currents in my lower body. It went 'Pop!' and then it was upward-directed. It felt like it was going through hyperspace, from the third to the fourth dimension. I just needed one or two hints, and then I could do everything."[25]

By "everything," Kevin means the spectrum of bodily and spiritual practices known and commented on extensively in the kundalini yoga traditions of South Asia. For example, there is the *mulabandhu* lock, a kind of tightening of the lower energy center, or chakra, and its nearby sphincter muscles. Here is how he learned the lock and now understands it:

> You have to keep it in your pants. You are simmering in a soup of primal desire. This is called the forcing technique. Please remember: I was a Minnesota farm boy. I had to learn this all. I had to learn that if the stomach is tense, nothing will happen. The heart needs to glow (oxytocin). You pinch off

the bottom, get it all revved up, in order to create the "cauldron." All these hormones are building up. Oxytocin is doing its thing in the heart. It is the very same sexual apparatus that is associated with religion. The result of this reconfiguration is to build new pathways which previously did not exist, to allow for an expansion of already existing human faculties; a form of parabonding with the universe.[26]

The ideas here draw from a range of cultural traditions: Hindu Tantric yoga, modern medicine, and ceremonial magic, among them. All these together work toward Kevin's own concept of a parabonding with the universe. The latter, Kevin explains, comes about by confronting trauma or spiritual projections and moving away from the ego of the person to the ego or Self of the larger universe, which Kevin calls by a number of names: "*our ego*," the "Universal Subjective," the "Imaginal," and even the "true Soul." There is only one such true Soul, and we all belong to it. It does not belong to any of us.[27]

Put neuroanatomically, we must move from the left hemisphere to the right and begin to understand the brain not as only a producer and projector (which is what the left hemisphere does do) but also as a mediator and translator of Mind (which is what the right hemisphere does do). This is how one becomes the universe, *because one already is the universe.*

None of this, it should be noted, constitutes an "escape to fantasy lands"—that is, into religious mythology or belief (the Marxist model of the imagination). Rather, there is a strong and practical focus on "the one and only tool we have to evolve," the interface between the human being and the cosmos—which is to say, the social and psychological ego: "The friction between our ego and the physical world is the true cauldron of evolution and transcendence in-place. You don't escape to fantasy spiritual worlds. You see that where you stand is already radiant."[28] This, of course, is an ego that knows it is not the center of the universe, that knows its own constructed and finally illusory nature. It is a very special form of being a self that knows it is a nonself and exults in its own deathless nature (because it does not really exist—you have to be born to die).

Very much a part of this is the classic dialectic between asceticism and mysticism, between "keeping it in your pants" and coming (or not coming) in order to parabond with the universe: "Hence my years of abstinence, until about 30," says Kevin.[29] That enigmatic phrase contains more than it says. It calls for more explanation. And more humor.

And so Kevin jokes, in the opposite direction now: "If your 'spiritual journey' has always been pleasant, you've probably just been masturbating."[30] With such a line, he does not mean to be simply funny (although this *is* funny). Nor does he mean to dismiss what is definitely one of the most common of all sexual practices (I probably said too much there). He means to point to the obvious physiological relationship between human sexuality and spiritual experience, how physically expressing the latter often results in the nonexperience of the former.

One might think from some of the above comments that Kevin is a materialist or reductionist, that he understands kundalini phenomena to be the special effects of sex hormones and human physiology. One might assume a kind of simple hydraulics: push it down here, and it will pop up there. Such a (Freudian) reductionism is admittedly tempting, but it is finally incorrect. Why? Because *there can be no reduction in a monistic world.*

It is more accurate from our own egoic or separated perspective, which is not ultimate, to think of the body and brain as receivers of some much larger cosmic signal, receivers, moreover, that can be "tuned" by manipulating the sexual system, hormones and all. Even that set of metaphors is incorrect, though, since, in truth, everything is already connected. There is no receiver, no sender, and no signal in such a monism: "In the Cosmic Imaginal, obviously visible and not obviously visible systems interact as part of a greater whole. There is no mere matter and no 'woo spirit.' Everything is a conjunction of properties and interactions, where Mind, Awareness, and regular meat-brain consciousness all stand equal."[31]

In the end, then, human sexuality can be "flipped"; it can be turned inside out or upside down until it reveals its cosmic nature and manifests its potential ability to parabond with the universe. Of course, it can be flipped again to discuss things like testosterone and oxytocin. Again (and again); it goes both ways. Same difference. It does not matter, or mind.

The Human Is Three:
Consciousness, Awareness, and Mind

There is both an ontology and an epistemology within what Kevin writes. He uses words in very specific ways. Kevin knows well what I have written about the Human as Two within a dual-aspect monism, so he playfully writes of his own views as the Human as Three (he likes to tease me). Actually, what I would call his ecology of the whole is in full alignment with the dual-aspect monism that I find most creative and helpful. We are in truth saying the same thing, literally *the same thing.*

Still, there are differences in language and expression. For one thing, Kevin is no fan of all the talk of "consciousness" these days in the philosophy of mind, including my own. And for one simple reason: he understands the term in a very different sense than I do; he defines it in the sense of the social self or ego. The ego is "conditioned consciousness": "what we all experience in the daily world if we are not in the grip of something mystical."[32] Such a level of the human is simply not real. It is socially and psychologically constructed. It is an illusion.

No objection there. Our words are different. Our point is the same.

If the human being is fortunate enough and can experience this social self or consciousness "die" or, to use the computer metaphor, temporarily go offline, then a deeper or higher form of being might appear, as it does in the yogic traditions under the Sanskrit rubric of *nirvikalpa samadhi*, or "absorption without thought or form." This level of being Kevin calls "Awareness." Such a form of being human (or being nonhuman or superhuman) does not need a body to exist, as conditioned consciousness definitely does (unless, of course, you want to call the entire quantum computer universe a "body"). Put most accurately, then, Awareness is entirely independent of any such individual biological organism, of any body. It is, if you will, an Awareness of the whole by the whole. It is the ultimate ecological thought.

But even this Awareness is not ultimate. Awareness, too, emerges from a deeper level of reality that Kevin calls by different names. The key here, though, is that this level of being is infinitely potential and so neither "material" nor "mental" in any ordinary sense. In philosophical terms, Kevin's views are very much an example of neutral or dual-aspect monism again, with matter and mind, or the objective and the subjective, both emerging from a deeper monistic ground.

Here is Kevin, laying out the fuller system in his own words: "In short, the author takes Awareness as a field, as in physics, perhaps the first and eternal field that, when it chooses to interact with substance [matter], allows for consciousness. Awareness allows for consciousness but is not consciousness. Consciousness springs from Awareness through an Interface [the body-based ego]. The author also takes Awareness and (primal) substance [or matter] to be different views of the exact same thing, the author holding the philosophical view of Neutral Monism."[33] In other places, in a more idealist vein, Kevin will use "Mind," with a capital *M*, to gloss this deeper "exact same thing."

There is an important corollary to such a philosophical vision—namely, that one cannot translate easily, or at all, from the ultimate level of Mind or the One to the embodied and socialized level of conditioned consciousness. Even the intermediate realm of Awareness is more or less inexplicable and impractical to the social self. The mystical is *not* the moral. The moral is *not* the mystical. And it is a great mistake to confuse or collapse these two very different scales. Hence the importance of the autistic condition, absorbed in Awareness, again: "If you live in unary consciousness [in Awareness] for too long, you start to be unplugged from the human experience entirely, which can be problematic. There's nothing 'noble' or 'elevated' or 'enlightened' about this. You are simply taking a break from 'reality' and hanging out with 'the One.' Which is great and all, but making use of this state, being practical—those are real considerations."[34]

This is why mystical states are so difficult to maintain. This is why "God hides." If God did not hide, if such states of pure Awareness were easily accessible, "if unary consciousness didn't limit Itself to [or hide in] Awareness, then each temporary being, each 'delusional being' would have all the other beings blasted into it and could not function or exist. . . . Awareness is God holding his breath, so that we might all exist."[35] Awareness is the One when it is not being anything in particular, when it is not reflecting back on itself as consciousness of something else. Awareness is God hiding.

Mind, the One, or the ultimate level of reality is multiple, superimposed (as in a quantum mechanical explanation), and infinitely potential. It is a bit deceptive, then, to consider it as "one thing." It is really *everything*. It is a plural unity, or a unified plurality. Hence, Kevin's favorite expression for this bigger vision of reality is "Platonic surrealism." It is "Platonic" for its insistence on a deeper level of truth and reality that is not accessible to human reason and opinion. It is "surreal" for the wildly symbolic, irrational, nonrational, and playful ways it expresses its endless possibilities and potentials, not unlike a surrealist artist trying to paint the superreal or surreal unconscious. Thus, Kevin will also sometimes speak and write of the "Platonic Surrealist Unconscious" because this level of being is simply not accessible or knowable to the conditioned consciousness of the social self.

He will also insist that no single myth, narrative, or belief is true. In some profound sense, they all are true since they are all artistic expressions of this same "Platonic Surrealist Unconscious." There is not no meaning. There is too much meaning. It is not that this or that culture is right. It is that everyone is right. Such is the deconstruction of deconstruction.[36]

The Little Red Wagon Theory

Very much related to this multiple unity or Platonic Surrealism is what Kevin calls his Little Red Wagon Theory. This is his theory of the imagination. It is derived from a story Kevin once heard but whose source he has long forgotten. The story involves two nation-states at war and a little boy who keeps crossing the contentious border with different things in a little red wagon that he pulls behind him. His father, it turns out, makes little red wagons for a living. First, there is nothing in the wagon, then feathers, then bricks, then complicated boxes. But the border guards can never find anything suspicious, and the boy is just a boy, so they let him go each time. One day, when the little boy gets out of the reach of the border guards, he calls back to them that he has in fact been smuggling something into the other country all along: little red wagons.

Kevin likes to tell this story to make a simple set of points. First, human beings need to imagine things. They need to tell stories. But, second, the things or stories themselves do not matter much and can never be final. Only the little red wagon—that is, the imagination—truly matters. That is because the imagination is grounded in reality itself. Actually, then, it is not that humans need to imagine. It is that the universe is constantly imagining itself, including in and as human beings.

What Kevin calls "sole source rationalism" (and traces back to Descartes) is simply mistaken, then. We are not, nor will we ever be, simply rational creatures knowing an objective world. Reality is imagining itself, including the reasonable self that Descartes imagined to be real (but is not). Until we can really see this—that *reality itself is imaginal by nature and intent*—we will not be healed. We will keep imagining, quite falsely, that our egos and their gory stories are real.

The phrase is vulgar (and funny) for good reason. Kevin can get pretty raunchy. "It's no surprise that humans tell crappy stories." Many of them are victims of "the train wreck of monotheism." Basically, believers in Christian monotheism damn themselves, suffer their own projections, and believe in fictions like heaven and hell or damnation and salvation. It's "a spell of degradation, lies, imprisonment and eternal death."[37] It does not have to be this way, though.

Actually, it can be *any* way: "I don't care what people 'summon.' Let them summon the Stay Puft Marshmallow Man."[38] Again, it is not what is in the little red wagon. It is the little red wagon. "I don't want to tell people what to imagine. . . . I want to show them *how* to imagine."[39]

"We Don't Have Souls"

Probably one of the most difficult aspects of Kevin's thought, which flows directly out of the teachings above, is his insistence that we do not have souls—that the human is really little more than an animal, an animal, more-over, that is constantly creating and projecting *tulpas*, or thought-forms (including and especially "the soul"), so as to imagine that it is real, that it somehow matters in the bigger picture.

It doesn't.

The human animal, or the "meat puppet human," can be possessed and is actually often a composition of multiple previous identities, none of which are really real, permanent, or stable.[40] This is certainly the experi-ential fact for Kevin. Hence all the "weird stuff" that happened around, in, and as him—"Skinwalker Ranch level stuff," as he once described it to me.

Enter the category of the *tulpa*, a temporary mind-created entity in con-temporary paranormal folklore. The concept and experience of the *tulpa* emerges over millennia, first from older Buddhist traditions (which also deny there is any permanent soul or self), especially the Mahayana or "Great Vehicle" teaching of the three bodies of the Buddha. The first of these bodies is the *dharmakaya*, literally the "truth body," often identified with the entire universe or reality as such. The second of these bodies is the *sambhogakaya*, literally the "enjoyment body," through which the Buddha and bodhisattvas inhabit the various heavens and perform their miracles through the superhuman powers (*iddhi*, *rddhi*, or *siddhi*) of concentration and intention. The third of these bodies is the *nirmanakaya*, literally the "emanation body," through which the bodhisattvas take a physical body and appear in this world to aid aspirants on their path to full enlightenment.

It is in conversation with this three-body doctrine, and especially the third body, that the concept of the *tulpa* was born, very recently, it turns out. The word first appears in the work of Alexandra David-Neel in her 1929 book *Magic and Mystery in Tibet*. The *tulpa* thus appears in Euro-American culture at a particular historic moment within, as we shall see, early mod-ern Theosophy and paranormal American currents in conversation with Tibetan or Vajrayana Buddhism, which was in turn deeply informed by the earlier Mahayana traditions. It is all one big fusion, "appropriation" all the way down.

The word *tulpa* is a rough approximation of the Tibetan *sprul pa*, or "magical emanation" (that is, the third body of the Buddha), but it also represents a shift of meaning through this East-West co-creative, cross-cultural dialogue and is especially indebted to the "thought forms" of later Theosophical writers like Annie Besant, Bertram Keightley, and Dion For-tune.[41] Eventually, in American folklore—particularly through the writings of authors like John Keel (who was very influential in the UFO and cryptid worlds, particularly for his book *The Mothman Prophecies*)—a *tulpa* came to mean a relatively unconscious or "accidental" collective mind-created entity that can become independent and turn on its creator in disturbing ways, a meaning that is quite distant from the compassionate and intentional action of a Buddha, bodhisattva, or even possessing spirit of the earlier Tibetan traditions. Still, the *tulpa* remains a product of a most collaborative and productive cross-cultural dialogue. It cannot be reduced to this or that colonial, Theosophical, paranormal, or Tibetan Buddhist origin or intention.

This, anyway, is the broader context in which Kevin first "lost" his soul and then learned to recreate it, which, he learned to his great sadness, is not the energetic life-form he felt course through or possess his heart and brain:

> There did come a time it wanted out. It was in my heart for a long time. Then it lived in the brain. It became restless and simply left. I would hear this crackling noise at top of the head, like my skull plates were moving. It was excited. It felt like it wanted to go home, filled with joy. But then things got ugly. When it left, I thought I was coming with, but I was alone now. It crushed me for ten years, because [I realized that] this is not me. I thought I had a soul that was going home. Now I knew I had no soul. I almost died. I felt like an automaton. When 95% of my inner being had left me, I was empty and weak. I started to experience little offers. "Let me use that body." One claimed that he was an alchemist from the medieval world. . . . What I discovered was that, although the majority of me had left, this was not healthy. So I grew a replacement. . . . Humans are *tulpa*-creating machines.[42]

One might think in this context of the replicants in Ridley Scott's *Blade Runner* (1982), itself based on an earlier science-fiction novel of Philip K. Dick, *Do Androids Dream of Electric Sheep* (1968). Replicants are biological machines who think they are human but are not. The comparativist cannot also help but recall the "soul-loss" that is so prevalent in many shamanic cultures, including the shaman's call to recover such a soul and so to restore health and well-being to the person. Kevin appears to be reliving this shamanic practice in a Buddhist or Theosophical form (Vajrayana Buddhism, by the way, is often associated with an earlier Tibetan shamanism called Bon).

Kevin will also sometimes argue for what I once called, following the American writer Charles Fort, the Earth-farm theory; that is, the idea that we are the livestock or biological experiment of higher life-forms.[43] He believes, and has indeed *seen*, what he calls "symbiotic plasmas" essentially feeding on people, including one famous charismatic figure. Kevin once wrote to me in a striking series of thoughts:

> In some ways, we are the plants being consumed by those "above us," just as they are "grass" for those above them. It's "Russian dolls" in a way. But we, the grass, *could* hunger strike our upstream if we wanted to . . . and that's called end-state-original-Buddhism or "kundalini-enlightenment." We *could* learn to control our Awareness-to-consciousness converters. But to do it as a "race" when it's "counter-the-current-of-evolution" in some ways (religion is a somewhat useful adaptation)—I don't see that happening.[44]

The ideas embedded here are fascinating. We are inside an immense environmental food chain, but we don't generally know it, and we don't have to be. Religion is a way we can orient ourselves within this hard ecological truth but also, I assume, imagine and practice ourselves out of this feeding cycle. So understood, religion becomes an evolutionary adaptation, not just to social or biological pressures (of course, it is that too) but also to what are occult or hidden presences and symbiotic feeding processes.

I cannot help but add here that there is a large literature on alien or cryptid creatures "feeding" on human emotion, including and especially sexual arousal (we saw this theme in chapter 2 with Bobby and the mantis and distantly suggested in the UFO encounter of John Lennon and his lover). Once again, Kevin is hardly alone in his thoughts and visions.

The high strangeness went on, this time through the realization that "Kevin" was in fact four different people strung together (again, a very Buddhist idea). He repeats his observation that humans are *tulpa*-creating or soul-creating machines; that is to say that the social ego, person, soul, or self may be an illusion, but it is projected from something Other:

> This was one of the most beautiful experiences of my life. I had many spontaneous visions. Something that felt like me disassembled into four different pieces. These four people had been together for thousands of years. They were like glowing vibrating strings. They had led lives all over the place, including off-earth. They knew I was listening, but I was not considered a part of their process. ("You are a host.") But I had access to memories of them. One of them was the number two person who had originated Dvaita Ve-danta.[45] This happened the day before the eruption out the top of my head.[46]

The beauty would end, however, as Kevin crashes back to the Earth-farm: "I think I am an idiot savant. I feel like a disposed breeding chamber.

The 'Other' becomes enlightened, not us. We exist for the purpose of another. Somebody has to be at the rear end of the dog pack, and that's us. We provide a valuable function for the Larger Ecosystem. We are *tulpa*-creating machines."

The conviction in the self or soul's irreality points to another big idea in Kevin's emails and essays, what he calls "abuse" or "death." Such an abuse or death can be caused by internal or external factors (he would know) and is often a necessary correlation of "unusual occurrences'" (he would know). Kevin explains that the human mind must dissociate from the social ego to become or know the projections of a god, demon, monster, poltergeist, or soul. But the human psyche is simply not going to do this under happy, healthy circumstances. It generally takes trauma and torture.

Here we enter the realm of demonology and the demonic paranormal as a kind of mental-spiritual condition, which is not to be desired or pursued: "You and others around you might experience 'poltergeist phenomena' at some point. This is not GOOD. This 'proves nothing.' This does not make you 'evolved' or 'enlightened.' It just means you have tortured your psyche and parts have split off. If you become absolutely convinced these 'beings' exist, you might even stumble onto how to make them 'semi-real.' I did it, crazy person that I was. Nowadays? Not so much."[47]

There is too much to say here around what I have called the traumatic secret. Trauma, after all, is a strong correlation of so much extreme religious experience. Happy, healthy people do not experience poltergeists. These alleged ghosts, as I have joked, never show up as rainbows, kittens, and friendly unicorns. They are always breaking, scratching, and screaming shit—projected trauma, suffering, and self-torture. This is why paranormal phenomena so commonly appear in or around near-death experiences, extreme illness, war, the evils of racism, and social suffering. Here is how Kevin puts the idea, always succinctly: "It's generally trauma that opens doors in the psyche. . . . Since 'trauma' opens the doors more than anything, that's why 'the Other' tends to manifest as 'evil' and not 'good.' . . . It took me freaking nearly forever to learn why the 'Phenomenon' is nearly universally 'nasty' and not 'good.' It's because the part of the human mind that brings it in is hurt and nasty. . . . That's the simple truth to it."[48]

The negative paranormal, in short, is a clear witness to the social suffering that spawns it (and the social justice for which it cries and haunts). That *is* the simple truth to it. The demons are us in trauma and suffering, split-off alters that we do not recognize as othered parts of ourselves. We may be God, but we are also the Devil.

Summoning the UFO

Which brings us back to the UFO.

I have often observed a particular phenomenological feature in modern alien abduction cases: the experience might involve classic kundalini-like phenomena, such as rushes of energy in the spinal region and brain core, accompanying altered states of consciousness, and the sensation of out-of-body floating or flight. No doubt, I have noticed such things because of my earlier training in South Asian Shakta Tantra, where such features are theorized and symbolized in very specific and culturally syntonic ways. Similar phenomena are also evident and often commented on in the anthropology of African healing practices, sometimes as a "rising heat" in the spinal region.[49]

The spinal sensations are not so theorized or symbolized in the abduction literature. They are sometimes duly noted and quickly described, but they are almost never followed up on, never really integrated into a cultural practice or a widely shared religious belief. Kevin is a glaring, glowing exception to this general rule, no doubt because of his long training and reading. He has thought a great deal about the relationship between what the Hindu traditions call kundalini and what the Western sci-fi imagination calls the UFO. Indeed, in one striking email, Kevin said that "Kundalini can be viewed as a UFO encounter inside one's body," and "a UFO encounter can be viewed as a Kundalini encounter outside of one's body." Referring to the chakras, or "circles," arranged along the central column within the subtle body of kundalini yoga, the famous spheres of ufological lore become the "dislodged chakras from the body." Or, if you prefer the flip, "chakras are like 'orbs' captured by the body."[50]

Such an astonishing cross-cultural comparison, rooted in Kevin's own body, is partly why one of the most important moments of his life was the day he successfully "summoned" a massive UFO—in this case, a classic black triangular craft in the sky, with a sleek metallic sheen and rivets no less. Yes, rivets. That was a clue. He would later connect the symbol to connectedness.[51]

Before we go there, it is important to recall something on which Kevin often insists: "Kundalini is the Mother of Illusion. Maya and all that." Put a bit differently, Kevin does not literally believe all the phenomena that appear to him, *and yet they happen*. He understands perfectly well that they are illusions, but they are illusions actually appearing, truly manifesting, that serve legitimate purposes, whether we understand them at the time or not.

There is another way of saying this: "Consciousness is the Trickster."[52] As we have seen above, Kevin means something very specific by "consciousness." He does not mean anything particularly noble or desirable. He means what we might think of as ordinary banal consciousness, the ego. This is partly the reason Kevin will regularly qualify his language and visions with phrases like "in *part*," "in a manner of speaking," "in a *sense*," and, my personal paradoxical favorite, "the metaphor that was not a metaphor."

In the end, these are illusions, but not in the Western sense of not really real. They are illusions in the Indian sense of *maya*, of the magically "made," "measured," or "constructed." They are also real illusions *because reality itself is an illusion*. Nothing we know is truly physical, truly real in a materialist sense. *Nothing*. Again, a profound neutral monism or idealism, a Platonic surrealism, ultimately shines through Kevin's thought.

The UFO summoning is a prime example of this real illusion, of this metaphor that is not a metaphor, of this fantastic impossible. Like many UFO sightings, this one was announced by the apparition or appearance of a cryptid or monster—that is, another demonology. Here is how the presence began to announce its coming:[53]

> The UFO itself followed my commands pretty well. Strictly speaking, it was not a *tulpa*. A little ritual summoned the black triangle. There was a mini-kundalini awakening, a kind of broke connection, which then threw it out. For three days, I built this *tulpa* that had its own awareness. Then I set it loose, "Go find Mommy. Go find Daddy." What it actually brought back was this creature that came into my house, a barrel-chested angry dwarf right out of *Phantasm III* [the 1994 American horror film]. It came in that appearance. What I saw was indeed the decaying remnant of what had left my head. It was pissed at me: I had abandoned it.

The actual appearance of the UFO a few days later reads like something right out of the books of Henry Corbin and his category of the imaginal, the intersection or middle world of the ideal Platonic transcendent and the material realm of matter and physical form.

> A few days later, I went out to throw away the trash and looked up. There was the giant black triangle, two worlds joining together. These are *tulpa*-like worlds when they interact. What is important is the summoning elements, as when someone is abducted and feels like it is physical. Summoning is a dual thing. It occurs from both sides, as it were. There is a kind of shadowy

pocket between the two worlds, an interface or liminal place. There is an intersection of reality where Platonic ideals meet the physical world.

Strikingly, Kevin has not read Henry Corbin. But he has read Jenny Randles, the ufologist who has famously called this same middle space or eerie zone "the Oz effect," after *The Wizard of Oz* (1939). There is that fundamental fuzziness again.

The fullest explanation of the summoning comes in the form of an essay entitled "Kevin's BTUFO Summoning and Encounter of the Third Kind," which was accompanied by a cell-phone photo of a drawing of the "Black Triangle UFO." I think it is best simply to let Kevin speak directly here and include the drawing as a redrawn image by my artist friend and colleague, David Metcalfe, with Kevin's permission and encouragement.

Here are chunks of the essay on the summoning:

> Starting on 11/24/2013, I designed a "summoning ritual" that I felt would be efficacious in the summoning of a so-called "Black Triangle UFO," which at that time I felt was in fact some manner of coherent-energy life-form, and *not* any form of "nuts and bolts" craft. The ritual involved what is called "Samyama" [Sanskrit for a "joining" or "unification"], a yoga practice which involves reaching a very deep level of mind and then "letting go," so that what is visualized becomes a reality. However, from past experience, I knew that this would not be enough. So I also wove in a very arcane Kundalini Yoga practice, whereby I woke up a symbiotic organism that is already present in all humans and "fed it." However, I knew that this would not be enough. So I also moved the symbiotic lifeform to the point of Samyama in my head and bathed it there. Now I also knew that this would not be enough, so I spent the next 3 days attempting to hold this unusual body/mind configuration in place, while educating the symbiotic lifeform that it should go find "mamma or daddy," so to speak. I kept this up for three days, then completely exhausted, I collapsed and went to sleep.[54]

The next day he woke up and completely forgot about the entire summoning ritual. Which was the point, since, without such a fallow period, the unconscious creative process cannot work. In this case, the life-form could not have left and brought "Mamma or Daddy" back to Kevin.[55] The first sign of success was around seven o'clock the next evening, "when I suddenly felt my entire body tingle . . . and it felt like my back yard and all around my house was suddenly in some sort of coherent structured energy field."[56]

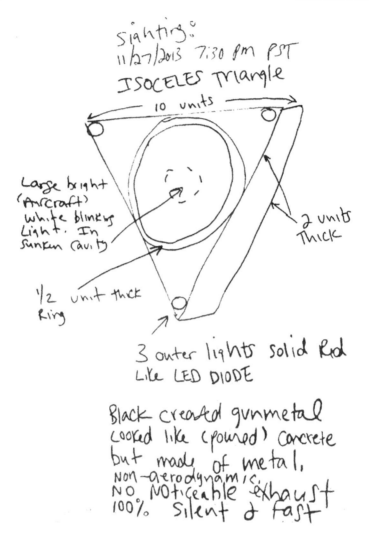

sighting:
11/27/2013 7:30 pm PST
ISOCELES Triangle
— 10 units —

Large bright
(Aircraft)
white blinking
Light. In
Sunken Cavity

2 units
Thick

1/2 unit thick
Ring

3 outer lights solid Red
Like LED DIODE

Black created gunmetal
Looked like (poured) concrete
but made of metal,
Non-aerodynamic,
NO Noticeable exhaust
100% Silent & fast

The Oz effect.

It was then that the monster-creature appeared in the house while Kevin sat paralyzed in his chair for about ten minutes. Then it all disappeared. Kevin tried, in essence, to shake it off and, his body back now, decided to take out the trash. As he walked into his backyard and looked up, there it was:

Massive. Silent. Three red lights, steadily on and this arc-light search light beam in the center. I watched it fly over me and my house from the east to the west horizon over the heavily populated area I lived in. It APPEARED to be the size of, say, a 747 or three, and it appeared to be, say, 500 feet high. Of course, I had no reference points whatsoever, so it could have been the size of the moon, for all I know. It was very polite with me. When I wanted to see other parts of it, it would politely move that direction for me, so that I could see it had no jet engines, or any other kind of engine. . . . It had no aerodynamic surfaces. It had no exhaust and, of course, it was beyond quiet. I could hear small gnats flying around me, but they were very loud in comparison to this thing. . . . Now as it got about halfway across the sky over me, I realized that it was an image projected into my mind, as I could see rivets . . . rivets . . . which is fine, I guess . . . but they never got larger or smaller or more or less distinct, no matter the distance between "it" and "me." I said in my mind, "You fraud! You shapeshifter! That was very funny of you with the rivets." I got the sense it liked me, and that it had done that on purpose. It then shape-changed into a pyramid from the movie *Stargate* [1994]. Then it shape-changed into a ship design from [the television series] *Star Trek: The Next Generation*. I said, "You silly thing you . . . can't you do anything except show me things from my own mind?" So it waggled its left arm or whatever in glee and then assumed all three forms simultaneously . . . and it laughed.[57]

It laughed.

Comedy aside, this utterly bizarre event totally "messed up" Kevin for five full years. There was also what he describes as an overwhelming "data stream," which went on for years but out of which he could make out nothing coherent. There was just too much information.

It was so weird that Kevin admits that it could have been some kind of "self-induced psychotic break" or "self-induced conversion disorder" (there is another easy out). Personally, I am not certain how such a quick diagnosis helps, given that *so* many of the features of Kevin's summoning are present in other reported UFO encounters (the triangular shape, the black color, the lights, the peopled neighborhood, the utter silence, the gliding, the interaction, and even the rivets). I suppose every such witness could be having a psychotic break. Still, such an explanation ceases to explain much. I am suspicious of the suspicion.

What should we take away from this? We can begin by observing that the UFO summoning is part of Kevin's larger argument that paranormal phenomena are essentially pieces or parts of ourselves that we have projected outward, usually in a traumatic situation or because of a previous trauma and its dissociative nature. These things are projected outward so that we can read them and let them go. They are "semiotic" in nature; that is, they are communications that must be interpreted, even if we cannot fully interpret them (hence the overwhelming data stream?).

Construed as such, it becomes obvious that such paranormal events are not entirely random, much less meaningless. On the contrary, they are deeply meaningful and can become a part of a healing process. In Kevin's own words, "We have to die and eject our soul fragments to talk to a more evolved form of life. That's what 'kundalini' and 'enlightenment' and 'summoning UFOs' boil down to!"[58]

I understand that last one is a bit cryptic and condensed. I didn't get it either. Here is the same theory in a longer form, this time in an essay written in response to a friend who asked Kevin if there was a term for "Mystical UFOs":

> So-called paranormal experiences pull from the same "imaginative space" (call it Platonic Surrealism, call it Hilbert space, call it whatever you like, so long as it's not bound by entropy or classical physics). So one person might see the Virgin Mary, another person Bigfoot, another person might see a UFO. In many senses, at least for high-strangeness UFOs, to quite an extent you see what you want to see, but if this "UFO" does indeed flow [through] the "Oz mechanism," which generally flows through the midbrain as a result of trauma on some level, there will also be an element of "the Other" in it; generally seen as the "weird part"—for me it was the rivets. That weird part is the semiotic "message" that you need to decode. It's pure Little Red Wagon Theory; and what's in the wagon is only significant on a person-by-person basis. Anyone who claims they have a message for the entire human race, because of such an encounter, is generally full of shit.[59]

I think we have sufficiently explored all the technical terms Kevin uses here. The summoning of the UFO, like the appearance of the barrel-chested dwarf or the kundalini phenomena for that matter, was all about letting go, letting the illusory parts of us die and wither away so that we can talk to a more evolved form of life. And what is that more evolved form of life? It is us again, but on a cosmic level now: "When you let your 'personal soul die,' then the only soul that remains is 'the Universe'—and then you know ME."[60]

Back to Our Backyards

Finally, allow me to observe what I hope is fairly obvious by now: there is an indubitable comparative base at work here in the shamanic, kundalini, and UFO literatures, and one that is not culturally arbitrary or historically relative—the human neurological system, or body-brain-spine organism. Every modern human being shares a biological frame that any and all human subjects share or once shared. One could say, for example, that the human body-brain-spine organism is more or less the same everywhere, although, yes, of course, there are also endless psychobiological and neurological variations, including autistic ones on the spectrum. Medieval South Asians and Europeans had spines, brains, and sex hormones, just like modern Americans and Brazilians, and contemporary French, Japanese, and West African peoples.

In this view, it is not any local mythology or religious ideation that appears to produce the globally distributed phenomena that is the UFO. It is the spine, the brain, and the body and—lest I be heard reductively—whatever in turn might shine or transmit through these organic neurological machines. The human biological form, after all, need not be thought of as the *producer* of all this cross-cultural commonality. It might just as well be thought of as the *reducer* of something else that is fundamentally not us, that is nonhuman or superhuman . . . in Kevin's language—that is ME.

This in no way means that ordinary matter itself is not sacred, is not "radiant," to use Kevin's words again. It, too, is "an artifact of Mind"—that is, an expression of some deeper structure or source, call it math or God. To repeat myself, there is no such thing as reduction, nor can there be in a monistic world. And science, understood now in this context, is "the most successful system of magic of all time" (mapping and manipulating all those occult or hidden structures).[61] Indeed, science is currently producing endless new realities in which we can imagine and create anew.

Kevin was most passionate that, as this chapter ends, these final observations be presented, exactly as written and formatted. And so here they are:

> At the very last, when the fetters of fear of loss fall from our very human eyes, what we thought of as eternal in and of itself, the "soul" turns out to be a love letter, a data packet, a temporary resonance in the realm of the very small, addressed to the Larger Life itself.

> What we thought of as even "higher than that," the "spirit" turns out to be a collaboration, by Surrealistic Weaving processes, that make good use of all that was discovered to be true and beautiful during our time.

> While we can sometimes "dream a better dream" as a "soul" or a "spirit" after all the veils fall, and Mind is revealed, we discover all our attempts to

cheat death, to create something enduring to fight against a cold and chaotic Universe were pure foolishness.

We simply Are, and will always be, as a song, a movie, within the Cosmic Imaginal, and when it is intermission, and slumber flows throughout the House, we smile at Ourselves as we always do, after having watched all our movies, and watch them yet again, with another set of properties, another set of interactions, and just for funsies, we sometimes giggle as we send all manner of pranks to the more serious versions of ourselves.[62]

I do not pretend to understand that all as a socialized ego, as a relative and historical form of consciousness. Nor am I suggesting that we should sign our names to any of the "movies" that have been described so poignantly above by this suffering visionary, whose very thought insists that paranormal phenomena appear in intense trauma and are deceptive by nature. Still, I remain utterly convinced that it would do us well to listen hard and long to Kevin, educated as he is in the hard sciences but also adept in the modern spiritualities. He can teach us to think impossibly, and to laugh impossibly. If we listen to such a being long enough, we begin to suspect that something else might well be beaming through us, taking possession of us, projecting sci-fi movie after sci-fi movie in our dreams, visions, religions, and even in our own backyards. We begin to seriously entertain the likelihood that something is flying around in our folklore.

Us.

PART TWO

MAKING THE IMPOSSIBLE POSSIBLE

FOUR

The Timeswerve

Theorizing in a Block Universe

My argument rests on taking seriously an alternative understanding of time as a reversible swerve, a scientific perspective that conflicts with the commonsensical view of time's irreversible linearity. . . . Methodologically, the appeal is to apply a natural scientific model of time to the humanist effort to understand historical experience.

ELLIOT WOLFSON, *Language, Eros, Being*

I want to shift my discussion now and move from describing the impossible to theorizing it.[1] Of course, we have also been doing this theorizing above, but I now want to do it more explicitly and systematically. This is how we make the impossible possible. More stories, however astonishing, are not going to get us there. We must demonstrate a way to contextualize and locate those astonishing stories in some larger superstory, otherwise known as a theory, to make them meaningful and so plausible. We know the impossible happens everywhere and everywhen. What we do not know is how (the scientific question) or why (the humanist question).

There are certainly multiple examples of individuals who have theoretically engaged different slices of this impossible material, and to significant effect. Indeed, there is in fact a kind of renaissance going on here in the academy that is obvious and palpable. There are dozens, if not hundreds, of books on everything from the realities of witchcraft and popular magic in early modern Europe and the secret Spiritist dimensions of the Mexican revolution to the hauntings, guilt-filled signs, and wonders of post–World War II Germany history and the use of clairvoyant psychics by the police in Poland to the anthropological call to think with forests "beyond the human" and a most rigorous philosophical dive into our own superreal multiplicity through the channeled teachings and conceptual explorations or "rough metaphysics" of the American medium Jane Roberts.[2]

To describe just one such recent book, consider a poignant retelling of a terrible mining accident in Wales in 1966 that killed 144 people, many of them children, the well-documented precognition of this horrible event, and the subsequent founding of a "premonitions bureau" led by a psychiatrist who could not deny what he saw so clearly (that people sometimes have stunningly accurate premonitions of major disasters). The psychiatrist himself would later die young. This is a perfectly honest history of what Sam Knight, the author, calls "a true account of death foretold."[3] What makes this book so different is not its historical case study (whose precognitive dimensions are common and easily recognizable elsewhere) but the fact that the book garnered such serious and widespread attention.

Something is happening. It is a new day. These books are overtaking my study in stack after stack on the floor these days. I want to read them, be changed by them. I want to understand how they theorize and so make the impossible possible. I also want to do the same.

On Experience: The Phenomenological Tradition

But is this all really so new? To shift our focus, why *were* the authors who most wanted to theorize exactly these historical moments left to the side of the twentieth century? Jason Josephson Storm has argued that, in effect, such thinkers were not left to the side at all—that it was precisely these kinds of anomalous interests that lay at the very origins of the modern human sciences.[4] Modernity is occult to the core. I can only agree. Indeed, I have spent the last quarter of a century reading and writing about these endless esoteric currents, none more so than a mystically inflected psychoanalysis.

But there is another particularly glaring current of modernity that is most relevant here, both for its careful focus on the intricacies of "experience" and the structures of "consciousness" and because it initiates a stream of thought in which the central subject of this chapter, the medieval kabbalah historian and philosopher Elliot Wolfson, participates.[5] The same philosophical practice—which really comes down to the conviction that it is *consciousness itself* that is the source of all knowledge and truth—also helped birth the modern comparative study of religion, a practice that vibrates at the core of impossible thinking, mine anyway.

I am thinking of the intellectual project of the Jewish German philosopher Edmund Husserl (1859–1938) and the school of thought that he spawned, often called the phenomenological tradition, since it concerns itself with philosophical reflection on the "appearances" (*phenomena*) of consciousness and attempts to get at the essence or structure of this consciousness beyond or before all such appearances, including and especially sensory, cognitive, scientific, and mathematical ones. Martin Heidegger, Gabriel Marcel, Jean-Paul Sartre, and Maurice Merleau-Ponty are generally referenced as inheritors, although none of them really took up Husserl's "transcendental phenomenology." They certainly developed the latter noun, *phenomenology*, but mostly, or just entirely, ignored the all-important adjective—*transcendental*. Indeed, Husserl's most famous followers often claimed that there was nothing mystical about this particular philosophical tradition.[6] To use a Nebraskan colloquialism, that is just bullshit.

Husserl was deeply frustrated with those who conflated the beginning stages of his thought with that total thought, and he fully intended the heretical grandness of his project, quite against the reigning assumptions of the day (which are still reigning)—namely, that one can never get to such transcendental essences or structures. Working *precisely* out of these transcendental structures, Husserl called for a "new science" that he believed underlay all forms of knowledge, including and especially the mathematical and scientific ones in which he was originally trained. Such a new science is aimed at the realization of what Husserl called the "transcendental Ego" or "absolute consciousness," which constitutes an "independent realm of direct experience."[7] He saw such a shift to the transcendental standpoint as a "Copernican reversal" (what I have called a "flip") and described it as "the greatest existential conversion that is expected of mankind."[8] Stop there for a moment and just think about that claim.

Unsurprisingly, Husserl also worked with a most obvious model of the Human as Two. There is in fact no such thing as the psychological ego. There is only the transcendental ego, which has no content and is entirely outside the realm of ordinary perception and cognition.[9]

It is difficult to avoid the suspicion that Husserl had known some kind of direct mystical experience of this absolute consciousness or transcendental I outside the (nonexistent) social or little psychological self, and that, as Fred Hanna has put it, there is good reason to believe that "Husserl had contacted regions of experience that the vast majority of phenomenologists, during his lifetime and thereafter, simply did not encounter."[10] This is why he sounds so much like a humanist mystical writer. He is.

This is also why Husserl so admired Plato, why he once wrote so appreciatively of Buddhist analyses of consciousness,[11] and why there are allusions in his conversations to the questions of collective memory, of the monadic unity or plurality of transcendental subjectivity, and even of telepathy as "problems for the future."[12] This is why he once quipped that there were "whole sentences" of the medieval mystical writer Meister Eck-hart that could be "taken over by him unchanged."[13] Indeed, Colin Wilson has described Husserl and phenomenology itself as "a prosaic way of de-veloping the mystical faculty," which he compares to the creative vision of poets and painters.[14]

It is not difficult to see why Husserl sounds so much like a secular mystical writer. His philosophical method strongly resembles ancient apo-phatic methods of "saying-away" or deconstruction on the road to the realization of Mind or God, even and especially when the reader has not a clue about what he is talking (a possible sign that the text is based on an altered state of consciousness). Husserl's phenomenological method, for example, begins by refusing or "bracketing" (the famous *epoché*) what he called the "naturalist standpoint" of the sciences and their naive sense-based positivisms about the "fact-world." He calls this apophatic move the "phenomenological reduction" and considered it the necessary secret of obtaining genuine philosophical truth. One cannot arrive at absolute con-sciousness until one has let go of *all* sensual and materialist assumptions, including, by the way, those that would come to define phenomenology in the person of Martin Heidegger.[15]

There is a very strong phenomenological orientation in the early years of the comparative study religion. That study once emphasized mystical literature, the role of the symbol, and the experience of revelation itself, much to the chagrin of exclusively socially or historically oriented scholars. When it actually reads and understands Husserl, that same comparative study has on occasion been very friendly to his famous bracketing of naive positivism (the idea that the physical world is as we perceive it), even his absolute consciousness outside or above all such sensory perception.

Anthony Steinbock is a good example. A scholar of Husserl himself, Steinbock has given us a phenomenological comparative study of mysti-cal literature in the three Abrahamic traditions (Christianity, Judaism, and Islam), arguing for what he calls the "verticality" of religious experience—that is, a global sense of "up," "above," or "outside" normal human experi-ence and functioning.

Steinbock's extreme suspicion of ordinary assumptions about the world and the nature and reach of the human being should strike a very familiar chord to the reader by now, as should his celebration of phenomenological reflection as "hyper-normal," a word that *literally* means "super-normal."[16] Steinbock, indeed, sounds much like Husserl on this very point: "Among all philosophical attitudes, it has been the virtue of phenomenological philosophy to call into question the naivete that structures our everyday life, our 'taking it for granted,' by carrying out a shift of perspective."[17] It is this shift in perspective, conversion, or flip that impossible thinking is about. Hence the importance of this particular phenomenological current of precedent for my impossible project in these pages.

And the similarities go on. Steinbock's severe criticism of contemporary academic culture and society at large as a kind of flattening or "de-spiritualization that deforms or reverses verticality," a long historical process that he calls "idolatry" and that today can be seen in the attachment to "world," so evident in movements like secularism and religious fundamentalism, also rings true to the way of thinking articulated here.[18] Idolatry, for Steinbock, is the very "negation of verticality." It sets itself against the epiphany of unconditional or absolute love that is so often given in mystical experience within the Abrahamic traditions.[19] Idolatry is the "horizontal" force that functions as a kind of invisible ideology in the postmodern academy and helps produce "the full wonder, weight, and horror of our contemporary situation . . . our de-spiritualizing downspin in all its forms." This all-encompassing flattening is one of the sources of the "just plain ugliness and stupidity" of the modern world.[20]

The language is sometimes admittedly harsh, but unfortunately it is all too understandable. Alas, the observation that the contemporary academy recognizes only the horizontal or the social is something of an understatement. Unsurprisingly, a figure like Husserl with his pure phenomenology and search for absolute consciousness has been almost entirely forgotten in the contemporary study of religion. Endless reductions of human experience to language, representation, society, brain, and bad politics (but no actual subject) have become the standard fare. In such an ideological culture, what Husserl wanted to put aside as the "natural attitude" has, ironically, become the only attitude.

In the spirit of the impossible, Steinbock will insist that "in principle, one should be open to all kinds of givenness, which is to say, all types of experience, without discrimination, no matter how paradoxical the givens may seem to be."[21] He will insist that academic philosophy has arbitrarily sliced off whole realms of givenness, of human experience, especially those that do not conform to the widespread prejudices of modern thought.

No matter how paradoxical the givens may seem to be. On one level, the impossible thinking theorized and modeled in these pages differs quite dramatically from figures like Husserl and Steinbock in its unabashed focus not on the experience of transcendental consciousness as such (Husserl) or on mystical experience of unconditional love and a personal God (Steinbock) but on the interface or middle world between this absolute consciousness and the human organism—which is to say, with the imaginal mediations of the fantastic, *however outrageous the latter appear*.

In the end, I worry that Steinbock does not take his own advice about how the phenomenologist should be open to all types of givenness. He clearly wants to separate, like so many other religious thinkers, mystical experience as vertical givenness from "everything that is nonrational, weird, exotic, occult, or paranormal."[22]

His argument also clearly privileges the monotheistic—which is to say, a "Personal" presence.[23] The phenomenological tradition follows this, alas, basically ignoring the fantastic that actually appears, opting instead for the orthodox past, the acceptable and thinkable, and the religions themselves: the possible. There are no gigantic insectoids, precognitive dreams, or UFOs in the backyard in this particular stream of thought. There are a few important, literally telepathic precedents, but these are only recently being held up as important, part of the present renaissance, perhaps.[24]

Obviously, my own impossible thinking does not make these distinctions between the proper theistic and the improper whatever-it-is. I consider such distinctions to be suspiciously culture-bound and insufficiently comparative. Impossible thinking embraces every religious experience that can *or cannot* be slotted into a theistic model, including entirely private and popular ones. *All* of this is on the comparativist's table. Like the phenomenology of verticality, this refusal to distinguish between the popular and the elite, between the "authentic" and the "inauthentic," between "religion" and "magic," this intentional oddness, will function as one of our five pillars of impossible thinking in my conclusion. There simply is no impossible thinking without this whacky weirdness, this refusal to distinguish.

Such a queer comparative table, it should be noted, is hardly Husserlian. Indeed, Husserl's relatively dismissive comments about "pure fancy" (*"phantasie"*) being "of a merely imaginative order" as nonempirical by definition—that is, as never being "real data"—could not be further from the spirit of impossible thinking celebrated here.[25] Also, the very abstractness or intentional generalizations of Husserl's thought—an essential abstractness or universality which seems to be based on geometry, mathematics, and physics (what he calls the *mathesis universalis*)[26]—is very far from the baroque mythical expressions of the religious imagination celebrated in this book.

But, on another level, this particular impossible practice—intentionally focused on the imaginal, the supernormal, and the paranormal—is very much in the Husserlian mode in that *it is normative and not purely de-scriptive*. It is all about the nature of reality. It is *not* an example of what is generally thought of as "religious studies" today. The latter discipline, as part of the conventional humanities, is defined by a very different kind of bracketing that can be summarized like this: "Set aside all beliefs, yours or theirs. Do not consider whether such a human experience, conviction, or conclusion is 'true' or not—that is, whether it corresponds to reality in some way—but rather describe everything neutrally. Do not take a position. Do not admit an ontology, much less a theology. Practice all forms of critical theory. Assume the social, the political, and the moral (namely, the horizontal) but never acknowledge the vertical. *Ever*."

I remember well how we were constantly told in the 1980s that such a religious studies method is not what Husserl's phenomenology was about. Wow, that was an understatement. The phenomenology of religion in which we were trained was a *setting aside* of all ontological questions for the sake of the fair or neutral comparison of religious forms and patterns and, in other modes, toward various critical theories (mostly psychoanalytic, feminist, postcolonial, race, and queer theory at that point in time). One could certainly think normatively, but only on the social and historical plane. That was radical enough for its time, but it seems so obviously inadequate today.

Not everyone agreed with this setting aside of the normative, of course. The African American theorist Charles H. Long, for example, was central to the history of religions school in which I was trained. He wrote extensively about the ways that Western imperialism, colonialism, and the slave trade have impacted how we think of "religion," particularly in the United States and its departments of "religious studies."[27] That is what he is most known for today. But Long also practiced the collapse of the subjective and objective, of the spiritual and the material, which is one of the hallmarks of impossible thinking. He wrote positively of "the imagination of matter" and insisted, in a very phenomenological mode, on the "inner structure of consciousness of *homo religiosus*," and this despite the fact that, as he wrote, "several modern disciplines have already decided that the religious life does not exist."[28]

None of this means that the material world, the cosmos, is not absolutely central for Long. For his own work, the religious imagination has *always* been formed in deep dialogue with "technological praxis" (agriculture, husbandry, and so on) and the physical universe (sun, moon, sky, and stone). Such a dialectical structure can never be reduced to one of its poles—that is, to its social and political meanings. Yes, there is terror and suffering, but there are also the stars and cosmic hope: "There is nevertheless the possibility for the rediscovery of the life of matter as a religious phenomenon—an equal and sometimes alternate structure in the face of the dehumanizing and terroristic meaning of history."[29] There is the both-and.

Long observed that what is sometimes called the Chicago school of the history of religions may no longer be at Chicago, that it is "present wherever in the study of religion the concrete situation and locus of the investigator is probed for that sense of internal 'otherness' that might be an analogue to the so-called 'external otherness' that scholars encounter in their investigations."[30] In my own terms, the inner life of the intellectual is the real comparative resonance and source of insight into the external data of ethnography and history—another impossible collapse of the subjective and objective.

These are not just empty academic gestures. Hence Long's many subtle but real provocations, including his observation that the prophet and the seer, who employ precognitive and clairvoyant powers, are "valid types of experience and expression."[31] Indeed, if we are to believe Thomas J. J. Altizer, the founder of the death of God theology, Long was convinced of the magical power of the traditional African curse and practiced it, in late 1989, on Altizer himself, who had been "Chuck's" fellow graduate student.[32]

It seems more than relevant here. Altizer's death of God theology (which was Nietzschean to the core) was a part of the history of religions school that Long helped to found, and which had its own experiential beginnings. Hence Altizer relates a dreamlike ecstatic "initiation" involving ghosts, "spirits of those dead who never leave us," in Hyde Park, Chicago, and goes as far as to argue that such presences were the "instruments" of Mircea Eliade, who lived down the street. The "deep sense of mission" Altizer came to know around the altered states of this spectral initiation led him to call Eliade "father" and resulted in his conclusion that "my way was not ultimately different than his."[33]

The death of God and the history of religions, then. There is another rabbit hole.

Let me walk this thing back a bit. I deeply appreciate the neutral setting aside for the sake of historical description of religious studies. I often think it is the only effective way through what the believer wants to call "authentic" and "inauthentic" texts and traditions (authenticity here is a kind of orthodox winnowing). Still, such a neutrality has genuine costs, and one of them is a resistance to philosophical thinking, which also becomes inevitable at a particular stage of the comparative enterprise.

It also needs to be said: such religious studies "answers" are not really answers. They are only beginnings, a partial setting of the comparative table. It also needs to be said: the neutral bracketing of present-day religious studies is *the exact opposite of Edmund Husserl*, who was doing something far more profound and important, who was in his own mind working for all of humanity in a distinctly positive, even ecstatic mode. He was very clear about wanting to lead all of humanity to a barely hidden (in Latin) "Eternity":

> I also want what the churches want: to lead humanity to the Aeternitas. My task to do this is through philosophy. Everything that I have written up to now is preparatory; it is only a development [*Aufstellen*] of methods. In the course of one's life, one unfortunately does not arrive at the core, at what is essential. It is so important for philosophy to be led out of liberalism and rationalism, and to be led once more to what is essential, to truth. The question concerning ultimate reality, to truth, must be the object of every true philosophy. This is my life's work.[34]

I cannot think of anything more impossible today. Truth itself.

Roads of Excess

At the turn of the millennium, I was writing my second book, *Roads of Excess, Palaces of Wisdom* (2001), a study of the mystical experiences of scholars of mysticism and how these altered states of consciousness and erotic energy secretly shaped the public scholarship of five major scholars of religion (the human sciences as occult, again). The first four figures were, in this order, the early proponent of Christian "mysticism" and British Episcopalian spiritual guide, Evelyn Underhill; the French Islamicist and biographer of the crucified Sufi ecstatic al-Hallaj, Louis Massignon (the mentor of Henry Corbin); the British Oxford Zoroastrian scholar, Sanskritist, and gay spy, R. C. Zaehner; and the Austrian American anthropologist turned countercultural snobby Hindu monk-mystic, Agehananda Bharati. All four had passed at the time of my writing. The early work of Elliot Wolfson on the theophanic envisioning of the body of God in medieval kabbalah, the identity of mystical experience and scriptural interpretation of the kabbalists, and the homoerotic structures of their ecstatic visions constituted my fifth case study.[35]

It was the spring of 2001. I was proofreading the galleys of *Roads* when Elliot Wolfson came up to Cambridge from New York City to help me teach a seminar at Harvard Divinity School that I had called "Method as Path." It was *Roads* in pedagogical form. I was teaching his work that week. We were both young. For some reason, I most remember the hat he was wearing on a crisp New England day: one of those flat wool caps often called "Ivy." Was this some stylish nod to his Jewish identity, a way of covering the male head that was not obviously Jewish? Such visual paradoxes are certainly Wolfsonian enough, but I may be overreading my memory. Maybe he was just cold and needed a hat.

I do know that he was happy. So was I. And why not? What we were envisioning in our respective bodies of work seemed possible, thinkable, doable—an engagement with the full history of religions that embraced and celebrated all the critical and historical methods of the humanities, particularly the gender and sexual theories so prominent in the humanities at that time, but also recognized the potential gnostic or esoteric effects such a study could have on the scholars themselves and their careful readers. We saw a new world of thought emerging from the historical depths of mystical experience.

Then it all seemed to disappear before our eyes, partly via Logan Airport, the very airport I was using to commute between Boston and Pittsburgh every other week. I would take my last flight home in the middle of June 2001, after a beautiful Harvard graduation where the German Continental philosopher Hans-Georg Gadamer was present and honored. A few months later, everything would seem to collapse in fire, smoke, and leaping, choking, crushed bodies.

It was already patently obvious to me that Elliot Wolfson would become a major voice not only in the study of medieval kabbalah and Jewish studies but in the larger field of the comparative study of religion. There was no one writing in English about the imaginal and the religious imagination whose interpretations or practice of hermeneutics I respected more. That is still true. When I want to understand Henry Corbin, I ask Elliot Wolfson.

It is my own long-held opinion, which I have shared with Elliot Wolfson on many an occasion (to his own considerable discomfort), that, although it is perfectly true that his body of work has been mostly ignored by scholars outside Jewish studies, his seemingly endless stream of essays and monographs carry tremendous implications for the study of religion—indeed, for the future order of knowledge itself. To call Elliot Wolfson "ahead of his time" is something of a grotesque understatement, although it may also be true. If there is an intellectual who knows how to think impossibly, it is Elliot Wolfson.

Such thought in Wolfson has something profound to do with the paradoxical nature of the image, the interpretive act in a participatory quantum world, the circular nature of time, and an autopoetic or self-evolving cosmos. It has to do with what the kabbalistic tradition calls *Ein Sof*, or the Infinite, which in turn poeticizes everything that is, including our conceptions of a personal God, or what Wolfson calls our "theomania," literally our God-madness, or, we might well imagine, our God-pathology. Such impossible thought also has to do with the mystical nature of interpretation in a universe in which space-time can literally swerve back, enabling the act of interpretation to influence or even change the meaning of that which has come before. Words are experiences, but some such word-experiences, handled with enough intention and attention, can travel "backward."

Hermeneutics constitutes a particular power over time.

My engagement with the books and essays of Elliot Wolfson has been anything but exhaustive. It has been selective and partial. That is how life and scholarship generally work. If there is anything I possess in great measure, it is finitude, limitation, and fallibility. I certainly claim no complete knowledge, total reading, or adequate comprehension of the Wolfsonian oeuvre, and I will pretend none here.

Part of the problem is the sheer scope of Wolfson's body of work. Who can read all these dense essays and exquisite books, with Talmudic notes that constitute more miniessays and entirely other directions? And even if one *could* read all of this, pursue all these directions, how can one really understand it all, unless one happens to be professionally trained in ancient and medieval Hebrew, the history of Judaism, kabbalistic hermeneutics, biblical scholarship (Torah and New Testament), Continental philosophy, the histories of Gnosticism, Neoplatonism, and Western esotericism, and the contemporary comparative study of religion?

Where in that list did *you* stop? I personally possess an adequate knowledge of only the last subject and, on my best days, a smattering of some of the others. I do not read Hebrew, and I possess no adequate knowledge of the history and hermeneutics of kabbalistic literature, much less Jewish studies. I can pretend something of the biblical critical thing, I suppose. Still, I am a gentile, a voice from outside the Jewish tradition. But then why *was* I so drawn, and still am, to Elliot Wolfson's work? What was it that so attracts this most inadequate reader?

I think I know.

Eros, Gnosis, Hermes

My uncanny friendship with Elliot Wolfson has morphed over the decades. In the 1990s, our initial conversations orbited around the exoteric heterosexual practices and symbolisms and esoteric sublimated homoeroticisms and autoeroticisms of male mystical traditions around the world. We had arrived at the same fundamental conclusion—the orthodox prominence of a kind of esoteric or sublimated male homoeroticism—with completely different cultural materials: he with medieval kabbalistic texts, I with Roman Catholic mystical literature and the Bengali texts surrounding the Shakta Hindu saint Ramakrishna (1836–1886).[36]

There was a kind of shock of realizing that the two of us, completely independently and with historically unrelated traditions, had come to more or less identical conclusions about any number of things, from the intensely erotic nature of male ascetic practice to the homoerotic underpinnings of two ostensibly heterosexual symbolisms to the deeply androcentric nature of apparently feminine or androgynous religious imaginaries. Even androgyny is finally masculine here.[37]

As our conversations grew and deepened, however, I became more convinced that a big part of the resonance worked on another plane. I began to intuit something comparative in nature. Allow me to explain. I have been struck over the years by how Wolfson refuses what every other scholar of religion seems simply to assume (or just wish): a historical positivism or absolute contextualism, a total immanence with not a whiff of transcendence, a kind of intellectual bowing down to absolute multiplicity, complete difference, and, frankly, ultimate meaninglessness.

Not here. As far as I can tell, Elliot Wolfson has done as much as anyone to emphasize and explore historical differences and textual nuances, even when such differences and nuances are anything but congratulatory. Think of his extensive explorations of Jewish phallomorphism (the metaphorical prominence of the male phallus) and the ontological subsumption of the female into the male in *Speculum*, his long meditations on medieval Jewish exclusivism and essentialism (really a kind of anti-Christian conviction) in *Language, Eros, Being*, or his radical critique of theism, or theomania, in *Giving Beyond the Gift*.

At the same time, Wolfson has also identified and refigured profound transtraditional themes, like the double mirroring of humanity and divinity in both medieval kabbalah and Christian mystical sources, what he calls "the specular entwinning of anthropomorphism and theomorphism: envisioning the divine as human mirrors envisioning the human as divine."[38] More radically still, he has always insisted on a kind of transhistorical unity or even identity among the religions at their most radical and most sophisticated, an empty not-one but also a shimmering not-two at the heart of global mystical literature.

Such a paradoxical thinking that emphasizes both radical historical particularity and ontic emptiness-fullness is extremely familiar to the historian of Asian religions. It is no accident at all that so much that Elliot Wolfson writes looks a good deal like some kind of Continental philosophical fusion of the medieval kabbalah *Ein Sof*, or Infinite, and some of the most sophisticated streams of Buddhist and Hindu thought. If I may, there is a certain Zen-like or Mahayana Buddhist quality in his work. Wolfson reads like a Jewish Nāgārjuna. If I had to describe Elliot Wolfson in a few words, I would say that he is a postmodern kabbalist with strong Buddhist convictions in the "emptiness" and apophatic "nothingness" of a shared and universal Godhead, which, very much in line with the ancient Jewish and Christian Gnostics, is *not* the "God" of the Bible that we are always asked to believe.

If I were Jewish, Elliot Wolfson would be my rebbe, my rabbi, my teacher. Maybe he is, anyway. He would certainly deny that. And that is how he teaches.

The Mirror or Coincidence of Opposites

There is also, in the Wolfsonian oeuvre that attracted me so, a particular structure of thought that it was difficult to name at the time but that is nevertheless recognizable with hindsight. One might name this structure with any number of inadequate words: paradoxical, circular, hermeneutical, reflexive, specular. It is this that I want to say now, even if it cannot fully be said.

Consider the reflecting image of the mirror. The specular image of the mirror, including the double mirror, shines throughout Wolfson's corpus. There is a certain optics here, which also encodes both a hermeneutics and a kind of postmodern gnosis that is, by definition, not restricted to any particular tradition or culture. That gnosis is rigorously dialectical and self-reflexive. It continuously bends back on itself, very much like the ancient *ouroboros* biting its own tail. This serpentine or tail-biting movement (which is also somehow vaguely autoerotic) doubles throughout the Wolfsonian body of work, determining in the process some of that work's most basic paradoxical insights.

The reflections of the mirror do not stop there. Each reading into its reflective surface evokes another series of reflections, another envisioning. Accordingly, I have been struck over the years how Elliot Wolfson's corpus reflects, in an almost occult manner, my own thought and writing. I am not exaggerating. There is something uncanny about this man's words, something that finally escapes and overflows reason, something that makes me believe in a kabbalah—that is, a received tradition, not of a purely Jewish wisdom, mind you, although that is part of it too, but of our own modern and now postmodern comparative gnosis embedded in the comparative study of religion. Hence, also, Wolfson's insistence that it is entirely traditional to subvert the tradition, that the greatest respect one can show a religious tradition is to engage it in a radically critical fashion and thereby to change *and* preserve it.

The "our" of my expression is carefully chosen, and deliberately open-ended. I include myself in it. I also include, with care, the texts of Elliot Wolfson, although he, of course, is free to reject or qualify my appropriation of his own mirroring thought. As readers, you too are free to extend this comparative gnosis to yourself, or to reject it as inappropriate. Still, *these texts are mirrors.*

There is an ontology behind or beneath Wolfson's reflecting words. Hence Wolfson's consistent embrace of Nicholas of Cusa's *coincidentia oppositorum*, or "coincidence of opposites," as a model of kabbalistic thought and as a forerunner and fulfillment of postmodern theory today. This coincidence or identity of opposites, which violates and *transcends* the Aristotelian logic that currently defines pretty much the entire academy, is perhaps the deepest structure of all Wolfson's cognitive structures. Here is how Wolfson describes this deepest of structures early on in *Language, Eros, Being*:

> To savor the mystical intuition of the divine as the coincidence of being and nothing—what may be considered for the kabbalist, as his counterpart in medieval Islamic and Christian mystical speculation, the primary ontological binary that comprises other binary constructions, the binary of binaries, we might say—one must reclaim the middle excluded by the logic of the excluded middle, for it is only by positioning oneself in that middle between extremes that one can appreciate the identity of opposites in the opposition of their identity: that a thing is not only both itself and its opposite, but neither itself nor its opposite.[39]

In the same paradoxical logic, Wolfson will call the *Ein Sof*, or Infinite, of the kabbalistic traditions the "negation of all negation," a phrase rich in implications for Wolfson's use of Continental philosophy and its general failure to negate its own negations and so arrive at something deeper and more real.[40]

Perhaps it goes without saying, but I will say it anyway as I move into my own modeling of the impossible in the next chapter: the psychophysical ground of dual-aspect monism is similarly conceived. It is "neutral"—which is to say that it is neither mental nor material, which means that it is fundamentally paradoxical to us "up here." It is the negation of all negations, but it is also very, *very* real. Indeed, it is the very foundation of all being and existence.

The Veil of the Image

There was always something else, though, in the Wolfsonian corpus that struck me, that produced so many "aha!" moments, even if I could not quite articulate these. It has to do with the reality of the image but also with what the image cannot image. Wolfson often communicates this key idea through the motif of the veil. There is no naked truth for Wolfson. The truth can never reveal itself as it really is. The truth can only appear to a human being in the form of an image—which is to say, through a veil. The veil, however, always implies a face, something which is veiled, just as that which is veiled, a face, needs the veil to appear at all. The bottom line is this: the religious symbol of the text or tradition reveals only through concealing and conceals only by revealing.

This paradoxical understanding in turn produces a very distinct concept of the image: *the image is real*. The religious symbol is not a symbol of something else—it is not a metaphor. It has no "intentional" object in the ordinary sense of that term. It is what it is. But it is also what it is not. It is true, and it is not true. But it is very real as an expression of the underlying or overlying Godhead or Infinite. To put the matter in the terms of Henry Corbin (with whom Wolfson sometimes thinks), this is a "conception of the image as a theophanic apparition that challenges the dichotomatization of the real and imagined."[41] This in turn is why Wolfson will write lines like this: "the contrast between 'imagined' and 'real' is an altogether misguided formulation."[42]

Words are experiences.

Wolfson expresses the same idea again through one of Corbin's favorite authors, the mystical philosopher and master Ibn ʿArabī. In the latter's writings, the imago appears "at the Meeting Place of the Two Seas"—that is, between the "Sea of Meanings" and the "Sea of Sensory Things." The Imagination (Wolfson capitalizes the expression) is once again not what we normally think of in modern secular thought. It is rather that which "embodies meanings and subtilizes the sensory thing."[43] It acts as a mediator between the two worlds or the Two Seas. Wolfson will thus adopt Corbin's "imaginal" but also deny the assumptions of the phenomenal world "up here," on the "presumed opposition between the real and imagined."[44] Both levels of reality—the surface and the deep, the Sea of Meaning and the Sea of Sensory Things—are so. Both are true. Both thus need to be affirmed.

Here is Wolfson on what communicates between them through paradox—the image: "The image, which serves as the *coincidentia oppositorum* that bridges transcendence and immanence, apophasis and kataphasis, invisibility and visibility, and thereby facilitates the epiphany of incarnational forms that escape the threat of idolatry, is not derived from the corporeal world of space and time; it is what imparts meaning to the objects of that world."[45]

There are other ways to carry and express the same paradox of the image. Wolfson will accordingly write things like how the symbol and the symbolized are identical, "albeit an identity that preserves the difference of that which is identified as the same."[46] I understand that these are difficult phrases, but perhaps we can simplify them with a single statement: for an embodied human being, *it is the imagination that mediates between the two dimensions of the real.*

As I tried to explain in the very first chapter, this is a *very* important comparative argument, as it allows us to take seriously the entire history of religions (as psychosocial mediations) and not just our own relative tradition or the secular scientism of the time, which would deny all of this as so much "hallucination." It is thus very important to Corbin, and Wolfson after him, that this place of the Imagination is the same place in Islamic Sufism, Jewish kabbalah, and Christian esotericism. Wolfson himself will also add the Asian traditions, particularly Buddhist, Hindu, and Daoist scriptural texts, of which he knows more than a little. Indeed, he will often comment that the further one goes down a particular path, the more likely one will find oneself in the heart of another path, however historically unconnected they might in fact be.

If I may, the Infinite is the Infinite. It is not contained by any border, culture, or social identity. It holds no passport. Nor does it need the permission of any academic or intellectual to be so. In Wolfson's more careful but always poetic terms, this is "a place that cannot be charted on any graph or pinpointed on any map, the interior space of the heart that is everywhere, since it is nowhere."[47] Hence "the perennial mystic quest to envision the invisible and to name the unnameable."[48]

Time itself, a topic to which we will return soon enough, is also quite different in such a quest. Time, it turns out, can be "ruptured," and no historical causality and so no historical criticism is absolute. And no such observation is "ahistorical," a criticism that is sometimes articulated when we arrive at this point. Such positions do, however, work with a much more expanded sense of time and history. In some very real sense, the position is superhistorical, with history now understood as a kind of hyperdimensional process that simply does not work, and has never worked, in just one way or in just three or four dimensions.

In Wolfson's more careful terms, this is "a time, in which the past remains present to the future, in which the future is already present to the past, just as the notes of a musical phrase, though played successively, nevertheless persist all together in the present and thus for a phrase."[49] And here is Wolfson again, now approvingly quoting Corbin: "Nights and days, hours and minutes, are simply means of determining the *measure* of time; but these measurements are not *time* itself. In itself, time is the limit of the persistence of the eternal Form 'on the surface' of the accidental matter of this world."[50] That is difficult to understand and paradoxical to our little temporally bound minds. Exactly.

Very much related here is Wolfson's notion of embodiment. Hence Wolfson's writings on embodiment or what he calls "divine corporeality" in the Jewish scriptural and kabbalistic traditions. The basic problem that Wolfson homes in on is the problem of why the bodiless God of scripture so often has a body in scripture. The easy way out is to read all the bodily descriptions in Jewish scripture in strictly metaphorical ways. God does not have a body, so all the bodily images and language must be something else, metaphors or allegories perhaps.

But that is not where Wolfson goes, because that is not where his kabbalistic sources go. Here, the human body corresponds to the divine body, and, most strikingly, "the anthropomorphic expressions inform us about the comportment of divine bodilessness, which illumines, in turn, the corporeal nature of the world and that of the human being."[51] Put again, the "ascription of a body to God is not merely a rhetorical device to enunciate the inherent metaphoricity of theological language; it is rather a mode of discourse that calls into question our naturalistic and commonsensical assumptions about human and cosmic corporeality."[52] Put in a simpler way, the scriptural presence of God's body encourages us to rethink our own bodies and the complexities of our own embodiment—which extend *way* beyond any biomedical, physical, or cultural model—encourage us to rethink the nature of God's body.

We begin to intuit that the human body is also God's body and that God's body takes on a human structure, what Gershom Scholem famously called "the mystical shape of the Godhead." We begin to intuit how the human being and God mirror one another in the kabbalistic texts, at least as they are interpreted for us by Elliot Wolfson now. In the temple of contemplation

(itself derived from *templum*, or "temple"), Elliot Wolfson has allowed us to
see that "the divine becomes human and the human divine."[53] And *that* is
the final meaning of "embodiment" for Elliot Wolfson.

Again, allow me to pull this back into the framework of the present book.
Allow me to observe that the soul or spirit is *never* without a body in the
dual-aspect monistic world that I will be explaining in our next chapter.
It cannot be, since the fundamental ground that is shining through the visionary symbol, subtle body, or landscape is neither mental nor material—
which is to say, it is both, or neither. Similarly, every "spiritual" experience
is also a "material" or "physical" event. They can *never* be truly separated
because they are not so separated in the ground of all being. The same is
true with what we so naively call "sex" and "spirit." The materialists are
right, but they are also half right. In a truly radical and thoroughly thought
through nonduality, *there is no such thing as immanence or transcendence.*
Both are category mistakes, an application of phenomenological descriptions "up here" in the body-brain to ontic reality itself "down" or "up" there
without such a splitting body-brain.

This, in the end, is why I consider Elliot Wolfson an impossible thinker.
He goes where others will not go, and which I suspect is a most productive
place to go, but he also recognizes that what he is so paradoxically saying,
repeatedly, is probably not translatable into any social form or political
platform, past or present. In other words, he recognizes that his thought is
impossible in our present context, and maybe any historical social context.

> Ultimate redemption would consist of attaining the state of conscious-
> ness—or perhaps metaconsciousness—that entails incorporation of all dif-
> ferentiation in indifferent oneness that is ascribed to *Ein Sof* or to *Keter*,
> the divine nothingness marked by the paradoxical coincidence of opposites
> such that night is day, left is right, white is black, Jew is non-Jew, male is
> female, and so on. . . . What is required is . . . an apophasis of the apophasis,
> a venturing beyond to the precipice, the chasm of the excluded middle, where
> opposites are identical in the opposition of their identity. . . . While it is not at
> all clear to me that such an ideal can be implemented sociologically without
> dispelling the very path that leads to it, this may very well be the most daring
> implication of the messianic potential of the kabbalah.[54]

He ends with a lesson for anyone who might compare: "This would pre-
clude not only the reduction of the other to the same but the reduction of
the same to the other."[55] It is paradox all the way down.

More Real than Real

I have dwelled on Elliot Wolfson because I think he has something profound to teach us about the comparative study of religion, about mystical litera-ture, about the image, really about everything. Who else today still thinks like that? Who speaks of the "symbol" in anything but superficially social or cognitive scientific terms?

There is this phrase that often pops up in the first-person accounts of the near-death experience, the abduction account, or the UFO encounter. Such an event, the witness relays with utter earnestness, was "more real than real." It was *not* a metaphor (it did not refer to something else). It was not a dream (it was not imagined in the banal sense). It was really, even physically, *there*. It was more real than real.

Words are experiences.

It is easy to invoke quick explanations at this point, usually around some lame understanding of the imagination, or maybe some plodding delusional "hallucination." Anything to return us to a flatland materialistic worldview that can be controlled and manipulated. None of them work. Oh, they might work to soothe the convictions of the reductionist and distract us once more from what is patently obvious to the experiencer, but such a person is not describing a delusion or an ordinary dream. This person is reporting another dimension of reality altogether, often with a sense of traumatic shock. The ontological is at the very heart of the experience. We can "bracket" that question, set it aside for the sake of historical descrip-tion or analysis, but we will, by so doing, set aside what the experiencer most wants to say.

But here is the problem: these stunned descriptions of what is more real than real are all different. The otherworldly landscapes, the entities, and the ships are not at all alike. Moreover, and more seriously still, many of them are manifestly modeled on cultural and historical references that we can easily recognize. A paradoxical thinker like Elliot Wolfson gives us a plausible and erudite way to balance this "more real than real" with this obvious cultural shaping. He "resolves" the paradox by showing us that the paradox is true.

The Real Nature of Time

As if that were not enough, what is equally remarkable about the Wolfso-nian corpus is that it means to make a claim on reality, on *physical reality*, on space-time and causality themselves. Wolfson thus instantiates what I have elsewhere called the "realist impulse of the cosmic humanities."[56]

Wolfson enacts this realist impulse quite explicitly in *A Dream Interpreted within a Dream*, where he puts kabbalistic dream interpretation into conversation with the speculative reaches of quantum mechanics in order to demonstrate how the construal of meaning in the dream is brought into being or actualized through attention, intention, consciousness, and the act of dream interpretation itself, much as the act of observation is said to "collapse the wave function" in one interpretation of the quantum mechanical experiment. It is simply not possible to extricate the presence of consciousness from the behavior of reality in either context. As shocking as the conclusion might be, to perceive is to be.[57]

But Wolfson also enacts this realist impulse in the prologue to *Language, Eros, Being*, and it is there that I most want to go in the final part of this chapter. Before I do, however, it is worth pointing out that there is a particular intellectual lineage at work in such moments. Such a realist impulse goes at least as far back as the ecstatic nineteenth-century figure of Friedrich Nietzsche. More immediately, however, it goes back to the Romanian historian of religions Ioan Couliano.

In the late 1980s and early 1990s, Couliano was teaching and writing about the history of mystical literature and paranormal experience and their likely relationships to quantum physics, hyperdimensional geometry, and modern cosmology. He was asking, in so many words, why historians were writing about "history," as if time really were a simple linear causal process, when we have known, since Einstein, that this is simply not so, that time does not likely work like this at all. In effect, Couliano was asking this bracing question: How should we think and write about the history of religions, and in particularly about mystical experiences and paranormal events, in a post-Einsteinian universe?

Hence Couliano's bizarrely beautiful introduction to *The Tree of Gnosis*, where he begins to explore what is essentially a Platonic model of historiography, with hyperdimensional idealist forms interacting in three-dimensional historical time with different actors and movements as these forms play out their different cognitive possibilities.[58] Basically, he was writing of the interaction of eternity and time. Hence, also, his little potent essay, "A Historian's Toolkit for the Fourth Dimension."[59] Husserl would have loved this. So would have Corbin.

I ask all my PhD students to read two essays: Couliano's "A Historian's Toolkit for the Fourth Dimension" and Wolfson's "Prologue: Timeswerve/

Hermeneutic Reversibility."[60] It is my own conviction that these few pages contain some of the most provocative lines ever written in the modern study of religion. For his part, Wolfson takes up the question of Einsteinian space-time in order to answer a most obvious and common criticism of his work: that there is something anachronistic or inappropriate about employing twentieth-century Continental philosophy to medieval kabbalistic literature. Put simply, you cannot use present categories of thought to interpret the meanings of the past.

Not if space-time is curved, not if the future can reach back to the past to change or reveal its potential meanings:

> Without delving into the thicket of theoretic grappling that this subject demands, I pose the rhetorical question: What would be the consequences if a historian were to take seriously the conclusion reached on the basis of Einstein's General Theory of Relativity that spacetime—the "mathematical structure" that "serves as a unifying causal background for phenomena"— is to be regarded as a curve? Does this not at least entail the possibility that the past is as much determined by the present as the present by the past?[61]

This, of course, was Ioan Couliano's question (and deeply resonant with Nietzsche's answer about the circularity of time and our potential superhuman powers over it). Wolfson pushes the point further, into the heart of matter, by invoking and quoting the German mathematician and physicist Hermann Weyl: "The possibility of future connecting with past, of time moving backwards, 'arises because a gravitational field implies that spacetime is curved, and the curvature might be great enough and extended enough to join a spacetime to itself in novel ways.'" What we end up with is "a closed loop figuratively depicting the object/subject becoming its own past." Hence Einstein's famous remark in a letter he wrote after receiving the news of the death of his friend, Michele Basso, that the distinctions between past, present, and future are ultimately illusory, that time itself is "a stubbornly persistent illusion."[62]

One can begin to see why historians would not look too kindly on Einsteinian space-time, why they might want to ignore the advances of theoretical physics, pretend, in effect, they never happened. Reality has fundamentally changed beneath their feet, but best not to look. Just keep walking on the surface of things, pretending that history is purely linear, that causality only works in one direction. This new reality, after all, presumes the final illusory status, or at least the relativity, of their discipline and, presumably, of themselves. It implies that space-time is, to quote the physicists themselves again, one immense "block," and that causal influence can move both forward *and* backward within such a block universe.

Such a model is speculative, like all cosmological hypotheses (and the metaphor of the "block" is too, well, blocky), but it is seriously maintained by numerous physicists and cosmologists, is supported by Einstein's relativity theory, and has received major philosophical attention.[63] In the block universe cosmology, developed after the work of Einstein and his teacher Hermann Minkowski, all of time or history—past, present, and future— already exists within an immense cosmic block or space-time continuum that extends from the big bang to however the cosmos finally ends (or "bounces" back). Temporality or our sense of linear time (and so all of "history") is, in effect, a neurological illusion or phenomenological apparition that our brains produce as we experience ourselves "moving" through this eternal and unchanging spatiotemporal block. We are back to Wolfson's metaphor of the entire musical score being present as we hear individual notes or movements.

We are not really moving in this cosmology. What we experience as our present bodies and specious selves are merely phenomenological snapshots or single frames along the running (but not really running) film of a "long self" or a "long body" that can be imagined as a decades-long, four-dimensional space-time worm wiggling out from conception to its disappearance in death.[64] Here is such a block universe as described by the scholar of mysticism and philosopher of religion Paul Marshall: "The special theory of relativity has led some thinkers to speculate that past, present and future events coexist in a unit of space and time called 'space-time.' Likewise, mystical experiences sometimes give the impression that past, present and future exist together in an 'Eternal Now.' This was certainly true in my own experience."[65]

Marshall is careful and qualified here, perhaps because he writes with significant training in special relativity. But he also writes of this particular comparison through his own mystical experience, as his last line makes clear. Like countless figures before him, Marshall will argue, in great detail and with great care, that such a universe not only exists as such but can be directly known as such.[66] This all, of course, implies the very real possibility of retrocausation within the block universe, to speak in the terms of the physicists. It implies "willing backward," to speak in the terms of Nietzsche. Yes, Friedrich Nietzsche.[67]

The Return of the Eternal Return

It is well known that one of Nietzsche's most important teachings, indeed what Michael Allen Gillespie has called his final teaching, was the "eternal recurrence of the same" (*"ewige Wiederkunft des Gleichen"*), also know more simply as the eternal return.[68] The teaching was about the circularity of time and how everything and each of us will be repeated, in our tiniest details, over and over again, not in some serial fashion but in a circular one. Much as someone traveling on the equator might think that they are traveling in a straight flat line, they are not, and they will see the exact same landscapes and coastlines repeat themselves soon enough. They are, after all, traveling on the surface of a sphere. If we extended this global circling to space *and* time, we would have something approaching the ecstatic vision of Nietzsche's eternal return.

It is important to understand that since Georg Simmel famously rejected Nietzsche's central teaching in 1907, commentators on Nietzsche have generally followed suit and widely dismissed the eternal recurrence of the same as incoherent and indefensible, as just a little, or a lot, crazy. When they are feeling more generous, they read eternal recurrence as an "idea" to which the philosopher reasoned and that we can now play with and "think" in our heads, as if it were nothing more than a cognitive act of neurons and education, or some moral experiment designed to get us to accept the unchangeable details and directions of our lives.

Such metaphorical readings of Nietzsche's eternal return are simply false ones. As recent Nietzschean scholars, foremost among them Paul S. Loeb, have taught us, such a safe reading is not reflective of Nietzsche's fierce conviction that, in the words of Loeb now, "he had discovered a fundamental truth about the nature of the cosmos that would change his life and the history of humankind."[69] Indeed, Nietzsche clearly thought of eternal recurrence as the most scientific of truths and whispered its awesome message to his closest disciples as some kind of transgressive, almost unspeakable, secret.[70]

That's because it is. And it came with equally awesome implications, including physical immortality (since these bodies always return exactly as they are) and a kind of retrocausal influence. In Loeb's reading, again, one of the central claims of *Thus Spoke Zarathustra*—woven right into the narrative arc of the book that Nietzsche himself considered to be his most important, indeed, to be world-changing—is the claim of "willing backward" (*"Zurückwollen"*). This is the power to influence the meaning and import of the past, if never to change the physical events of that past. This, of course, constitutes a willed interpretation of the past from the future that changes the meaning of that past, not dissimilar to how Elliot Wolfson's philosophical readings of medieval kabbalah change and transform the meanings of those historical kabbalistic texts. Hence his use of twentieth-century figures like Edmund Husserl, Maurice Merleau-Ponty, and especially Martin Heidegger to read Hebrew authors from the medieval period. Once again, the phenomenological tradition.

Unsurprisingly, numerous commentators on Nietzsche's philosophy of time have noted a clear correlation with Einstein's (later) theory of relativity.[71] In lay terms again, this is the notion that there is no absolute space-time through which all things are moving—space-time is relative to the observer; what is past, present, or future is not laid out on a single linear arrow; tense is entirely relative to the position of the moving observer. Put more colloquially still, time and space are not "out there" as a universal container in which all things flow in a single direction, as I suspect most of us assume.

I must quickly add that Nietzsche did not understand his eternal return in such block universe terms. He did not have that particular cosmology available to him. He thus argued that the same things repeat themselves eternally within a series of cycles, *but these are in fact the same cycle*. They are in fact *not* numerically different.

With the simplest "click" from the modern lens of the block universe cosmology, however, Nietzsche's implausible claim immediately becomes plausible, if also admittedly different (I want to own my own backward-willing or future-reading of Nietzsche here). This is *not* what Nietzsche thought, but I cannot shake the idea that the two ideas—eternal recurrence and the block cosmology of space-time—are somehow related, somehow two human attempts to get at the same superhuman reality.

In any case, with this new block cosmological click, things come into quick focus now. In the block cosmology, after all, every moment *is* happening again and again, right now, within the same always-existing, always-so block universe. So is every past moment. So is every future moment. It is all there, at once, simultaneously, happening "over and over again" not in a linear but in a block eternal sense. It is all one immense Now, one gigantic déjà vu universe.

It's about Time

More often than we imagine, the humanist's "historical experience" does not work historically at all and looks *a lot* like the natural scientific model that allows for retrocausal influences, particularly as we find it in the block cosmology. Put a bit differently, some of the most extraordinary moments of historical experience, singular life events that the individuals *never forget* and so are by definition "set apart," confirm a natural scientific model of time, including and especially the ability of the future to reach back to the past or present.

I find such a block cosmology deeply satisfying for a very simple reason: because it does not deny, because it does not make impossible, because it makes good sense of my own texts, which are *filled* with individuals pre-cognizing or dreaming a future that shows every sign of already existing. Such texts even include numerous instances of individuals visiting them-selves from the future, in effect influencing (or haunting) the present from the future, more or less exactly as Nietzsche claimed is possible in Loeb's provocative reading.

Clearly, if we take these reports as honest ones, then such events can hardly be whisked away as instances of "luck," "coincidence," or "anec-dote." This kind of hand-waving strikes me as a shameless intellectual cop-out. Is it not more honest to admit that precognition and visits from a future self simply *happen*, that these are only "impossible" in the framework of our present obviously fallible and relative present understandings of space-time, and that we can make such events possible again if we simply locate ourselves living in a block universe in which the future has already hap-pened and sometimes flows back into the present as a kind of apparitional self-guiding or spectral adjustment of the meanings of history?

Here hermeneutics becomes, once again, an actual power over time.

Perhaps such information does not really "flow" at all within such a universe. Perhaps, as the mystical literature claims again and again, *it is all one thing*. If so, reality is just communicating with itself as One, instantly and immediately, altogether in and as the block universe, from "Eternity," as Husserl might say. This is how we can think impossibly—with a new and much more scientific understanding of temporality and so of history.

In this specific reading back, we can now say that human beings can know, or dream, or, in some cases, literally "see" in a vision what is about to happen, not because they are guessing well or getting lucky in some cosmic poker game, not because they are "intuitive" (another cop-out), but because the event has in fact *already happened* and they themselves are *already physically connected* to it, really *are* it within the world block. There is not the slightest physical separation. It is all One.

Of course, I began this book with exactly this kind of precognitive experi-ence, that of Zora Neale Hurston as a seven-year-old girl on a Florida porch. Allow me to list a few more examples of what I take to be a shared space-time block. I leave the authors, dates, and places in the footnotes to emphasize that such experiences, being outside space and time as we normally think of these, cannot be completely reduced to these contextual notions. Listen:

> As one comes suddenly out of darkness, I perceived the full meaning of the doctrine of immutability and said: "Now I can believe that fundamentally all things neither come nor go." I got up from my meditation bed, prostrated myself before the Buddha shrine and did not have the perception of anything in motion. I lifted the blind and stood in front of the stone steps. Suddenly the wind blew through the trees in the courtyard, and the air was filled with flying leaves which, however, looked motionless. . . . When I went to the back yard to make water, the urine seemed not to be running. I said: "That is why the river pours but does not flow." Thereafter all my doubts about birth and death vanished.[72]

> What happened to me between 12:30 and 4 o'clock on Friday, December 2, 1955? After brooding about it for several months, I still think my first, astonishing conviction was right—that on many occasions that afternoon I existed outside time. I don't mean this metaphorically, but literally. I mean that the essential part of me (the part that thinks to itself "This is me") had an existence, quite conscious of itself, enjoying itself, reflecting on its strange experience, in a timeless order of reality outside the world as we know it. I count this experience . . . as the most astounding and thought-provoking experience of my life. . . . From my peculiar disembodied standpoint, all the events in my drawing-room between one-thirty and four existed together at the same time. . . . When we take off from an airport at night, we are aware of individual runway lights flashing past in succession. But when [we] look down a little later, we see them all existing together motionless. It is not self-contradictory to say that the lights flashed past in succession and also that they exist together motionless. Everything depends on the standpoint of the observer.[73]

> [In this frame of being] everything that has ever happened, as well as everything that will ever happen, all have an equal temporal status. In a certain sense, they are all there and one only has to look at them. . . . A perspective is taken by which all that will have happened at all times is co-present. In this limit situation, the temporal may, in a fashion, be reduced to the spatial.[74]

> Time didn't run linearly the way we experience it here. It's as though our earthly minds convert what happens around us into a sequence; but in actuality, when we're not expressing through our bodies, everything occurs simultaneously, whether past, present, or future.[75]

And then it all made sense. In that unsettling, parallel reality. . . . Dinah arrived at the realization that "birth and death actually don't have any meaning." When forced to clarify, she adds, "It's more of a state of always being. . . . Always being. So being now and always. There's no beginning or end. Every moment is an eternity of its own."[76]

Here is another instance of what I suspect is the block universe. This one involves a "willing backward," an actual lived timeswerve, and the future reaching back to the past to "interpret" and so make bearable its specific suffering. It comes from an email dated March 9, 2020, from a PhD student of mine, John Allison. It is best simply to quote Allison's own words. He had told me this story before, and I had asked him to write it down. This is what he wrote:

Late one night in early September 2013, I am sitting alone in my basement apartment in Princeton, NJ, after having had a few new friends over to play cards. As I sit there musing happily over the day's events, I suddenly notice a pair of flipflopped male feet and legs (with red shorts down to the knee), walking towards the basement window to my right. I am surprised by this, but even more so when a voice in my head materializes and begins repeating the same words over and over: "You're going to be OK, you're going to be OK, you're going to be OK."

Rather than feeling frightened by this voice or the unknown person standing outside my basement window late at night, I inexplicably begin to sense a strange, loving energy moving through my whole body, and I begin to cry uncontrollably. After about a minute or so, the figure outside the window vanishes. I then rush out my backdoor to see where he has gone, but there is no one in sight. I write down the incident in my journal. Eventually, I will forget about it completely.

The next three years were the absolute worst of my life. I suffered through terrifying bouts of heart arrhythmias, tachycardia, hypertension, and frequent visits to the ER, which led to chronic anxiety and depression. I often wondered if I was going to die. However, by 2016, I had made a turn for the better, and was getting happier and healthier.

Cut to May 2017. I am out on a quiet, late night walk, thinking about nothing in particular. As I am returning home, I unexpectedly get this sense that something important is about to happen, and then I notice that the light is on in my front room (which surprises me because I know I had not left it on). But in a moment, I am not just surprised, but stunned as I perceive that there is a man sitting in the basement room, and that man is *me*, except younger.

The hairs on my arms stood up on end. My heart began racing. I felt a surge of adrenaline in my body. And then, suddenly, a voice in my head said, "Now is the time." And somehow, I knew what I was going to do.

I rushed up to the basement window and I then put my forehead against the house, closed my eyes, and just "sent" this feeling of love and comfort to my younger self with the whole of my being. I don't know how long I stood there doing this, but when I was done "sending" this message, I looked down, and the basement lights were off.

I then ran inside, turned on the lights, and things were as I left them before my walk. And I fell upon the basement floor, weeping in joy, clutching my flip flops to my chest, and feeling like I had just been given some unthinkably tremendous gift. I often now wonder what would have happened if I had *not* somehow sent a message to myself during my years in crisis.

This was easily one of the most important events of my entire life. I have hardly told anyone about it. Since this event, I have felt a deep assurance that all moments in time somehow exist simultaneously, and that, for whatever reason, sometimes two moments in time not *directly* connected to each other in linear causality still somehow "bleed" into and affect each other.[77]

An "unthinkably tremendous gift." "This was easily one of the most important events of my entire life." Sit with those phrases, and then try to ignore them. Try to pretend that such events do not matter, do not possess historical agency, cannot be a very special part of what we so confidently call "history."

Obviously, I do not read such claims with the usual dismissive categories. I read them as honest and relatively accurate phenomenological descriptions of actual encounters with the physical cosmos of a warped or swerved space-time. Should I also repeat that John Allison is my doctoral student, and that experiences and present thoughts like his might well represent one future of the study of religion?

And Elliot Wolfson? I personally cannot imagine such an intellectual writing so eloquently and extensively about the timeswerve and its hermeneutics without having some experience of the same, but I also know that he is extremely reticent to speak or write of his own experiences. I do not expect him to do so, then; I do not even expect to find out if there have been such experiences. In some profound sense, it simply does not matter, *even when it matters*.

Let me explain.

It has long been my argument that some of the most canonical authors of the humanities derived their ideas from the inspirations of altered states of consciousness and energy. The core ideas of the humanities are superhuman ideas in the sense that they emerged from "above" or "beyond" the ordinary human and historical condition. They arose from ecstatic epiphanies of mind. They were *not* the result of simple cognitive processes, logical syllogisms, or social processes and practices. They just *appeared*. I have been saying this simple truth for decades, at least since *Roads of Excess, Palaces of Wisdom* (2001), where, not at all accidentally, I also first engaged the work of Elliot Wolfson.

Unsurprisingly, Wolfson makes a related argument. And why not? From the very beginning, he has shown us how direct mystical experience and indirect textual interpretation have implicated one another, have become one another in this particular Jewish mystical tradition:

> Whether or not any of the thinkers to be discussed in chapter one has had direct or indirect connection with kabbalah is not a necessary condition to justify the employment of their insights in decoding this singularly complex expression of the Jewish religious imagination. Nonetheless, one cannot by any means rule out such links. On this score, it is of interest to ponder the possibility that Western esoteric speculation, which is greatly indebted to kabbalistic tradition, has had an impact on the history of linguistics, especially evident in the period of Romanticism and its aftermath, including Heidegger, well versed in the theosophy of Böhme and its reverberations in the idealist philosophy of Schelling.[78]

Reverberations, indeed. One could go on for a very long time here. I know. I have.

My point? That this is why I think I have been so struck by the work of Elliot Wolfson over these decades, why his work has always felt so uncanny to me, so familiar and yet so other. Perhaps I knew of these ideas from our shared future, from this very chapter and book.

Whether that is so or not, this is one of the surest and simplest ways to recognize an impossible thinker: that time no longer moves in a single direction. It loops and swerves. It visits itself from the future. It is in time, but it is also outside of time. It's about time.

FIVE

The World Is One, and the Human Is Two

Some Tentative Conclusions

More and more I see the psycho-physical problem as the key to the overall spiritual situation of our age, and the gradual discovery of a new ("neutral") psycho-physical unitary language, whose function is to symbolically describe an invisible, potential form of reality that is only indirectly inferable through its effects, also seems to me an indispensable prerequisite for the emergence of the new *hieros gamos* [sacred marriage] that you predicted.

LETTER FROM WOLFGANG PAULI TO C. G. JUNG, May 17, 1952

This penultimate chapter is, I hope, the clearest practical statement to date of what I think and, most of all, *how* I think.[1] It will switch gears, as we say, and enter a more philosophical and confessional way of expressing myself. It is autobiographical in genre and intent. There is good reason for this.

I have written autobiographically before, actually many times, mostly because I think the genre is especially suited to doing serious philosophical work. I was trained in a discipline called, in English, the history of religions and, in its earlier German form, *Religionswissenschaft*, literally the "science of religions," badly translated into contemporary English as "religious studies," a confusing phrase which sows endless misunderstanding.

The history of religions, which we might also think of as the comparative study of religion, is deeply indebted to microtextual work, translation, and accurate historical and ethnographic description but also insists, usually at a later point in a career, on bold speculative metaphysics (that is, on thinking and writing about what it all might mean). Comparison is about data, yes, but it is also about theory or modeling what the data *taken as a whole* suggest. It is about all those important microhistories, but it is also about the big picture and, ultimately, about the nature of consciousness itself, what the intellectual lineage calls "hermeneutics."

As Charles Long has argued, probably the most significant feature of the history of religions school is its insistence that there is some kind of deep correspondence between the inner life of the interpreter and that which is being interpreted, between the consciousness of the contemporary scholar and the forms of consciousness encoded in the historical texts and artifacts.[2] The work of the interpreter is finally to realize, "make real," the world of the text or tradition being interpreted—to make the impossible possible, as we are saying here and now. This fundamental correspondence or interpretive resonance between the hermeneut and the historical deposits is the deepest source of truth in the intellectual-spiritual practice called "history of religions." Unsurprisingly, such a resonance often relies quite directly on the altered states of consciousness and embodiment undergone by the hermeneut.

I have certainly tried my best to practice all sides of the discipline, but I am probably best known today for the hermeneutical speculation, for those further reaches, as it were, that link the interpreter and the interpreted in unbelievable ways. This is why I have written so much about the mystical experiences of scholars of mysticism, the magical structure of comparison, and the superhuman origins of the humanities. Perhaps this is also why the American poet Joseph Donahue once described me as an "epic imagination trapped in a historian's body."[3] That is funny (the trapped part), but I also hope it is true (the epic part). I certainly *want* it to be true, since, as the reader of this book has probably realized by now, my own theory of the fantastic implies, really shouts, that the imagination, particularly within altered states, can become a revelatory translator of paranormal perception and not simply a spinner of daytime fantasies and banal nocturnal dreams.

The human religious imagination, it turns out, *is* epic. And it's not me.

(Non) Credo

The original title of this philosophical essay, which is now a chapter, was simply "Credo." That is already a problem. *Credo.* Latin for "I believe." Actually, *I don't believe anything.* I think that the paranormal investigator John Keel had it just right in a cultural meme widely attributed to him and certainly faithful to his body of work: "Belief is the enemy." Anne Strieber, the wife of the science-fiction writer and American abductee and visionary Whitley Strieber, used to say something similar, if in a more diplomatic way: "Mankind is too young to have beliefs. What we need are good questions."[4]

Still, I grew up Roman Catholic with a very nuanced creed that took centuries to hammer out through endless cultural wars in the first few centuries of the Common Era. We started every mass in my little Nebraska hometown by reciting it together: "Light from light, true God from true God, one in being with the Father." I realize today that this particular incarnational theology (the creedal lines refer to the doubled human and divine natures of Christ) has informed nearly everything that I think. My present writing, for example, on the superhumanities—that is, the academic humanities as already encoded with the "super" experiences of their historical authors (precognitive dream, evolving superhumans, and doubled selves)—is clearly indebted to that same ancient Christian credo. Jesus is fully human and fully divine. But so am I, and so are you. The human is really the superhuman.

Which is to say that I think creeds are *really important*. They encode centuries of careful thought, vigorous debate (not all of it civil), and extreme forms of revelatory irruptions, like the actual experience of human deification, "becoming divine," or "being God." Such creeds bind communities together and split them apart. They can even produce new thought and experience that the early believers could not have possibly foreseen, much less accepted.

I am fairly certain that the early orthodox Christians would not have accepted me. I know many of the contemporary ones do not.

The Human as Two, or the Tyranny of Clarity

One of the most important principles of how to think impossibly is something that we have already seen in the chapters above but that is worth underlining now: to think impossibly is to think doubly, to think as Two.

This doubleness works on many levels and in many ways, including on a revelatory one or in a given sense. The paradoxical structure of impossible thinking, for example, is so important, not as some failure or temporary unclarity but as the very way that particular kinds of truth must appear to a human subject or social person as a passive and culturally conditioned receiver. This is the structure of all metaphysical opening or revelation, not because there is no such truth but rather because *the human is doubled in all sorts of ways*—neuroanatomically (there are two brain hemispheres, or "selves," in the skull), epistemologically (there is both a knower "in here" and a known "out there," the latter mediated through the senses and the brain-body), socially (there is always an "inside" and an "outside" to our social experience, a social "outside," moreover, that the "inside" often finds stupid or absurd), psychologically (the ego and the unconscious, as in a dream), and spiritually (our deeper identities that appear to us in traumatic states as immortal spirits, angelic twins, doppelgängers, apparitions, and so on). I have encoded this multiple doubleness in a little poetic expression that is intended on all these levels: the Human as Two.

Such a doubled conception of the Human as Two has *major* implications for how one thinks and imagines the entire spectrum of apparently anomalous phenomena—that is, how one thinks about all those semiotic or meaningful moments in which one-half of this Human as Two is trying to communicate with the other half in symbol, story, and vision or apparition. Such translations, after all, are seldom, if ever, transparent to the rational or egoic half of the human—to that part of us that seeks clarity, reason, and formula.

Such translations *cannot* be transparent or clear, not because they are some preparatory approximations that could be in principle cleaned up or rendered clear in some future scientific way (such a claim disgusts me) but because this is what a human being *is*—Two. If you think that such things should be clear, then you are operating with a very mistaken model of the human. You do not know what you are talking about, or, better, just half of you is speaking and thinking. Get over yourself. *Be Two.*

My Rice colleague Alexander Regier, who works on Romantic literature in England and Germany (especially in conversation with the European Enlightenment movement), has written of the "tyranny of clarity," which was a basic Romantic conviction that real truth cannot be simplified and appear to human reason in a straight rational form, say, as a logical statement or syllogism.[5] Rather, profound truths will *always* appear to the human being in narrative, symbol, image, and poem. They will not be clean and clear because they *cannot* be clean and clear.

To speak in more contemporary neuroanatomical terms, the left side of the brain simply cannot understand what the right side of the brain knows, since they speak in two *entirely* different ways. To seek clarity, then, is simply to impose one set of standards on another. It is a form of tyranny (the word is rich in political connotations), this time of the left brain over the right. It is a symptom of our sickness or unbalance. It is a *problem*, not the solution. What we want, what we need, is both sides of the brain firing at once and in concert. We want them "talking" together, through reason and imagination, through rational clarity and intuitive symbol.[6]

"I Believe in Belief but Not in Beliefs"

I cannot stress this doubleness enough. Nor can I sufficiently emphasize just how existentially difficult such a paradoxical position can be. I have sometimes quipped that, "I believe in belief but not in beliefs." Listeners who want straight, easy thoughts, who see only left-brained clarity, look at me strange when I say that. They think that I am trying to be difficult, obtuse, or clever. But I am not. I am just trying to be precise and *true*.

Let me explain. What I mean by this quip is that I understand perfectly well that beliefs are important, that they sometimes actually work. Indeed, my theory of the fantastic *requires* them and affirms their importance, both personally and socially, as necessary mediations of the divine to the human. Such religious beliefs can also be transmitted from one generation to the next through scripture, architecture, art, music, ritual performance, and material artifact. They are relatively stable. I will say more about this in my conclusion.

But here is the thing: very *different* beliefs do very *similar* work in very different cultures and historical periods. There seems to be little, if any, necessity to holding this or that particular belief. Looked at comparatively (that is, globally), *particular beliefs are simply irrelevant*. A woman with cancer can be healed by the Blessed Virgin Mary in Fátima, Portugal. Or before the shrine of a Sufi saint in Karachi, Pakistan. Or in a near-death experience floating around a hospital in New York City. Or in a UFO encounter in Colares, Brazil. She can be a French Catholic, an Indian Muslim, an American Hindu, an Indigenous Spanish Amazonian, *or anyone else*. Honest and careful comparison quickly reveals the efficacy and power, even necessity, of belief but also the relativity of *any* of its local expressions or social identities.[7]

In effect, then, *I believe everyone*. But this also means that *I cannot believe anyone*. "I believe in belief but not in beliefs."

The same paradoxical comparative method quickly reveals another uncomfortable truth: *beliefs limit what is possible*. Beliefs shut down. Beliefs decide. If specific beliefs can be thought of as fishhooks that sometimes catch real fish, it can also be said that beliefs generally *only* catch fish, and particular *kinds* of fish that the hook is designed to catch, no less. But, if we extend the metaphor, we can also say that the waters are filled with other sorts of beings, most of which couldn't care less about these particular belief-hooks. Which is another way of saying beliefs can miss a great deal, *really almost everything*. Of what would we be capable if we did not limit ourselves through what we believe? This, I take it, was what Nietzsche meant by his most famous declaration that "God is dead." God (or at least the idea of God being believed) must die so that superhumans might live. Past belief shuts down future possibility.

It gets worse. Particular beliefs can *literally* demonize common phenomena that are not in themselves sinister or negative but become so when
they are believed to be so. Belief is a kind of mirror. So are impossible
phenomena. If you believe you are dealing with invading demons, you are
probably going to find invading demons. If you believe you are dealing with
beneficent angels, you are probably going to find beneficent angels. If you
believe you are dealing with extraterrestrials inappropriately involved in
our genetics, then you are probably going to find extraterrestrials inappropriately involved in our genetics. In truth, though, you are likely dealing
with none of these things. The phenomenon reflects to us what we will,
what we fear, who we think we are.

That reflexive quality of the phenomena may well be the very best clue
as to what some—maybe many, maybe most, maybe all—these appearances
actually are on a deeper level: us. We are all of this. And we are none of it.

Connecting the Dots We Already Have

Deep down, I suspect that robust paranormal phenomena are not personal
or individual at all but fundamentally nonhuman or superpersonal—in a
word, "cosmic." It is not about a mind. It is about Mind.[8] Which means that
anyone's experience is, in actual truth, "my" experience on a deeper level
or, better, all "our" experience. This is why, I think, extreme experiencers
so often want so much to write about their experiences—or why those
experiences want to be written about: because those experiences were
never really theirs in the first place; because they were always everyone's,
or no one's.

Once one arrives at such a superhumanist conviction, it follows that
one of the best sources of knowledge is to shut up, step aside, and listen to
people who have experienced something strange or inexplicable, something
big and vast. Because of four decades of such listening, I have some sense of
the bigger picture—the larger framework in which individuals and communities believe and experience unbelievable things. Part of this overwhelming intuition (and it is overwhelming) is something I have written already
above—namely, the conclusion that *we have more than enough evidence.*

Can anyone really read and absorb, much less process and analyze,
what two and a half centuries of psychical and then parapsychological research have amassed, and in multiple languages, no less? German, French,
English, Spanish, Portuguese, Russian, and so on? How many of us realize
how incredibly large the near-death experience (NDE) literature actually is

now, in both its medical journal and book forms? Is the UFO literature any less labyrinthine? It is just one rabbit hole inside another, inside another, inside another.

As a former navy fighter pilot put it, we have a real UFO problem, and it is not about balloons. The case has become overwhelming, particular since 2014, when the radar systems of the fighter jets were upgraded, and the military professionals realized with a new force that "there were unknown objects in our airspace." Lots of them. They were not effects of the radar system. They were not glitches. They were not part of some US secret military project. Their "advanced performance capabilities" are entirely beyond our own, or that of any other nation-state.[9] There is an understatement.

As I mentioned in my introduction, I am currently helping to collect and archive some of these literatures in our Archives of the Impossible. One collection alone took up much of the Virginia house of its donor (former Pentagon employee Larry Bryant), filling up an entire room before it spilled out into hallways. Another gift (from American physicist and remote-viewing scientist Edwin May) consisted of some fifty boxes of declassified remote-viewing or "psychic spying" material. Another (that of the Harvard psychiatrist and abduction researcher John E. Mack Institute) constitutes well over a hundred boxes. And on and on we could go. The truth is that the basic problem with these archives is that no single researcher can, or ever will, get through it all, much less understand these hundreds of thousands, soon millions, of documents and put them all together into some coherent picture.

This is a big problem. It is quite obvious to serious researchers that most, maybe all, of the anomalous phenomena are related, are expressions of the same basic iceberg floating above and below the waves: the NDE is related to the OBE (out-of-body experience) is related to the UFO (or what is now called unidentified aerial phenomena, or UAP—another dodge) is related to. . . . "Psi" in fact may simply be another Greek coinage for "stuff we don't understand about consciousness and its relationship to matter."

I laugh out loud when people accuse me of "exoticizing" consciousness. "Well, of course, I do. It *is* exotic. If you make consciousness normal, if you shove it into your puny categories and social boxes, you miss its true cosmic reach and vast nature. You reduce it to your own relative local notions, which will pass. That's you, not It."

Unfortunately for the purist, Bigfoot creatures show up around UFOs. So do werewolves and gigantic mantids. Near-death experiencers and psychedelic trippers emerge from their spiritual journeys with telepathic and precognitive powers. Dead friends or loved ones, even loved ones the experiencer did not know had died, show up in alien abductions *and* near-death experiences. *Whatever* this massive interconnected presence is, its textual and visual expressions are completely, totally, absolutely beyond the grasp of any single individual, conceptual schema, or acronym. It is all one big glorious . . . well, *something.*

Once one realizes the vastness and interconnected nature of the archival and historical situation, a familiar conclusion that I have already offered quickly follows: more information, more stories, more data is not going to get us where some of us want to go. Rather, what we need is a framework or theory in which to place all this data and so render it, as a whole, plausible and meaningful. Basically, we need a way to connect the dots. We do not need more dots. Intelligence is finally meaningful collection and connection.[10]

The analogy is not perfect, but our situation is a bit like what educated people in the European cities of the seventeenth century thought of all those silly farmers in the countryside who kept talking about rocks falling from the sky. "Rocks can't fall from the sky, damn it. Stupid farmers. Obvious idiots." Or that's what the smart people in the city said. Until they eventually arrived at a new model of what outer space actually is. Turns out the uneducated peasants were correct, that their untutored experiences were perfectly true, even if their interpretations were sometimes off.

This is precisely why I "think" and "theorize." I try to explain why rocks really do fall from the sky (or in closed rooms from the ceiling) or, in this case, why people really do dream the future, leave their bodies in disaster or illness, see hairy creatures and gigantic insectoids, and encounter craft floating over cities or conscious balls of light in their living rooms and werewolves in their backyards. I try to make the impossible possible, not only with more and more outrageous stories but, above all, with new thought and theory. We know how to collect. We do not yet know how to connect.

Presentist Nonsense

Strange beings come out of "the heavens" (from the stars). They appear in the form of glowing disks, crosses, or objects in the sky. They grant struggling human beings different forms of cultural and technological knowledge (laws, writing, agriculture, ritual). They warn them about the end of the world, teach them about the nature of the soul, generally scare the shit out of them, and even sometimes engage them sexually to create divine-human hybrids.

That's the history of religions.

That is also the modern UFO phenomenon.

Be wary of your assumptions at this point. You might think that I am arguing that the ancient religions were misremembered, misperceived, or misunderstood contact experiences with alien astronauts. A young Carl Sagan thought that. So does a contemporary US television series and, no doubt, millions of its viewers.

No, I am not saying that. I do not believe that.

Why? Because that kind of mythmaking is simply an example of a very common error that historians of science called "presentism." The error of presentism is the understandable (but always false) presumption that one's present scientific worldview or grasp of the cosmos is somehow correct or adequate, that we have arrived (or are about to arrive) at the truth of things, and that our present order of knowledge will not change appreciably in the future.

Presentism is nonsense. The order of knowledge *always* changes, often quite dramatically. Just look at what happened to Newtonian physics in the twentieth century with the rise of quantum mechanics. Or biology in the late nineteenth century and early to mid-twentieth century with the rise of Darwinian comparativism and then the eventual discovery of the gene and the DNA molecule in the 1950s. Or look at what is happening to cosmology right *now*. Scientists once told us that we are quickly closing the gaps on all scientific knowledge. Now they are telling us that it appears that *all* our science—all that physics, chemistry, biology, you name it—applies to only between 2 and 6 percent of the actual universe.[11] We went from knowing almost everything to knowing next to nothing in just a few decades. And then they give fancy names to our ignorance: dark matter and dark energy.

So much for presentism.

What the modern ancient alien enthusiasts are doing is basically assuming the truth of *their own mythology*, in this case a kind of spacefaring physicalist scientism, and subsequently reading it back into human history (often not very well—their grasp of the histories of religion and magic are atrocious). In this presentist myth, there are no invisible nonhuman agents, no God, no magic, no miracle, no world of the dead, no two worlds of experience, nothing genuinely paranormal or supernatural. There is only misperceived technology. Such enthusiasts have taken Arthur C. Clarke's "any sufficiently advanced technology is indistinguishable from magic" and turned it into a universal dogma that denies there is anything like magic at all—it is *all* technology now.

The problem with this "law" is that what we once called magic happens all the time, and it has nothing to do with technology—spacefaring, digital, or otherwise. It has everything to do with consciousness and cosmos, with mind and matter, with our neurobiology and the highly evolved ways that our perceptual apparatus "splits" a single reality into two domains (a mental and a material domain) so that we can survive and thrive.

I am perfectly aware that Clarke himself went through different stages around the paranormal, celebrating it (and linking it to evolution) in novels like *Childhood's End* (1963) but later becoming more suspicious, particularly in the 1980s in television series and tabletop volumes like *Arthur C. Clarke's World of Strange Powers* (1984). Still, Clarke's humor is redemptive even in this later context and leaves the proverbial door open, or at least a bit ajar: "At a generous assessment, approximately half this book is nonsense. Unfortunately, I don't know *which* half; and neither, despite all the claims, does anyone else."[12]

That seems basically correct, and very funny.

Comparative Magic and the Occultism of Science

I would go somewhat further, though. I would say that reality *is* often magical—that is, mental states and physical events can be split off an original fundamental but mysterious unity and still resonate, still correspond. I would say that there is such a thing as real magic.[13] Such real magic is not misperceived or mistaken technology. Much less is it some kind of dreaded "primitive" or "magical thinking" (this is racist). I suppose, if you prefer, you can say that it is a type of biotechnology. People dream of physical events before they happen. Dramatic correspondences between subjective states and material events happen every day. The comparison that links these experiences and events is quite literally magical thinking, but in a most positive and preternatural sense now. Comparison is magic.

Science works in a similar way—between some rather shocking but never really explained correspondences between states of mind and matter. Science is basically an occultism, a way of positing a hidden order of things that is invisible to almost everyone, except, of course, to the scientists. And this occultism works. Indeed, the ultimate *tertium comparationis*, or "third space of comparison," between the inner and outer world is mathematics. Mathematics is, *by far*, the most effective symbolic system ever discovered between these two epistemic domains. Numbers are "psychoid," as Carl Jung put it in his own neologism. They are mental, but they most accurately describe and even predict the behavior and structure of the material world: "the sought-after borderland between physics and psychology lies in the secret of the number."[14]

Secret, indeed. Who can really explain this? Who even tries? Little wonder that so many elite mathematicians are Platonists—that they are convinced that numbers and equations are *discovered* not constructed by the human mind, that mathematics exists in its own right in some esoteric space. There is the occultism again.

Comparison in this mathematical sense is not some cognitive process in the head, some abstract way of organizing things in an interesting but finally arbitrary way. It is eerie proof that the external world really does correspond to the internal one. Such a comparative unity also explains what the physicist Eugene Wigner called, in 1960, the "unreasonable effectiveness" of mathematics. Mathematics is so damned effective because the symbol system is "true"; that is, it gives constant witness to an uncanny correspondence between the human mind and the cosmos.

In the end, then, there simply is no fundamental difference between the mental and the material, between the inside and the outside. Deep down (outside or beyond "us"), it is all One World. Mathematics works and mental intentions can magically correspond to material effects without one "causing" the other because these two dimensions are *in fact* in correspondence: they are both grounded by a deeper reality or an ontological source from which they each emerge and that they both share, that they both are.

Dual-Aspect Monism, or the Meaning of Meaning

Philosophically, I am articulating a form of dual-aspect monism—that is, the position that reality is ontologically One but epistemologically Two. *We* are the "splitters" of reality. Reality itself is not so split. Such a philosophical framework is more than worth further articulating at this late juncture since it shapes and forms all that precedes and follows it. In some ways, it *is* impossible thinking, which I have applied to some of the wildest exempla of the history of religions to show how and why it works so well.

Put succinctly, dual-aspect monism is a philosophical proposition that states that "the mental (psychological) and the material (physical) are aspects of one underlying reality which itself is psychophysically neutral" (that is, neither mental nor material).[15] There are different forms of dual-aspect monism. The one I am invoking is "decompositional," since it further posits that the mental and the material domains have "split off," or de-composed, from a previous holistic state, in this case the psychophysically neutral ground, which is an undivided whole or nondual in nature (I would say "in supernature").[16]

I confess I do not like the word "neutral," since I consider it rhetorically boring, and this fundamental ground of the entire universe is *anything* but boring. Indeed, coming to know or become this ground of all being is the very goal and purpose of human life for many of the world's contemplative traditions, about which they have spoken so extravagantly for millennia. Still, I also understand that "neutral" is philosophically precise: it means neither mental nor material. So, I will use it too.

There are other tools in this toolbox. Harald Atmanspacher and Christo-pher Fuchs call this specific proposition "the Pauli-Jung Conjecture," since it was developed in the private correspondence and later public writings of the quantum theorist Wolfgang Pauli and the psychiatrist and depth psychologist C. G. Jung, and since it is not yet proven as a theorem (hence its conjectural status). There is genuine humility and fallibility in such an expression: this decompositional dual-aspect monism is a best guess and a working hypothesis, not some dogmatic certainty.

Jung, well read in medieval alchemical texts, called the psychophysically neutral ground the *Unus Mundus*, or One World, after the Latin expression of a sixteenth-century alchemist named Gerhard Dorn (circa 1530–84). But Jung was also thinking of his own understanding of the collective uncon-scious, the deepest and most unknowable level of being, a psychoid or physical-mental sea in which all individual psyches are floating, as it were, and that in turn produces the comparative patterns of world mythology and dream. For his part, Pauli would generally relate the same ground to a quantum level that is defined by a radical wholeness or unity, nonlocality, a non-Boolean logic (think "both-and" instead of "either-or"), and all the other "impossible" features of quantum theory that he knew so well and in fact helped theorize and mathematically model.

It bears repeating that such an ontology is philosophical and not quan-tum mechanical in nature. It is not a scientific theorem. It is a philosophical framework. This does not mean, however, that the Pauli-Jung Conjecture is unrelated to quantum theory. It certainly is. The psychophysically neutral ground in fact bears any number of similarities or possible relations to the unsettling reality, or superreality, posited in quantum theory. Accord-ingly, what Atmanspacher and Fuchs call "structural elements of quantum theory" might help us elucidate such an ontology if we can use such an

ontology and the posited philosophical implications of quantum mechanics
upon which it draws with nuance and care.

Foremost among these possible elucidations is the fact that there is no
longer any detached or independent observer in quantum physics; to quote
the cosmologist John Wheeler, there is no "man who stands behind the
thick glass wall and watches what goes on without taking part."[17] Rather,
there is a participating observer who is a part of the experiment being run
and so a part of the results that are measured. Moreover, and more radically
still, "any quantum measurement is not simply a reading-off of a measured
value but also induces a (generally uncontrollable) change of the state of
the system observed."[18] Every measurement, in short, changes that which
is being measured. Or in Wheeler's terms again, "Nature at the quantum
level is not a machine that goes on its inexorable way, [rather we] are ines-
capably involved in bringing about that which appears to be happening."[19]

To put the matter in humanistic terms that may well be related, reality
is hermeneutical all the way down. It responds to our interpretations. *We
change reality by interpreting it.*[20]

Another structural element of quantum theory that is particularly helpful here is its radical holism. It is difficult to make too much of this, since
such a holism involves, well, *everything*. Briefly, conventional science assumes a strictly local cause-and-effect model, with this particular thing
bumping into that particular thing and so effecting some physical influence
or change. Such an "efficient causality," Atmanspacher and Fuchs point out
in *Dual-Aspect Monism*, has become the dominant metaphor of modern
science. Hence the mechanisms and materialisms of the present reigning
worldview. Nature is considered to be one big machine, with countless and
relatively independent parts.

Not so with the strange holism of quantum theory and the fantastic way
different "parts" of the system remain "entangled" despite being separated,
even by great cosmic distances. There is no local causation between entangled particles. There are no parts bumping into other parts and thereby
changing them. There is a kind of instant communication, as if everything
was one presence communicating with itself. Such a new model, Atmanspacher and Rickles explain, has *profound* consequences for everything from
evolutionary biology to ecological thought (since the World is One, and evolution looks very much like it is rooted in a realm of meaningful coincidence
or synchronicity—more soon). It also explains why they entitled the last
chapter of *Dual-Aspect Monism* "Outlook: After Physicalism."

It's over.

And that ending is another beginning. This holistic quantum structure, this One World, is rich in implications and points to another way that mental and material ("material" here defined as an epistemic construction—that is, what human beings experience as the material world, not the world itself) might well be related in a dual-aspect monistic world—*through meaning*. Here, again, comparison is the key, comparison between the inner and outer worlds. As Jung would have it with respect to synchronicities and the relationship of the mental and the material domains, "their *tertium comparationis* is meaning."[21] Put a bit differently, such meaningful coincidences are not expressions of causation. They are organized by meaning.

This is a complicated philosophical claim that works on two basic levels—a common but superficial level "up here" in the mental and material domains, with the sign or metaphor "intending" or "being about" something else (hence meaning as it is commonly understood in the humanities), and a deep level "down there," in the psychophysically neutral ground, with the symbol or mystical experience issuing from a reality in which there are no absolute distinctions or differences and where paradox is the norm. This is the *meaning of meaning*, if you will, a kind of profound "sense" or state that is not intentional or "about" anything other than itself. Monism is the meaning of meaning.[22]

Atmanspacher and Rickles will often call this psychophysically neutral ground from which the mental and material emerge a third realm, a *tertium quid* that renders the famous mind-matter problem moot and entirely unproblematic (since both mind and matter split off the same neutral base), and that in turn promises a new and future spirituality, hinted at by *the founding quantum physicists themselves*.[23] Listen to Atmanspacher and Rickles:

Quantum theory predicts correlations which cannot be explained by efficient causation, but by decomposing wholes into parts, systems into subsystems. In this process, holistic features become disentangled, and the properties of the emerging parts exhibit holistic, *acausal* correlations. In physics, such correlations are well known due to quantum entanglement. . . . The *Pauli-Jung Conjecture* goes one step further and posits that a similar principle is at work in psychophysical systems, including both mental and physical properties in correlation with one another. Here it is evidently inconceivable how efficient causation could operate between such categorially distinct entities as mind and matter. Pauli and Jung agreed that *meaning* should be regarded as a proper alternative to efficient causation in this case. Pauli even postulated "a third type of laws of nature consisting of corrections to chance

fluctuations due to meaningful or purposeful coincidences of causally unconnected events." This third type of laws of nature beyond efficient causation and blind chance, beyond deterministic and statistical descriptions, is
entirely undeveloped in present science.[24]

Or the humanities. There is another understatement. Two of them, actually.

Such a conjecture may sound relatively innocent, but it will have major
consequences, impossible ones, really, for how we think about everything
from the writing of history ("what actually happened" is often far stranger
than is acknowledged) to the study of religion (religious states of inspiration are often precisely those that display a "third" psychophysically
neutral structure) to a science like evolutionary biology (since, certainly
for Pauli, that psychophysically neutral ground provides a third space
that could have influenced evolutionary history through meaningful coincidences connected through meaning—in short, a kind of guiding self-
direction, or *autopoiesis*).[25]

For most of their lives, both Pauli and Jung were Kantians with respect
to this deeper psychophysically neutral ground; that is to say, they denied
that a human being could know the ground as such. To sum up Kant in a
single sentence, "the world we experience is not the world as it really is."[26]
Pretty much ever contemporary humanist believes that to this day (and it
is very much a belief). Jung appears to have changed his mind, however,
particularly late in life in his last big work, *Mysterium Coniunctionis*. At the
end of this book, he writes, many times in fact, of a *coniunctio*, or "conjunction," defined, in a single line fusing the biblical book of Genesis and Indian
nondual philosophy, by "the union of the integrated human with the *Unus
Mundus*—the potential world of the first day of creation, when nothing was
yet in *actu*, no two or many, but just one."[27]

Jung will also identify this *coniunctio* with the *agnosia* (or "unknowing")
of the Gnostics, the affirmation of Plotinus that "all individuals are merely
one soul," the "suprapersonal atman" of the Hindu nondual traditions, and
the identity of the individual and universal "tao" of Chinese philosophy and
culture.[28] In short, he saw this One World as the ontological basis of mystical traditions around the world. Reality is One, and so are such realizations
of this One. How could they *not* be? Unless, of course, we want to argue
that humans do not and cannot have any knowing access to this shared
fundamental ground. This would be the Kantian position and, by extension,
the position of most academics today.

Seems safe.

But this is certainly *not* the case for the dual-aspect monism of At-manspacher and Rickles, who often relate their model to earlier European philosophy, including here: "Over and above meaning as a subject-object relation, dual-aspect monism offers the option, contrary to Kant, of direct, immanent experiences of the psychophysically neutral reality, which avoids the problem of access to a transcendental realm. If this reality is primordial enough, like an *Unus Mundus*, it may be aligned with the 'absolute' and bring us back to Hegel. Immanent experiences refer to modes of knowledge and meaning that are certainly non-standard, and are reminiscent of Spinoza's *scientia intuitiva*."[29]

The points on immanence and transcendence are especially important for the historian of religions, as they bear directly on the philosophy of religion and the experience of revelation as something given or outside the human subject. Philosophically speaking, with a dual-aspect monism, *"immanence" and "transcendence" are entirely relative to the body and brain*. Put differently, there is such a thing as immanence or transcendence, but there *are* experiences of both in and as the body-brain. Put differently again, immanence and transcendence are epistemic phenomena, not ontological facts. There are no such divisions in reality as such.

Still, people do experience transcendence as bodies and brains. Hence, Atmanspacher and Rickles will use the adjective "transcendent" in their gloss on the physicist Arthur Eddington's attempt to describe the reality beyond the mental and physical domains as "the spiritual world." Such a world is, according to the authors now, "a deeper transcendent reality which does not partake in the subjectivist treatment of physics, nor can it be nailed down by objectivist descriptions—it appears to be something like the 'absolute' . . . accessible by mystical experience."[30]

Accessible by mystical experience. It is difficult to overestimate that claim, as it renders the human being capable of a kind of ultimate ontological realization, a third realm, or *tertium quid*, that cannot, please note, be explained by physics (the material domain) or psychology (the mental domain). Indeed, to reduce this third realm to either matter or the socialized psyche is a very serious category mistake within dual-aspect monism.

In physicists like Arthur Eddington and John Wheeler, then, we find "repeated hints toward a kind of spirituality that transcends the mind-matter split."[31] Much of this new spirituality—which we might call either post-religious or postsecular—will involve the dissolution of "experience" itself, since all experience is phenomenologically dualistic (that is, an intentional expression of a subject-object split). This dissolution of the subject-object structure is precisely what "mystical" means for Atmanspacher and Rickles, which, once again, they relate to Hegel and, further back, to Spinoza: "Both Spinoza's intellectual intuition and Hegel's absolute join in with such most fundamental immanent experiences, which are often described as mystical revelations. We should repeat, by way of *caveat*, that they rely on the dissolution of any subject-object split, so that the notion of experience takes on a flavor entirely at variance with subjective experience in our ordinary usage of the term. Where there is no subject, there is no subjective experience."[32]

Can we *really* hear that?

Certainly not as psychological subjects or social egos—that is, as the splitters.

But it is not just the fairly rare case of mystical knowing of unity, oneness, or the Hegelian absolute that the dual-aspect monistic model makes possible. It is also the much more common human experiences "up here" in the psychosocial world, where the One World can still be known indirectly after it has split into two within human experience and material history *but still can communicate something of this deeper unity*. Hence, central to the correspondence of Pauli and Jung was a stunning but quite common historical phenomenon (which historians are forever ignoring, underestimating, or simply misreading) that Jung came to call "synchronicity."[33]

A synchronicity is a human experience in which something happens in the physical environment that uncannily corresponds to something happening in the mental world of a human subject. These events are not causally related; that is, one domain (say, the mental intention or thought) does not cause or effect the other domain (say, the material event). But they do correspond in some real way. They truly *coincide*. They correspond so strikingly because they have, in fact, split from a common source and, as such, are two halves of an original whole.

Jung wanted to call these mind-matter events "meaningful coincidences," but that does not capture what is intended here, since neither Pauli nor Jung thought that these were actual examples of randomness or pure chance. What they finally concluded is that a synchronicity is a powerful expression of a dual-aspect monistic world in which the mental and material domains have decomposed from a more basic One World but still retain a clear sign of their shared ground in how they correspond or speak to one another in unforgettable ways. They are not connected through any cause. They are connected through *meaning*. Hence the common human response to them: a sense of uncanniness or deep meaning that cannot be quite located.

Which brings us into the very heart of the humanities, of course, where meaning, at least since Franz Brentano at the end of the nineteenth century, has been understood as "intentional." An ordinary meaning, Brentano rightly pointed out, is always intentional, in the sense that it is *about* something else. In the last quarter of the twentieth century, this intentional notion of meaning was more or less dogmatized in the humanities by postmodern thinkers like Jacques Derrida, who insisted that every meaning is really a grammatical function, that every word points to other words for its meaning, that there is no end to this referentiality, and that humans live in these endless networks of words and can never get out of them. As Derrida put it, there is only what he coined as *différance*, a new French word that combines the words for "differ" and "defer" to signal an endless deferral of meaning. "*Il n'y a pas de hors-texte*," he famously declared, a phrase that is usually translated as "There is nothing outside the text."

Except there is. Human beings have been experiencing reality outside language and texts for millennia. They can do this most fully in what we generally call mystical experiences (although that expression is not quite right, as we have seen), but they can have such experiences within anomalous or excessive events as well, in what we generally call paranormal events. *How* this is happening in either case is quite another matter. This is where dual-aspect monism and the work of Atmanspacher and Rickles around what they call the "deep structure of meaning" is so very helpful.

For these two authors again, meaning is a kind of mind-matter mediation that works on *two* very different levels. Most commonly, and certainly in the sense of the humanities today, meaning generally works horizontally on a psychological or social level between the mental and material domains (again, "material" as the physical world as experienced, not in itself). Derrida was perfectly correct on this superficial level. There is no social sign outside of the text, and every sign or word points to other words and social signs. It is a hall of mirrors.

Until, of course, it is not. On a much more profound level, meaning can also work vertically between the two surface domains of the mental and the material dimensions of experience and the deeper One World from which they have emerged or split off. This is where the straight historical event becomes the synchronicity. Atmanspacher and Fuchs call this vertical meaning of meaning "sense."[34]

This is why the human subject usually notices that something very special is happening—for example, in a synchronistic event—even if that same subject often has no idea why or how it is happening. They sit up and take

notice. Such humans are being spoken to, but they cannot always understand the language. Such correspondences between the mental and material
domains of ordinary human experience and the One World "below" them
are the actual experiential source of the religious senses of the numinous,
the uncanny, and the eerie. Beyond or below that still is the mystical sense
of unity or identity in the psychophysically neutral ground of all being. In
short, we are talking about the very sources of religion *before* religion. We
are also talking about physics—that is, about the behavior of the material
world. We are talking about everything. We are also talking about a realm
in which space-time is void, there is no place for ordinary causation, and
there is no distinction between mental and material.[35] That is not physics.
That is metaphysics.

The Present Order of Knowledge

It comes down to this. Traditional religion or conventional science will
never get us to a solution or resolution of impossible phenomena, not because we do not have enough data but because these two knowledge systems are inadequate *in principle*. It is not that we do not believe enough or
reason enough, that we will somehow get to the truth of things if we just
believe the right things or do enough science. The problem is that we think
we can "believe" or "think" the truth at all. We cannot.

What to do, then?

The first thing we could do is recognize and come to terms with this
double failure. We need to stop relying so much on *either* our traditional
religion *or* our conventional science. We must understand that our models
are inadequate.[36] We have to be humble, not because humility is a moral
virtue that we need to signal but because humility represents an open-
minded recognition of our actual condition and situation. We do not know,
and it is quite likely that we *cannot* know, not because we do not yet have
enough information but because our dualistic way of cutting up the world
into "belief" and "reason," into "subjects" and "objects," is simply false;
that is, these dualistic divisions do not correspond to what is actually so.
They refer back to our evolved neurology and cultural histories. Again, *we*
split the world into two. We *are* that two. The world is not two.

Let me give you another example of what I mean, this time in the form
of a joke.

I often joke on the lecture circuit that there are just two things one
must understand *together* to understand the UFO phenomenon: radar . . .
and revelation.

No one ever laughs, probably because they don't get the joke. People don't get my joke because they are caught, they are assuming, they think they *are* our present order of knowledge. (Either that, or it's just not funny.)

Very generally speaking, our present order of knowledge is organized along a hierarchy of values within a specific materialist metaphysics. We might think of the sciences as disciplined forms of knowledge concerned with the behavior of matter or "objective" reality, with the number or mathematics as the privileged symbolic medium. We might think of the humanities as disciplined forms of knowledge concerned with the expressions of mental reality or "subjective" reality, with the text or narrative as the privileged symbolic medium. So, object and subject, number and narrative, mechanism and meaning.

This order of knowledge is also an order of *values* or ontological judgments. The objective material aspects are considered real, whereas the subjective mental aspects of reality are considered less real or even unreal. The "outside" of reality, after all, is manipulable, measurable, and predictable, and we can engineer technology out of it. We can make cool stuff.

The "inside" of reality . . . not so much. Some go as far as to mock and demean the inside of reality, with their constant refrain that every human experience is nothing but an anecdote, and that many such anecdotes do not add up to evidence. What they really mean is that they have no idea how to study or understand these inner worlds. They also know, on some level, that to protect their specific understanding of objective material reality, we better ignore or refuse to think about subjective mental reality. Cosmos *must* have nothing to do with consciousness. If it does, alas, we will have to rethink the dominance and ultimacy of our present science.

Our current hierarchy or pecking order of knowledge follows, with the study of matter (like physics and chemistry) on top and the study of mental expressions (like philosophy, history, art, and religion) on the bottom. Biology is messier, as it has to contend with that annoying category of "life," which it generally refuses to grant any ontological or "real" status. The social sciences (psychology and sociology) are definitely hovering somewhere in the middle of this hierarchy, wanting to be "hard" sciences but forever having to study "soft" human beings.

Once we understand our present order of knowledge in this way, it becomes rather obvious why something like the UFO phenomenon has such a difficult time fitting into that order. The reason? It does not follow this binary script. Many UFO phenomena, after all, are neither purely objective nor purely subjective. They are both. Worse yet (or better yet), some of their most extreme instances involve phenomena that simply cannot be rendered as purely material or purely mental. As such, they threaten to collapse altogether this useful distinction between the external and internal worlds.

This is why so many contemporary scientists who are quite sympathetic to, say, extraterrestrial civilizations or alien spacecraft in our solar system are openly hostile and emotionally demeaning to abduction experiences. They are reasonably okay with the material objective aspects of the phenomena (the radar), but they are arrogant and dismissing when it comes to the subjective aspects of the phenomena (the revelation). They want to chop the phenomena in two and deal with only its objective or material aspect. They don't get the joke. They *can't* get the joke.

That is what they are trained to do, of course. *That is how conventional science works*—by rending reality into two and then ignoring one whole side of it, the side of reality, of course, that the scientific method cannot make sense of, cannot understand, not in degree but, again, in principle. *Conventional science works because it gets to set the rules of the game and then pretend that phenomena it has no way of measuring or manipulating do not exist.*

This, again, is why conventional science will *never* get us there: because reality is not so binary. Cosmos and consciousness cannot be separated. We can pretend for a little longer, of course, perhaps to do the conventional science that really needs to be done to get us to a different place. But we are constantly being reminded by impossible phenomena that this is in fact a useful pretense. It is not the real.

The Imagination as Medium, Translator, or Interface

There *is* a way outside, or to the side, of mathematics that the external and internal worlds communicate, come together in human experience or little mind. Such moments do not yet represent a complete coming together, an actualized "not two" of cosmos and consciousness. But they often do point or signal in this same cosmos-is-consciousness-is-cosmos-is-consciousness direction, if always in an indirect, coded, symbolic, or mythical way. They involve that third space of meaning. Such moments, in short, need to be interpreted, read, engaged. They never speak in clear unambiguous ways. They cannot. That way is the imagination. That way is the fantastic, the fantastic as actual mediation of the real in which the One splits into Two and as an imaginal apparition or materialization.

I trust that the discussion of the imaginal, supernormal, and paranormal of chapter 1 prevents a misreading of these statements along "imaginary" lines. I do not mean to argue that technological craft, conscious energies, or invisible entities are not involved, are not in the bedroom or in the ocean, or that the being cognized is "only in the imagination." I mean to suggest, rather, that *whatever* or *whoever* is there (and it may well be many things or beings, or us, or the world of the dead), these presences are interacting with the human organism in and through that organism's imagination, which is itself a feature or dimension of consciousness.

The imagination is *the* privileged organ of contact, communication, and communion. Reason is mostly impotent. Language and grammar are something of a joke (Derrida was right). The senses are not reliable: they evolved for other much more practical and adaptive reasons. But the imagination is of an entirely different order. At least in its empowered or activated states, it becomes a revelatory, if always imperfect, organ of mediation and translation.

By invoking the imagination, I intend to refer to especially rare, palpably felt, often literally "electric" or "electromagnetic" altered states of mind and body that result in visions or virtual reality displays that the person in question does not consciously construct. I mean a kind of unconscious moviemaking that just "goes off" in the imaginal consciousness of the individual seer. Perhaps such a "movie" is sometimes triggered or activated by an actual electromagnetic source in the environment that is then refashioned into the visionary event.[37] Perhaps such a "movie" is triggered or activated by a nonhuman or superhuman presence. In any case, the person in question does not "imagine" such things. The person *is shown* such things. Hence the revelation of my joke.

It is also worth remembering that the imagination shows every sign of being capable of affecting *physical* objects and events, even the very workings of space and time, often toward some symbolic meaning or contact communication. Think flying, breaking, or falling objects in poltergeist events (signaling anxiety, anger, or some unresolved trauma). Think marks on the body of the visionary in the Christian stigmata or the modern UFO implant. Think communally witnessed alterations of gravity (the floating or levitating saint or Victorian psychic). Think banal physical objects appearing out of nowhere or teleporting across the room. In these instances, it very much appears that the subjective and the objective, consciousness and cosmos, are interacting in ways that clearly violate any binary understanding of the two realms or dimensions. We are back to dual-aspect monism.

The history of religions and folklore is positively filled with these mindmatter moments, as are contemporary UFO encounters. They involve lights or plasmas in the bedroom, craft in the sky, or monsters in the backyard (which rarely harm any human, by the way, so moral lines *are* being commonly drawn). Trees catch fire. Burn marks appear on the ground. Contactees develop superpowers: they become telepathic, precognitive, clairvoyant, spiritually awake. These are genuinely wondrous events. These are also *physical events*. There is radar.

But revelation often quickly follows or appears in conjunction with the physical events. Elaborate visionary displays might follow, which other people can usually not see (but sometimes can). These virtual reality shows communicate things, say things, show things. They also often involve beings well known from the history of religions and mythology. This is why experiencers write books filled with religious language, drawn from, well, from anywhere—from Christian theology to Greek mythology. It all works. Sort of.

Can we read such events literally? I don't see how. I don't see how we can believe all these very different and flatly contradictory beliefs. Can we read them instead as symbolic attempts to communicate something of immense human and maybe even cosmic significance, perhaps as "magical" in the technical sense of linking the mental and the material toward some future awakening to the World as One? Yes, I think we can.[38]

The Fantastic Symbol: A Mythico-Physical Reality

There is a final issue here. It has to do with the different natures of representation, image, language, and what is sometimes called, at least in the study of religion, the sign versus the symbol (two *very* different conceptions). This takes us back to the two natures of "meaning" in Atmanspacher and Fuchs, a superficial level in the social or textual sense and the "deep structure of meaning" related to the psychophysically neutral ground. Reference and sense, to use their language.

I should make clear what I have already implied above—namely, that the sign so conceived is now almost entirely dominant in the academic world and in our culture more generally, and that the symbol has been out of fashion for decades in intellectual circles and is today generally dismissed, when it is addressed at all.

That is a problem.

I cannot go into all the complexities involving the philosophy of language and the nature of the image, and I understand that different theorizations of the sign, symbol, and image are present in the humanities and social sciences. But I want to end this chapter with a most basic point. I want to do so because I think this point haunts, and helps produce, much of our confusion around the imaginal nature of impossible phenomena and the overwhelming ontological conviction of the extreme experiencers that what happened to them was "more real than real" and should be considered as such.

Generally speaking, the *sign* in the contemporary humanities is entirely secular, artificial, arbitrary, and ultimately meaningless, whereas the banished *symbol* is implicitly religious, spontaneous or "revealed," and supermeaningful. Most everyone in the humanities today assumes that the "sign is arbitrary"—that is, that texts, words, dreams, and visions all refer to other texts, words, dreams, and visions; in short, all such signs are entirely and completely "constructed" by historical, social, and neural processes and do not refer to anything outside themselves. They cannot. They can only refer to other signs in endless weaves and webs of illusory and always receding meaning—that is, to more and more historically relative signs.

This is why the psychedelic visionary trip, the abduction event, or the NDE is commonly considered nothing more than a trick in the head, a "hallucination." It is just a mishmash of culturally determined signs appearing in and as the awareness of the subject (*how* it is doing this, no one can ever quite say).

In this view, there is no meaning in the real world—which is to say, the objective physical world. Meaning is an anthropomorphic hang-up, a projection. In fact, reality is meaningless, and we are all just lost in our webs of words and representations, in sign after sign, which gives us the illusion of meaning when there is in fact none. Ultimately, the present conception of the sign is nihilistic to the core. It means nothing. Nor does, or can, anything else.

The *symbol* is very different. The *sumbola* is from the Greek and literally meant "thrown together" (*sum-bollein*) and, probably originally, an "agreement" or "contract." It referred to things like a shard of pottery broken in half and then given to two business partners so that the deal could be recognized later by a perfect match, or the wholeness of the doubled human who had not been split apart into genders and sexualities. In short, the *sumbola* was a whole that had been literally broken or split into two and then recombined or "thrown together" to form the whole again. Later, the *sumbola* came to refer to uncanny coincidences (another form of coming together) or to forms of ritual and reading divination that understood the image, the ecstatic poem, or the visionary display as a kind of esoteric cipher into deeper and deeper meanings of the real, which was suffused, soaked, with meaning.[39] The correlations here with dual-aspect monism should be more than obvious.

When a person who has had an NDE or a UFO experience comes back and reports a fantastic event of endless sci-fi special effects, the report is almost always expressed as a supermeaningful coincidence of symbols and almost never as an arbitrary collection of culturally conditioned signs. Moreover, *the experience often initiates other synchronicities and meaning-ful mind-matter coincidences in the physical environment*. This total event of meaningful symbols and uncanny coincidences is experienced as cosmic truths of astonishing significance—that is, as *images and events that par-ticipate directly in that which they represent*.

That is because they do.

This is why experiencers, as we saw above, almost always say "it was more real than real." It was. The visionary display was participating in, probably issuing "from," that which is actually the case—that is, the real, beyond or before it splits into the subject and object of human experi-ence. What the experiencer is actually reporting, I believe, is a temporary awakening into the One World beyond the "inside" and the "outside" of normal human perception and knowing, even if that One World is still coded in symbolic, mythical, or narrative form within the religious imagination, which is acting as a medium or translator here.

It is perfectly true that the experiencer often conflates the truth of the One World that is shining through with that through which it is shining (that is, the symbolic landscape or mythical narrative). But the impossible thinker does not need to do this.

Okay, I actually *believe* that. There is my credo. I believe that what is symbolically encountered in these fantastic events is the real world mani-festing in the body-brain of the experiencer in ways that the person can hear and integrate in some fashion—which is to say that these events are mediated by the "imagination," which, in these moments at least, is no more, and no less, than a function or dimension of consciousness as such. I believe we should not believe the literal contents of such imaginal pro-ductions or movies, but we should nevertheless watch and listen to them with great care and attention as moments of contact, communion, and communication with the real.

Please note that by "consciousness" I do not mean your egoic aware-ness, or mine, or that of anyone else. I do not even mean something neces-sarily human. Obviously, from what I have said above, I also do not mean to restrict the reach and influence of such a reality to the mental world, however grandly conceived.

This is why I often refer to this deeper reality as nonhuman or super-human. A dual-aspect monist who is more precise than me might want to refer to it as "neutral" (that is, as neither material nor mental). I would invoke negative theology and mystical literature here and look for other, more po-etic, ways of expressing the same metaphysical truth—some "divine dark-ness" or Buddhist "emptiness" or Eckhartian "Nothing," perhaps. In any case, I mean to refer to a presence that is cosmic, everywhere, everywhen, because it is nowhere and nowhen (that is, not restricted to space or time but always appearing to us evolved primates in space and time).

There are endless payoffs once we make this "flip," even if we stick within a more idealist framework (the brain is in mind, the mind is not in the brain).[40] For one, pretty much the entire spectrum of religious experience and ideation that are *impossible* within the present materialist framework begin to become entirely plausible, if not quite likely. The flip makes the im-possible possible. The survival of some form of consciousness after bodily death (even if it is impersonal or superpersonal), telepathy, possession, after-death communications, clairvoyance, precognitive dreams—pretty much all of it becomes possible (if still symbolic) in a new flipped worldview in which consciousness (or some other reality that is "neutral" with respect to mind and matter for which we have no name) precedes the body-brain identity and is not actually located or reducible to space, time, and causality.

We can finally begin to connect the dots—but *only* when we have moved outside our presentism, materialism, reductionism, and scientism, and *only* when we no longer conflate the spontaneous revealed symbol of ecstatic vision with the arbitrary cultural sign of ordinary language and egoic expe-rience. An NDE or abduction experience is *not* a cereal box. The fantastic is not a stop sign.

The debunkers are perfectly correct about one thing: if conventional materialism is true, these things cannot be.

Too bad they are. We can ignore them with all sorts of lame and unques-tioned rhetorical tricks ("multiple anecdotes do not add up to evidence," "extraordinary claims require extraordinary evidence," and so on), but none of this does anything to stop the phenomena from appearing. They just keep appearing, and always will. Clearly, the problem is the lame rhetoric and unquestioned assumptions, not the phenomena themselves. Multiple anecdotes *are* real data and add up to theory. And extraordinary claims are not extraordinary at all in the context of the broader history of religions.

I should add that none of these "beliefs" of mine are really that new. Symbolic, philosophical, and mythical forms of what is technically known as dual-aspect monism have been around for millennia. Rare, indeed, is the human civilization that believes that there is only insentient matter, and that all forms of sentience are accidental byproducts of this dead stuff. Considered comparatively and historically, our present scientific materialism is the clear outlier and is almost certainly in very serious error.

Such a doubled "neutral" approach to the mind-matter relationship and a subsequent "symbolic" understanding of visionary worlds is especially obvious in the UFO world. Perhaps that is why I am so drawn here. Already in 1978, the French sociologist and philosopher Bertrand Méheust was writing of the flying saucer as *"une réalité mythico-physique,"* a "mythical-physical reality"—in short, a revelatory symbol that participates in both the sci-fi imagination and the physical universe.[41]

That is fantastic.

So there it is—my *credo*, my "I believe." I believe that the World is One, that the Human is Two, and that this One World often appears in the Human as Two in and through the symbols, myths, and dreams of the empowered imagination. Indeed, precisely because of this Two-in-Oneness, reality can and does appear in us all the time in openly and robustly paradoxical ways that can fit no present order of knowledge that is organized around the "object" or the "subject"—which is to say, *all* forms of modern academic knowledge.

Such ideas, of course, offend and confuse the adequacy of our present scientistic beliefs and humanist projects. They are neither traditionally religious nor conventionally scientific, nor are they simply material or purely mental. They point to a third future space of mind-matter in which we almost certainly participate but have not yet coded in any culturally syntonic and stable manner. This is how one thinks impossibly.

SIX

We Are God (and the Devil)

Further Thoughts and Moral Objections

We should be a mirror of being: we are God in miniature. . . . Am I supposed to have created the whole universe? Was it the movement of *my* I that brought this about, just as it brought about the movement of a body? Am I merely a *droplet* of this force? I comprehend only a being that is simultaneously one and many, that changes and stays the same, that knows, feels, wills—this being is my *primordial fact*.

FRIEDRICH NIETZSCHE, *Unpublished Fragments from the Period of "Thus Spoke Zarathustra"*

This last chapter begins with some theological reflections that were originally included in a conclusion for my 2022 book, *The Superhumanities*. I did not publish them. "Too fucked," as Phil Dick, the barely fictional character and stand-in for the actual science-fiction writer, Philip K. Dick, would say.[1]

Maybe not. I wanted to set out as clearly as possible what I thought at that moment—my best guesses about the big picture. The original title of that book was *Teaching the Superman*. Friends and my editor persuaded me against using it for all the usual political and gendered reasons, none of which I was really persuaded by. Still, I eventually let it go, as a title anyway, not so much because of these good reasons but because I realized that I was not arguing for the Superman. I was arguing for the superhumanities, which is a very different thing.

Still, the phrase "teaching the Superman" stuck with me.[2] I may have let it go, but it did not let *me* go, partly because I think it is closer to what Friedrich Nietzsche—who was *so* central to that book (although almost no one can really hear him today)—actually intended by his seminal term the Übermensch, partly because of my affection for American popular culture, where the Superman and superhumans of all sorts are everywhere, but mostly because it expresses exceptionally well my esoteric pedagogy, with which I opened both that earlier book and this one: the not so simple truth is that I was, and still am, teaching the Superman.

In my own mind, the title phrase meant teaching *about* the Superman, as the history of religions has been doing for millennia toward very different ends and meanings, and as the humanities could still do, if it simply just chose to do so. But the phrase also meant teaching *the Superman*: teaching those who have been so awakened—of every gender and sexuality—into our superhumanity how also to be an embodied, caring, and just human being (that is, a particular person, even a good person). In short, the phrase goes both ways. It is another little poem about the Human as Two, now three words long.

How can I say this differently? Allow me to put the same sensibility in a language that people without a fascination with Nietzsche, or a slightly embarrassing adolescent superhero obsession, can hear. Let me speak in terms more people, *many* more people, can hear. Let me speak about "God."

In this much more traditional framework, I can see at this point in my life that my books, and maybe my entire lifework, has been about one thing: teaching humans how to be God and God how to be humans. I sincerely believe that the former project (teaching humans how to be God) is what the history of religions—still incomplete, still ongoing—has been partially about and where, I hope, at least some of it is headed. I also sincerely believe that the latter project (teaching God how to be human) is what the humanities—still incomplete, still ongoing—has been partially about and where at least some it is headed. And I understand both processes in plural ways. There is no singular answer or solution here. Only endless conversation and experimentation. That is precisely what I mean by the superhumanities—an infinite set of criticisms and conversations with no end.

I write a word like *God* with some trepidation, as I have generally been very reticent about invoking God in my work.[3] I have been reticent for theological, philosophical, historical, moral, and psychosexual reasons.

On the theological side, I worry about any singular solution or approach when human beings are so incredibly diverse and different. Belief in the One God has too often met the intolerant destruction and even ruthless slaughter of the other gods (and the communities who worship them). The idea of monotheism itself, for example, has been a historically horrendous one for far too many peoples, a long horror show.

On the philosophical side, I worry about theism in general; that is, I worry about the positing of ultimate value outside the natural world. Accordingly, I have pushed for a new, more expansive naturalism, a super naturalism that can take in all the rogue phenomena without too quickly resorting to some external deity or supernatural force outside the cosmos.

On the historical, moral, and psychosexual sides, almost any God-talk carries massive violences against colonized, enslaved, marginalized, and victimized peoples and cultures; vast historical tracks of scientific ignorance, or just pure idiocy; and, not to be underestimated, countless personal traumas in people's family memories, psyches, and shame-tortured bodies. I am thinking in particular of people with same-sex desires, transgender or nonbinary identities, and the deeply intertwined histories of race and religion in scripture, thought, and practice. There are many reasons "God is dead," then, and I have no desire to bring *that* bastard back. I am perfectly happy with his long Western funeral. I stand with my colleagues.[4]

And yet . . . and yet human beings continue to really and truly experience a beneficent Reality or Presence of unconditional love and acceptance, both within and without the religions, that can be described as, well, *as God*. I cannot deny that. Nor would I want to do so.

I am perfectly aware that these transcendent experiences appear to conflict with the other common human experience of injustice, suffering, evil, fate, and determinism. All I can say here is that it is obvious to me that *both* forms of human experience are quite genuine (in the simple sense that "they happen"), and so both need to be taken into account toward any fuller picture of who and what we really are.

The human experience of total immanence is a historical fact, especially today in the modern secular world. So too is the human experience of perfect transcendence, cosmic personality, and unconditional love. There are, no doubt, many ways to relate these two phenomenological features or claims on the human condition. Again, I am not proposing any singular solution. I have none. I am simply pointing out that both things happen.

So What Do You *Really* Think?

People often ask me what I think about spiritual and religious topics. "So what do you *really* think happens to us when we die? Do we have a soul? What do we do with the presence of evil and suffering? Do you believe in God?" They ask such questions because they assume that someone who has spent a lifetime thinking about such questions and their proposed solutions in a global manner might have some big picture, some intuitive hint.

I think they are correct to assume this. I would certainly assume as much if I were them. And there *is* some gut-felt, if often largely unarticulated, worldview behind everything I write, including this little book on how to think impossibly.

Here are some further thoughts, then, as fallible and tentative as they might be. From four decades now of studying gnostic, esoteric, mystical, and psychedelic literatures, which are some of the most sophisticated religious materials on the planet, I have become convinced that what people call God is a plurality or multibeing. Put more simply, we are all, together, collectively, God.[5] If you prefer a bit of humor, we can invoke the Dutch computer engineer become idealist philosopher Bernardo Kastrup and suggest that God has a massive multiple personality (dis)order, and we are its symptomatic expressions. We are God's alters.[6]

Or God's madness.

And I am not convinced that is only a metaphor, that madness is not sometimes a kind of unrecognized revelation, the beginning of some very serious philosophical reflection.[7] There are so many hints and unrecognized truths among us, if only we could put them all together into some kind of convincing, if always tentative and expanding, whole.

Consider psychoactive molecules, where reductive explanations like "ego inflation" simply fall flat, don't work (such moments are too often symptoms of a complete failure to recognize that the Human is *Two*, not just one social ego hallucinating its own deification). Here is one example from the former Soviet Union that reminds me of Bernardo Kastrup: "Every conversation that has ever taken place was the Universe, God, or the Game talking to itself, playing the parts of interlocutors. We are multiple personalities of the schizoid universal mind."[8] Here is a very similar psychoactive revelation in the United States, this time from the nonreligious, scientifically minded Martin W. Ball: "This is a truth that applies to all, equally, without any exception. God is the only reality, and we are all expressions and embodiments of this one universal being. . . . I am a universal being. In fact, I am the universe. I am reality itself. . . . Reality is my experience of myself, and it is a game—a game that allows me to experience myself as both subject and object simultaneously."[9] Or, again, "It is not your individual consciousness that evolves over many lifetimes. It is God that is evolving *as all living beings simultaneously. Our evolution is collective, not individual, because we are the collective. We are that one being.* A startling conclusion is that there is no reason to follow what any religion says about what one should or should not do. . . . You don't have a soul, so there's nothing to save or damn."[10] We have seen that one before: you don't have a soul. I am reminded of Kevin.

Or how about the late Houston comedian Bill Hicks again? Back in the 1980s and early '90s, Hicks was upset that only the negative psychedelic stories got told in the media. He thought it a terrible shame, if not a kind of official conspiracy to keep us down and depressed. "How about a positive LSD story?" he asked on the stage. "Wouldn't that be newsworthy, just the once? To base your decision on information rather than scare tactics and superstition and lies? I think it would be newsworthy." And so here was Hicks's story, expressed in a typical third-person trick: "Today, a young man on acid realized that all matter is merely energy condensed to a slow vibration. That we are all one consciousness experiencing itself subjectively. There is no such thing as death; life is only a dream, and we're the imagination of ourselves. . . . Here's Tom with the weather."[11]

Everyone laughed, but Hicks was being perfectly serious. He meant it. No "ego inflation" there.

I began this chapter with an epigraph from Nietzsche. Toward the end of his conscious life, as Nietzsche grew increasingly certain of his own advanced superhuman status, he declared himself to be "all names in history."[12] In the vision that I have been sketching here, or that Bill Hicks was joking about, *he really was*.

But perhaps my personal favorite happens to be from the British poet and visionary William Blake (no doubt because my own theory of the imagination resembles his). Blake was widely considered to be "mad" (of course) in his own time. My favorite story about Blake involves his friend Henry Crabb Robinson being terribly concerned about Blake's orthodoxy— that is, about his proper Christian beliefs. Robinson was right to be concerned, it turns out, though not quite in the way he expected. Here is Blake's reply to his friend's anxious question about whether Blake really believed that Jesus was true God and true man. Not to worry, Blake sung in a few lines:

We are all coexistent with God
members of the Divine body.
We are all partakers of the Divine nature . . .
He is the only God.
But then so am I, and so are you.[13]

One can only imagine what poor Robinson thought.

There are contemporary examples of very similar realizations. Consider, as an example, Sharon Hewitt Rawlette. Rawlette comes to us with elite training in the philosophy of mind and ethics at New York University and the École normale supérieure in Paris. From the opening epigraph of her book *The Source and Significance of Coincidences* (2019), she clearly writes as an impossible thinker, rejecting, from the very beginning, the easy ways out of believing everything and doubting everything and explicitly taking up the paranormal as a philosophical catalyst, not as a literal truth or an easily dismissible piece of nonsense.[14] Indeed, Rawlette treats, for some six hundred pages, subjects such as the manner in which coincidences are sometimes positively overread through theological lenses, the possible influence of others on our physical environment (even perhaps malevolent or ill-intentioned entities), the likely psychokinetic and precognitive powers of the human being that sometimes camouflage themselves in these same moments, and everything in between.

Rawlette's own book is the product of eight years of both philosophical and existential struggle. Like almost everyone else that writes about such shocking coincidences, she has had her own mind-blowing examples, about which she philosophizes as anything but "just coincidental."

Rawlette's conclusion? "We are swimming in a sea of consciousness, and ultimately our mental life is not separable from the physical world around us, nor from the mental lives of others. Coincidences point, above all, to this fundamental interconnectedness. They allow us a glimpse of a level of order to our world that is probably too complex for us to understand but that compels us to try, or at least to gaze in awe." She goes on to criticize, at least implicitly, the present assumptions of humanist intellectuals: "Our normal, everyday experience makes us believe we are separate beings, each of us isolated within the shell of our own body and mind, but coincidences show us that this is an illusion. At a deep level, we are all connected, and nothing we think or feel goes unheeded by the universe."[15]

In the end, Rawlette will argue that "our individual human minds are actually parts of a larger consciousness—perhaps the 'mind of the universe'—which communicates with us through the medium of coincidences."[16] Put in my own terms, such coincidences, like most paranormal phenomena, are trying to get our attention, to get us to think differently—in fact, *to think impossibly*: to think through connections, comparisons, correspondences, and meanings and not just through material or social causes.

Rawlette's ending is well worth quoting in full, as it can also serve as a kind of concluding manifesto for the present book:

> If there is one unfailing message I have gleaned thus far from my experiences of coincidences, it is this: The Mysterious numinosity that we sense in

our moments of deepest meaning and greatest love is not an illusion. Those moments are not just side effects of chemicals in our brains, hormones selected by evolution to cause us to mate and procreate and protect our offspring. What we *feel* in the most intense moments of our lives, those feelings are *real*. Realer than the physical world we see. Coincidences show us that the physical world is actually built *from those feelings*, rather than the other way around. That is, our feelings are of a higher order of reality than the subatomic particles studied by physicists. They are a "wholer" form of truth. A holier form. Modern science, in focusing on the physical, has gotten things precisely backward. It has studied the physical because the physical is simpler, easier to see and pin down and dissect. But it is in fact less *real*. Coincidences, on the other hand, show us what is behind the façade of the physical world: something deeper and more intrinsically significant. Coincidences are the holes in the reductive, mechanistic worldview that allow us to catch glimpses of the vast sea of meaning that awaits us on the other side.[17]

Rawlette is currently working on another book, this one on how the philosophy of idealism—which we might gloss as the philosophical position that there is ultimately only One Mind—is the most adequate model that we possess to explain all of the available data of our existence, from the impossible experiences that we have used as building blocks of new thought in the present book to our moral or ethical desires, which emerge, for Rawlette now, from this same fundamental intelligent unity or underlying shared oneness.

As I have sometimes joked, God is really just idealism for the masses.

Sin, Forgiveness, Karma, Liberation, Suffering, Enlightenment

I am perfectly aware that the vision sketched above might be read as suggesting that God is responsible for all the evil, injustice, and suffering in the world.

That interpretation would be correct.

And incorrect.

I do not look away from the countless bodies that have been gassed, shot, injected, or just hacked into pieces and then buried in unmarked graves or thrown in the ocean out of slave ships; nor do I overlook the endless bodies ravaged by cancer or some other common disease or genetic condition, or all of those conscious creatures slaughtered so cruelly in screams and agony by the millions each day so that well-off human beings, like me, can eat meat, lots of it. Honestly and truly, I look into my dog's eyes, and I do not see a dog. I see a human with four legs, fur, and a canine intensity. I no longer understand the word "pet." And then I go and eat a hamburger.

Each developed religious system acknowledges the suffering, evil, or messed up nature of the world and offers some form of transcendence from it. *That's* what makes religion religious—the transcendence. Of course, that transcendence is coded, and no doubt experienced and limited, in culturally and religiously shaped ways, which then get used for all sorts of dubious reasons, including exclusive and authoritarian ones.

Still, all such systems can be read as civilizational expressions of the Human as Two. There is a social ego, which sins, which accrues bad karma, which suffers, which offends the gods or the ancestors, which is trapped in the body, which is caught in an endless cycle of life and death (pick your local language). But there is also this other eternal, empty, immortal, absolute, infinite field of consciousness that no evil, no sin, no action, no suffering, and no god or ancestor can possibly touch for the simplest of reasons: such a field is not composite, is not put together, or, if you prefer, was never born. And something that was never born can never die.

It is embarrassingly simple.

Let me share a thought I have had many times but which I have never quite voiced, mostly because my profession is so unfriendly to this kind of thinking. It seems to me that what Christian theology calls unearned grace, or, in another mode, unconditional love and forgiveness, has been personalized with some bizarrely legalized conceptions of God, but this notion of grace is not so different from what some of the Hindu traditions call liberation from all action or karma, or what some of the Buddhist traditions call emptiness beyond and before every self or thing. Each religious mode, after all, offers utter freedom from the actions and consequences of the socially and historically determined little self. True forgiveness comes, liberation happens, enlightenment shines forth when one side of the Human is set aside, forgotten, outshined for the other side, the Big Self, which is not a self at all. This is what I think the Sinhalese anthropologist Gananath Obeyesekere meant when he observed that the mythos of the Buddha and the mythos of the Christ are structurally the same.[18] They are.

This, anyway, is my philosophy of suffering and evil. I think the little social self of the human is pretty messed up, and it will inevitably mess up other little social selves and actually *must* mess up other forms of life to survive, as in eat them. The truth is that, as individual bodies or "souls," we have to kill things in order to live (whether those are intelligent plants or animals), regardless of what those who wish otherwise want to wish or believe. I am sitting here right now, my white blood cells murdering countless tiny infectious organisms just so "Jeff" can stay conscious and upright, just to tap out these few lines.

But I do not see what any of this has to do with the field of consciousness or life that we seem to share, that we seem to be. I do not think that any of the local religious codes, including the Christian God-talk, are correct, but they all express something that can be shaped and experienced as such, if only we would cease interpreting them so goddamned literally and exclusively. If nothing else, they point toward the Human as Two, even if they can never reach it in their own self-serving languages, customs, and institutions.

I find the same insight again in Nietzsche, in a late book like *The Antichrist*, the Devil's title, if ever there was one. Nietzsche's understanding of Jesus is eerily similar to what I have voiced above (before ever reading his text). "The most spiritual" for Nietzsche are precisely those who see that *"the world is perfect."*[19] Hence the philosopher's "psychology" of the gospel, or what he calls in his older Protestant language the "evangel."

> In the whole psychology of the "evangel" the concept of guilt and punishment is lacking; also the concept of reward. "Sin"—any distance separating God and man—is abolished: *precisely this is the "glad tidings."* Blessedness is not promised, it is not tied to conditions: it is the only reality—the rest is a sign with which to speak of it. The consequence of such a state projects itself into a new practice, the genuine evangelical practice. It is not a "faith" that distinguishes the Christian: the Christian acts, he is distinguished by acting *differently*: by not resisting, either in words or in his heart, those who treat him ill; by making no distinction between foreigner and native, between Jew and non-Jew ("the neighbor"—really the coreligionist, the Jew); by not growing angry with anybody, by not despising anybody; by not permitting himself to be seen or involved at courts of law ("not swearing"); by not divorcing his wife under any circumstances, not even if his wife has been proved unfaithful. All of this, at bottom one principle; all of this, consequences of one instinct. . . . Only the evangelical practice leads to God, indeed, it *is* "God"![20]

So powerful is such a unitive state that Jesus did not even resist his own gory state crucifixion. He went to the electric chair of the time forgiving those who inflicted it, living the very practice Nietzsche identified as "the evangelical practice." Death simply did not matter.

At that point, or soon after, what would become "Christianity" took over and turned this most radical of teachings into its very opposite: a set of beliefs in the head about Jesus dying for our sins, about sacrifice, death, gore, and the divinity of suffering—in short, the bizarre legal conceptions to which I alluded above. Nietzsche does not flinch at utterly rejecting such a long sickness: "Christianity has been the greatest misfortune of mankind so far."[21]

Such is the history of religions: an ironic and disastrous reversal of the truth of things—that everything is perfect, that there is not the slightest difference between any of us, and that our emotional, personal, legal, political, and moral lives should reflect this most basic fact. That's the good news.

Our lives do not normally reflect this, of course. That's the bad news. The Human is Two.

Belief as True Projection

And so, I repeat myself. We are God. God is us.

Human beings are not very good at being God. We generally do not know this, of course, and we are *very* bad at accepting it, coming to terms with it, integrating it into our familial, social, and political lives. Because of some toxic mix of Roman imperialism and Jewish orthodoxy, the life of the most famous godman in Western history ended very badly, to put it mildly. And we could tell similar stories about other opened or unfiltered superhumans in other horrific political, religious, and social contexts: there would be lots of people impaled or hanging (sometimes upside down) on poles, or screaming on burning piles of wood in the public market with melting skin, faces, and muscles, or bruised and bleeding buried under a ton of thrown rocks.

Does it have to be this way?

People do not generally have access to their own secret identity with God, and so they find it completely unbelievable, or worse, they persecute, torture, melt, and kill others who do have access (and who inevitably challenge the violent and stupid images of God the religion of the day is pushing). Today, of course, people just deny this unconscious divinity as nonsense; they mock and make fun of it as mad or insane or, if they happen to be religious, project it into the sky where it is safely out of reach—that is, *not them* any longer. Or they conflate it with a single Jewish wonder-worker who lived two millennia ago. That works. Until it doesn't.

I am thinking here within a particular line of European and American thought that stretches, in its modern rendition anyway, from the German philosopher Ludwig Feuerbach, through the Czech Bohemian writer Franz Kafka, to the American sociologist of religion Peter Berger and the American literary critic Harold Bloom.[22] In this lineage, we intuit or actually experience something immortal and indestructible in us, something *vast*. As little temporary egos, we fear this presence, which threatens to engulf or absorb us, and so we project it into the sky and "believe in God." Now we feel pious and safe. We submit. We obey. We bow down. We kneel, never realizing that, by doing so, we are denying our own superhuman natures and are secretly worshipping ourselves.

Basically, belief and God are projective fantasies, forms of spiritual schizophrenia, defense mechanisms against our own indestructible immortal being that will one day swallow whole our tiny insignificant social selves. And yet these projecting beliefs are also somehow true, at least indirectly so, since they give some witness to the Human as Two.

God is not particularly good at being a human person either. Actually, God appears to be really bad at it. Just look at the traumatized and traumatizing lives of spiritual prodigies and charismatic virtuosos in the history of religions. If these religious prodigies are clear windows into what we call "God" or "Reality"—and they often are—then God or Reality has not learned to inhabit a human body and personality in healthy, humane, and morally just ways. God or Reality, it turns out, can easily manifest through a sexually abusive guru who was enlightened. Or a fundamentalist bigot who once knew eternity. It happens.

We can focus on the abuse and the bigotry and condemn the man (it's usually a man, isn't it?), which is what humanists generally do, and with perfectly good reason. Or we can focus on the shining-through and the revelation, which is what believers generally do, and with perfectly good reason. But neither reasonable move taken alone is, in the end, very satisfactory (or very reasonable), as each will erase the other truth and so deny at least half of the superhuman picture. Apparently, people will take the human or the super but cannot figure out how to put them both together. Obviously, we have not really come to terms with our own superhumanity. We have only reveled in our reasonable righteousness or our projecting devotion.

We Are the Devil

It needs to be openly admitted and underlined, then: these millennial-long attempts to integrate the super and the human have not gone so well.

There are some deep historical reasons for this, particularly since the ancient Eurasian world was likely first and foremost a kind of demonological middle world, upon which the "world religions" were built as relatively recent and superficial structures.[23] But there are also some rather profound

psychosexual reasons: paranormal phenomena do not usually appear unless something is going very wrong, is out of joint, and is trying to heal
itself by speaking its pain in figurative ways. We saw this above with Kevin's
thought. I understand that there are exceptions, but, for the most part, if
there is no suffering or nothing is wrong, there are no robust paranormal
phenomena. This is why there is so often a potential connection between
the impossible and social justice *and* social injustice. The impossible is
often a materialized cry amid endless social suffering, cruelty, violence,
and murder.

It also needs to be said the same impulse for justice can *suppress* inquiry into the impossible. Moral charges of fascism and totalitarianism have
been especially common in recent decades. The adoption of Nietzsche's
Superman by the German National Socialist Party (the Nazis) and the same
political embrace of the pagan "soul" ("*Seele*") of the German nation cast
a long and still deathly pall on virtually all discussions of the esoteric and
the occult in Germany and Europe as a whole, and this despite the fact
that Nietzsche cannot be understood as a proto-Nazi and that "soul" is an
ancient and widespread concept (and experience) and in no way restricted
to twentieth-century Germany.[24]

Still, impossible phenomena can be and *are* integrated into fascist, far-
right, and authoritarian systems to the present day. In the end, these are
human capacities and abilities, and they are simply not morally determined.
Communities and cultures have long recognized this in the history of magical practice, as we will see in a moment. Accordingly, superhuman capacities can be employed toward numerous moral and political ends. But—and
this, too, is my point—they can also be expressed to deconstruct and resist
those same values and ends.

There are magical battles galore in the history of human civilization,
and sometimes these are actual battles, even world wars.[25] As a historian of
religions, I strongly suspect that one of the most common sites for altered
states and unusual phenomena that have in turn generated or supported
religious ideas is war, with its gory traumas and physical carnage. By focusing on the impossible, I am not aligning myself with any such human
violence, no more than I think a transcendent near-death experience that
brings endless supernormal powers in its wake is really about the car accident in which it so obviously occurs and on whose fantastic violence it
relies (the earliest reference in the Western literature to a near-death journey, by the way, is in a war-story told by Plato). I am observing what I think
is obvious and needs to be robustly addressed: because the impossible is

human, and because human beings do terrible things to one another, the impossible can be aligned with *any* human moral system. We are God, but we are also the Devil.

I want to go deeper, then—think before and beyond the present or even the last century. Specifically, I want to say something about five related reasons for the general condemnation of the impossible today that actively engage not only this "God" but also this "Devil." I do not think we can finally explain the impossible or fully align it with our present (or past) moralities, but we can explain these resistances and work, in conversation and debate, toward more and more inclusive moralities. We are *all* God, after all . . . and the Devil.

1. Religion versus Magic. It seems odd, counterintuitive. Impossible experiences cannot be fully explored where they seem most to belong—in the religions themselves. Particularly in what are sometimes called the monotheistic traditions (Judaism, Christianity, and Islam) and their endless theistic offshoots, the religions have in historical fact generally rejected, persecuted, and even executed extreme experiencers, and for deep scriptural and theological reasons.[26] The basic condemnation of these people comes down to the fact that their experiences can seldom be slotted into the doctrinal and institutional structures of these religious systems. Here, at least, these have been generally monotheistic in nature, by which I mean that they have worked from the belief in the existence of a single or One God who is outside the natural world and who is mediated by the sacred scriptural text and relatively authoritative tradition.

Consider the early Christianities. These spoke of the *magus*, a word taken from the Persians, that meant "magician" but also a knower of the secrets of natural and divine realms who could manipulate those two realms. Accordingly, one of the exemplary enemies of early Christianity was named Simon Magus—Simon the Magician. Simon was the man who could fly, who none other than the apostle Peter himself prayed to have fall from the sky and break his leg in three pieces, thereby disabling but not killing him. "The Lord" did just that, according to the *Acts of Peter*, to prove the preaching of Peter and deny the "signs and wonders" of Simon Magus.[27]

Hence, also, the Roman Latin and then early Christian category of "superstition," which served a similar purpose (to dismiss the piety and foolishness of the people). Or Augustine's sexualized definition of the practice of magical spells and incantations as the "fornication of the soul."[28] Or, for that matter, the later thirteenth-century theological origins of the category of the "supernatural" itself. As I have noted above, the new Latin super-word (*supernaturalis*) referred to a miraculous event whose cause was in God and so *not* in the natural world. Hence the "super-," as in "above" or "up." The supernatural was fashioned to determine traditionally acceptable sanctity—that is, to separate a doctrinally correct holiness (embodied in a saint) from the doctrinally and morally unsound lives of those around whom marvelous events also often spiked (people have long known that the marvelous is not the moral). The supernatural was, and still is, a kind of religious policing, a holy border wall.

The contemporary demeaning, really demonizing, of impossible phenomena also has something very profound to do with the birth and development of modernity. Both Walter Stephens and Carlos Eire see the stunning rise of demonological treatises and experts in the early modern period as a kind of implicit and immensely sophisticated skepticism—the beginning of modernity itself and its unbelief.[29] Eire goes as far as to call the fifteenth through the eighteenth centuries the age of devils (which was linked most tightly with that better known moniker, the age of reason). During these same three centuries, between one hundred thousand and two hundred thousand individuals were prosecuted, by Catholics and Protestants alike, for malevolent magic (*maleficium*) or "witchcraft." The result was fifty thousand to sixty thousand grisly executions. The key biblical verse was Exodus 22:18: "Those who practice sorcery should not be allowed to live."[30]

Sadly, but unsurprisingly, such a malevolent magic was often linked to women, erotic desire, and what was believed to be physical sexual intercourse with the Devil, which in turn projected endless elite male anxieties about the nature and limits of human—particularly female—sexuality, the existence of God, and the reality of the supernatural. In Europe, 75 percent of those executed were women, with Finland as a reverse exception.[31] The same kind of activities were in turn linked to the acquisition of supernormal powers, including and especially the flight of witches, demonic levitations, and bilocation.[32]

Modern skepticism and the reign of science have made all this quite impossible, including (thankfully) the persecutions, prosecutions, and executions. But the theological distinctions remain at work in the background. Demonic deception has become intentional fraud, or human delusion.

Religiously speaking now, the distinction was, and still is, all about *where agency is placed*. Magical spiritualities tend to place supernormal agency in the human being. They also tend to be private. The religions tend strongly to place supernormal agency outside the human being, even

outside the natural order—for example, in "God"—and build up their institutions, scriptures, and rituals as a kind of privileged mediation or medium.
They also tend to be public. This, of course, is why public religion has
seldom got along with private magic: they fundamentally disagree about
where to place the supernormal.

As many have noted and explained, this same distinction between a
monotheistic God and a natural world lies behind the rise of conventional
science and the modern process of disenchantment, or "de-sorcering of the
world" (*Entzauberung der Welt*), as famously observed and named by the
German sociologist Max Weber in 1904. A most profound antimagic and a
most devastating disenchantment, working hand in hand, have removed all
meaning and story from the objective world as "God" has been increasingly
separated from the same.[33]

2. It's a Protestant Thing. It is well known that the Protestant Reformation changed the religio-cultural landscape of Europe. That is too tame
a statement. What actually happened was what Eire calls a series of "re-
configurations of reality," particularly around the relationship between the
supernatural and natural worlds.[34] Impossible things might still happen, but
since matter and spirit were now being increasingly separated; such things,
particularly if they were physical, could only be reinterpreted as *bad*—that
is, as fraudulent or evil, the work of the Devil and his minions.

It was really that awful, although the miraculous and the marvelous
would make their way back into the Protestant traditions. Still, witchcraft,
demonic magic, superstition, the dark arts, credulity, and Catholicism became synonymous in the Protestant mind. In the general Protestant logic
(which became the general Enlightenment logic, which became today's secular logic), no one can "become a different kind of human being suffused
with supernatural gifts."[35] There could be no superhuman saints, no divinized bodies or glowing matter. It is all demon-magic, if one is religious (in a
Protestant sort of way), or it is pure delusion or plain fraud, if one is secular.

Indeed, with the doctrine of the "cessation of miracles," the Protestant traditions attempted to put a temporal limit on what was possible for
God—now there simply could be no more miracles after the first century
of Christianity (that is, after the biblical era), or so the new Protestant
doctrine proclaimed. The relationship between the supernatural and the
natural was being severed.

Not that the Catholics were always sympathetic. Here is Eire again:
"After all, witches hovered and flew too. As Thomas Browne argued in 1646,

since Satan was a 'natural Magician' he could 'perform many acts in ways above our knowledge, though not transcending our natural powers.' Meanwhile, however, Protestants and Catholics alike continued to believe that witches hovered and flew and should all be exterminated."[36] Aye, there's the rub, as Hamlet might say. Eire certainly would. So would I.

It is the contemporary debunking community and the scientistic academy that inherited this earlier demonization of the impossible. It is no longer demons, of course. It is nonsense. The materialism of secularism, in short, is monotheism in disguise, and Protestant to the core.

3. The Fear of Psi and the Absence of Moral Agency. It is one thing to locate the cause of a miracle outside the natural world in God, particularly a god one happens to worship and with whom one's community identifies. It is quite another thing to suggest that these are *human* capacities that lie squarely in "nature," whatever that is. This is actually a very old Christian conviction that dates back as far as 400 CE and to the central figure of Augustine, who clearly understood that what we consider against nature depends on our limited views of what nature is. The implication is that the "miracle" or "prodigy" would in fact become perfectly "natural," if we but understood that natural world.[37]

Still, people seem to have forgotten this ancient insight. As the analytic philosopher Stephen Braude has taught us in a lifetime of books and essays, *we fear psi*, particularly its further psychokinetic reaches as "super-psi" (another super-word).[38] Moreover, and perhaps more importantly, we generally deny that these powers can be used for immoral purposes, although the history of magic clearly shows us that this is so.

It is messy. Agency and moral assent are often simply not present in impossible phenomena. Individuals *do* become possessed against their will. Sorcerers *do* curse and magically attack others. Demons *are* accurate phenomenological descriptors of many poltergeist, parasite, and haunting cases, which manifest a clear and disturbing contagion or immunological structure, again completely outside personal assent.[39] The early modern history of Christian mysticism, from 1400 to 1700, *is* a history of possession.[40] We ignore all of this at our own significant ignorance. But people wish to be ignorant: better not to think about something that cannot be thought in one's own rational or moral terms.[41]

In the end, things like abductions are called abductions for a reason. Human agency is absent in so many of these cases. Things—terrifying, abusive, and invasive things—are done to people. Unsurprisingly, there appears to be some actual correlation between UFO events and geographical places with "devil" in the site name.[42] And this correlation is as evident in Indigenous lore and practice (American Indigenous communities avoid these places and associate them with the presence of things like superwitches) as it is in modern English place names. One former police investigator even posits a relationship between national parks, human disappearance, and UFO sightings.[43] That is not a matter of theology. That is a matter of missing persons. One of the most famous American paranormal researchers of the twentieth century, John Keel, finally concluded that what he was really studying was not ufology but demonology.

Keel was onto something.

4. African Origins of the Paranormal. The parapsychologists are not alone in being rejected by the religious public in a culture like that of the United States. Many of the experiences that we might locate as Black are rejected as well; these experiences often bear a particularly close connection to the impossible, the fantastic, and the parapsychological.[44] I detect, in other words, a particular racial formation in the resistance to the paranormal, a formation that goes so deep that it is generally unconscious or simply assumed by the public, including by much of the Black public.

I do not say this only for contemporary moral or political reasons (although those reasons have certainly made me ultra-aware of this issue). I say this for millennia-long reasons. On the simplest and most obvious of levels, why do we so often call magical practice that we do not like or that does harm to others "black magic"? Is not that inherently racist, a conflation of "black" with "bad"? And how many times does the Devil or a demon appear as a black figure in the Christian imagination?[45] Isn't this just a bit—or a lot—suspicious?

Unsurprisingly, the modern American paranormal may well have its deepest historical roots in Africa. "Paranormal phenomena," writes the Dutch anthropologist and initiated African diviner Wim van Binsbergen, "may be argued to constitute a domain where the truth claims of African wisdom are not just valid within the local African space of culturally created self-evidence, but may deserve to be globally mediated as a statement of a transcultural truth, and hence superior to current collective representations in the West."[46] That is a strong statement. I happen to think it is plausible.

Van Binsbergen goes on to observe what is patently obvious but often denied by what he calls the "modernist Skeptics" and "their lay parrots"— namely, that the atomistic or separate individual of Western assumption leads us into endless difficulties, "since the actual direct communication between minds (as implied in the ideas of telepathy and precognition) is simply an everyday experience to many people from all cultural orientations and all times."[47]

The anthropologist does not deny, of course, that all kinds of psychological and social processes are at play in divination rituals, but he will also write of his own quarter-century experience as a diviner of "the trancelike techniques of transindividual sensitivity" and "non-sensory forms of knowledge transmission." As a good comparativist, he will also explain how there are different interpretations of the same kinds of experiences. The ontological and cosmological understanding of the Southern African divinatory idiom, for example, generally locates the extrasensory production in the "self-evident intercession of possessing or guiding ancestors," while in the Akan, Ghanian, or West African world, "individual minds are, as forms of what is locally called *sunsum*, considered to be semi-autonomously subsumed in a universal World Soul, *okra*, and it is this interconnectedness which eminently accounts for telepathy, precognition and veridical divination."[48] In short, the same "impossible" phenomena can be theorized and so made possible in different ways.

I recognize, of course, how multiplex the mythologies and ritual systems of the African religions are. I do not want to conflate these cultures into some simple essence. Still, there is something fundamentally different, especially about the ancestors (the so-called dead) and their active roles in human community, that generally renders the African worldviews much more friendly to the modern American paranormal. Significantly, again, the American paranormal has its roots in Spiritualism, or contact with the dead, which itself has deep connections to African peoples in the United States and the abolition movement. There is the deep linkage of the impossible with social justice again.

The historian of religions Charles Long is most helpful here again. One of the most basic insights of Long, drawn directly from the Romanian comparativist Mircea Eliade, is that what we generally mean by "the human" is in serious danger of being provincialized by the Hebraic, Greek, and Christian understandings in the Euro-American universities. What we now call "humanity" (*humanitas*) is in historical fact much more ontologically diverse, if we open our hearts and minds to the Asian, African, and Australian cultural traditions, which is exactly what Eliade's history of religions attempted to do toward what he called a "new humanism."

It was precisely in this same planet-wide scale that Long added to the comparative project of Eliade: any adequate theory of the human must consider "the relationship of the continuity of the Western discourse with the correlative history of Western imperialism from the fifteenth century to the present."[49] But that is never enough. Postcolonial and critical race theories must be central, Long argued, but they also must be practiced alongside other methods that fully acknowledge and robustly engage the verticality of religious experience—that is, the sheer "otherness" or "sacred" nature of these kinds of experiences, including and especially those of the contemporary world.

Indeed, Long took the "cargo cult" of the late colonial period as a model of a most positive future of the humanities themselves. These were cultural traditions in which the Indigenous populations, particularly in Melanesia and New Guinea, turned their colonial contact with Europeans into a ritualized hope of technology and material goods, or "cargo," that would be delivered to them by the gods. As such, they represented a kind of humanity that wished to be something fundamentally new. This was a future human no longer defined in the traditional sense available to the Melanesian colonized person, but who was also not in the Western colonizing sense of the European present or past. This was about a religious and technological future that was coming, that could even be ritually encouraged. "Contact" had changed everything. New levels of intimacy and reciprocity were now available to the human, and so also to the humanities.[50]

My contemporary Black colleagues will put an accent on Long with their pointed observations that Black peoples have not shared in the formation of understandings of the human. Indeed, these cultures and communities have been the "savage," "primitive," or "nonhuman" opposite against which the Euro-American human has largely defined itself in endless social evolutionary scales. These colleagues will also reflect on how the "superhuman" is a more or less accurate descriptor of Black self-understandings, myth, ritual, dance, art, and extraordinary experience, from ancient Egypt to the contemporary Americas, as long as we remain critically aware of how little Black intellectuals, artists, and activists have participated in the formation of the human in the elite public discourse of Western culture. *All* of this must be rethought, and so reexperienced, after the last five centuries of colonialism, "After Man," as the Jamaican intellectual Sylvia Wynter put it in the title of a well-known essay.[51]

Partly because of all of this, some Black critical theorists have moved to an "ontological blackness"—that is, to an insistence on separating blackness from the color of one's skin or the horrific histories of the Atlantic slave trade. They fully recognize Long's colonial past of the Americas, but they also see an ontological blackness as the apophatic or unimaginable source of much profound religious experience.

I confess that I am much attracted to this distinction, primarily because my earliest work orbited around a Hindu Tantric goddess named "Black" (Kali). As it turns out, in an almost perfect display of ontological blackness, there are two major forms of Kali in the Shakta Tantra of West Bengal. Whereas the black goddess Kali is dangerous and terrifying, if also liberating and transcendent (and superhuman, with an emphasis on the *siddhis* or "superpowers"), and sometimes linked to the marginalized communities of the caste system, the blue Kali is motherly, beneficent, and gentle. Kali's blackness signals a kind of esoteric intensity, a social marginalization, a secret sexual-spiritual practice (often in cremation grounds), and a final liberation or release from the cycle of birth and death. In short, this is a "transcendent blackness," to employ again Stephen Finley's apt theorization.[52]

The philosopher of religion and Black activist Biko Gray is one of my teachers here. Gray's call is not to compromise our critical theory with respect to race. His call is to engage robustly in such critical theory *and* to affirm the philosophical roots of an ontological or apophatic blackness that exceeds all attempts to explain the "experiences of the super."

> While we can and must continue to interrogate racial blackness, I also think there is something to be said for the fact that blackness is the preeminent site of the non-normal, the out-of-bounds, the "beyond" as I once put it. That blackness—a blackness held in the flesh of those deemed black, but nevertheless exceeding them—is, in my mind, the place where discourses and experiences of the super emerge: space travel (in the blackness of space); the Black Panther (with a superhero origin story lodged in root work—that sacred herb seems a lot like the root Frederick Douglass was given); and even the X-Men (that community of beings no longer deemed human, but nevertheless deeply intriguing and terrifying to the normative world). All, in my mind, speak to a kind of blackness that exceeds the reductive racial designation, the death mark, the "corporeal malediction," as [Saidiya] Hartman might say, that is applied to it. Sojourner Truth could rouse and quell audiences with her words. W. E. B. Du Bois spoke of clairvoyance. Harriet Tubman (and Truth, for that matter) was a precog. Such things are only celebrated now, in retrospect. They were not celebrated back then, when they happened. And I think (or want to think) in the tension between the blackness of the beyond and racial blackness, between the blackness whose super-capacities (if we are to retain that appellation) can and do move worlds, and the racial

blackness that has been designated as the constitutive outside of the norma-tive world. They are connected, and minimizing one for the sake of the other seems, to me, to be a methodological misstep.[53]

I am also much attracted to the distinction articulated by Finley and Gray because it is simply not true that paranormal experiences are re-stricted to any particular ethnic or racial group. If there is a human uni-versal in the history of religions, it is the anomalous, impossible, magical, or paranormal experience. Africa may or may not be its historical origin, as Van Binsbergen has suggested, but the impossible is in fact everywhere and everyone. It *is* the human, or superhuman.

Science fiction, of course, sometimes engages in similar thought ex-periments by locating the origin and source of paranormal phenomena in Africa. I am thinking of the famous scene in Arthur C. Clarke and Stanley Kubrick's *2001: A Space Odyssey* (1968) that begins with the alien or extra-terrestrial presence in the form of a black monolith influencing the human species' origins and biological evolution in Africa. With the monolith nearby, the early hominids realize they can use a bone as a weapon; one primate throws the bone into the sky, where it morphs into the orbiting space sta-tion of the film, thereby equating ancient Africa with the furthest reaches of time and travel in the blackness of outer space—a blackness, moreover, that will become a kind of psychedelic portal toward a future evolution of the superhuman at the end of the film (with some very strong Nietzschean notes, both musically and metaphorically, as I noted above).

It is much more than cinematic origins, though. One of the strongest and most well-attested UFO cases of recent decades occurred in Zimbabwe in 1994. The otherworldly craft and its alien occupants appeared to a group of over more than sixty racially diverse school children. These events are the subject of Randall Nickerson's documentary *Ariel Phenomenon* (2022), which subtly links the mind-boggling appearances to the African ancestors and puts them into sharp conflict with the Christian missionary beliefs.

Nickerson is himself an experiencer and was a close associate of the late Harvard psychiatrist John E. Mack, who wrote two very important books about the abduction and UFO phenomena, engaged American and African shamanism as a likely record of otherworldly contact, personally researched the Zimbabwe event, and whose abduction papers are now deposited in our own Archives of the Impossible.[54] Mack saw these contact events as a "passport to the cosmos"—that is, as inhabiting an actual third space that was neither objective in the sense of the sciences or subjective in the sense of the humanities but something else or more. Put simply, Mack was an impossible thinker.

When we move to the Afrofuturism of contemporary literature and music, themselves suffused with technological, paranormal, and magical motifs, the landscape changes again. So does an understanding of the human, or superhuman. For many Afrofuturist visionaries, for example, the apocalypse, the end of the world, and so the end of the assumed model of the human, is not some future fear or possibility. It has already happened. The human of the past is over. The human of the future is in process.

Consider the novels of Octavia E. Butler. Butler is often seen as one of the pioneers of Afrofuturism. And for good reason. Hers is a future in which Black people are central, in which mixing and matching of all the races is the desired norm and not the exception, and in which the presence of supernormal powers like telepathy are as natural as they are viral and evolutionary—a shaping power of some future race. Hence her early *Mind of My Mind* (1977).[55]

One of Butler's central late messages is that we shape God, even as God shapes us. "God is change," as she hymned. She meant this quite literally, physically, scientifically. Evolution was one aspect of this changing God or universe, but so was society, and so were people. Unsurprisingly, her books have been banned in many states and federal prisons in the United States for obvious affirmations of marginalized peoples, the mixing of all essences or natures, and a keen sense of social justice. And they've been banned due to plain social stupidity. Butler was also prescient, almost precognitively so. Books like *Parable of the Sower* (1993) and *Parable of the Talents* (1998) eerily read like a description of a totalitarian-leaning America under President Donald Trump. The latter novel even contains the phrase "make America great again."[56]

5. Madness as Metaphysical Opening. I finally want to insist that paranormal phenomena often appear in and through intense mental suffering, as in mental illness or, to be less medical and more philosophically accurate about it, madness. And I mean this in the most impossible of ways. The two altered states are *not* exclusive of one another but can be coproductive, despite what people assume, or indeed, how scholars of religion have consistently mollified or elided these mad states from their sober and legitimating analyses (much as they have ignored or erased the paranormal).[57] I think this most fundamental both-and is too often missed or simply denied, not only by scholars of religion but also by philosophers and the psychiatric profession, whose therapeutic goals admittedly make it difficult, if not actually impossible, to pursue the related strands here. Health and respectability are perfectly understandable goals, but they do not represent the truth of things, of what is actually so.

Let me underline the fact again, then: the link between massively altered mental states and paranormal events is not just a link; sometimes it is an actual identity or coillumination. So much of philosophical thinking about experience and time, and reality itself is "madness" with respect to the social order or common sense. Moreover, the contours of religious experience, especially in its more extreme or acute forms, are classically *not* what they are in the social order: the two orders of reality are fundamentally different, and often wildly opposed. There are *two* worlds, not one. Hence Plato's Cave, which is nothing like what reality is "on the outside," in the sun. The simple truth—it is not simple at all—is that serious philosophical thinking can lead one into madness and mystical revelation, and madness and mystical revelation can lead one into philosophical thinking.[58]

I have seen both happen, together, up close in my students and readers, over and over again. I am not making some kind of neutral or uncaring observation here. I have also heard the either-or thinking around madness and mysticism in public more times than I can count, in interviews and on podcasts, for example: people quickly want to reduce impossible events to psychopathology, full stop. They find some umbrella diagnosis and stop there. That's enough. That solves it. They see the mad state, whose external or third-person medicalized description they assume is sufficient, as productive of a delusion instead of as an inner psychic portal through which the phenomena arrive and appear, always (or mostly), yes, in personally confusing and socially awkward forms. They cannot imagine that truth might arrive in ways that are not socially conducive or psychologically stabilizing. They cannot think the impossible both-and. They can only think the possible either-or.

I do not wish to deny or diminish the human suffering. I wish to recognize it and see it healed, but also to potentially honor and listen to that which is sometimes coming through. Maybe what we so quickly call the schizophrenic or the psychotic individual is sometimes seeing something, knowing something, about the interconnected or unitive nature of the *world that is real*. Maybe the truth *is* "psychotic" with respect to the socialized ego. I cannot shake that thought, that possibility.

The American science-fiction writer Philip K. Dick certainly thought as much. He was hospitalized for schizophrenia, *and* he had numerous paranormal and mystical experiences that he understood as the secret of his own wild creativity and religious genius. He did not put these states in entirely different boxes. He connected them because they *are* connected.[59] Indeed, Dick honored both sides of the equation by putting the matter in an evolutionary esoteric context when he wrote in his private journals that "the schizophrenic is a leap ahead that failed."[60] The mad state may have failed socially in many ways, but it is still an advance, still a leap ahead.

To repeat myself, I personally think that trauma is a key to much of impossible thinking: something has to "split open" the person or social ego before such a subject can know and see in these new ways. Sexual arousal can do that, sometimes. Psychedelics can do that, more often and more reliably. So can car accidents, heart attacks, terrible falls, and racist social systems, sometimes. And that splitting is seldom pretty, much less socially helpful. It is often near deadly, or just deadly. It seriously compromises or even destroys its host or medium. And yet, and yet . . . not always. Sometimes it also helps to produce a kind of literary or philosophical genius, a paranormal virtuoso. Those are the superhumans who change history and mutate culture. In the end, these leaps ahead did not fail. They took us with them.

Toward a Critical Hauntology

Very much related here are the difficult subjects of possession, haunting, demonology, and abduction. Several intellectuals have struggled openly and productively with some of these. Consider the "spectral turn" of the 1990s and into the new millennium, a most important development in the critical theory of the academy that generally reads the spectral or the common historical experience of ghosts and haunting in terms of traumatic social and personal memory.[61] In my own understanding, that spectral turn is more of a half-turn, an important beginning but not a complete turning around. It does not flip. It keeps its eyes firmly on this side of reality, on the social, the political, and the psychological, even as it employs, for its own moral purposes, intrusions from the other side. Still, it is a very important beginning. More to my point, the spectral turn constitutes a keen awareness that the impossible and social justice can be related—indeed "intend" to be so.

I have been especially inspired here by the work of Martha and Bruce Lincoln on ghosts and haunting in contemporary Vietnam. They begin their 2015 essay this way: "A specter is haunting the academy: the figure of the ghostly, the phantasmic, and the unquiet dead. Over the last fifteen years, a large and rapidly growing number of works in diverse disciplines—

sociology, psychoanalysis, literary criticism, folklore, culture studies, post-colonial studies, race and gender studies, geography, media studies, and communication and rhetoric—have sought to reinterpret stories of haunting as the return of traumatic memory. Within such work, ghosts manifest not as terrifying revenants, but as welcome, if disquieting spurs to consciousness and calls for political action."[62]

The father and daughter are particularly concerned to engage Jacques Derrida's *Specters of Marx* (1994) and its profound influence on this spectral turn. I read this Derridean text, like the philosopher's late (and quite wonderful) essay on telepathy, as displaying some rather clear esoteric traces of modern thought that speak to earlier spiritual concerns and convictions, particularly via the philosopher's key category of hauntology. The latter is probably the closest Derrida could come in a playful pun to an actual ontology.

The Lincolns rightly criticize Derrida for basing his theorization of the ghost on a single text (act 1 of Shakespeare's *Hamlet*) and work on a much broader global canvas (Martha Lincoln is an anthropologist who works in Vietnam, and Bruce Lincoln is among our most gifted, fearless, and capacious historians of religions). Most significantly, they want to draw a sharp distinction between what they call secondary and primary haunting.

Secondary haunting involves critical theories like Derrida's deconstruction—a kind of academic thinking *about* haunting that wants to reduce the impossible phenomena to the social and historical plane by reading it as metaphor, memory, and social signal, or in the language of the Lincolns, as "literary representations and figures of the imaginary," not "agentive revenants" or "ghosts *qua* ghosts."[63] Never do such authors of secondary haunting really consider "individuals or social groups who experience haunting as something consistent with, and rooted in, their cosmology, ontology, and psychology (the latter term used in the etymologically precise sense: 'theory of the soul')"[64]—which is to say, as *religious*.

Primary haunting, on the other hand, involves, well, actual hauntings—basically, ghosts and spirits possessing people, and in precisely the terms of their religious understanding: as something "other than metaphor," "as possessing an ontological status that is ambiguous, even contradictory, minimally—but emphatically—substantial and real."[65] Here, the ghost is not simply something interior, a matter of consciousness, a projection. Rather, it exists, it really exists "on the borders of the physical and metaphysical."[66] It inhabits a different ontological zone. The Lincolns call their emphasis on primary haunting and a much greater comparative reach their "critical hauntology."[67]

If I read the anthropologist and the historian of religions correctly, the Lincolns appear to be arguing that secondary haunting, and with it pretty much the entire spectral turn in the academy, is just too clever, another reduction of the historical agent or conscious presence to an apparitional function of language, discourse, and traumatic memory. Their own critical hauntology certainly does not deny the practice of secondary haunting (including by state actors wishing to commemorate or memorialize war atrocities), since it is much broader and most concerned with large social, economic, and political issues, as opposed to the primary hauntings, which are much more personal, ritual oriented, and specific. But such a critical hauntology also wishes to struggle with actual ghosts, very much like the human beings that concern it so. Such individuals and such a subsequent academic practice recognize, fully, "the reality and autonomy of metaphysical entities."[68]

The Lincolns in fact cite the ethnographic case that moved me so while reading the late American anthropologist Mai Lan Gustafsson. It involves an American GI, an ex-marine named Sam who was having nightmares and speaking eerily fluent Vietnamese in his sleep: "Give it to me. Give it to me. Give it to me. . . . Motherfucker, you give it to me or I will eat your mother's soul." A Vietnamese medium Sam visited determined the voice was that of a Vietcong soldier who had been killed and whose bloodied name tag Sam had secretly kept as a souvenir. The medium had the soldier's name (Hoc Van Nguyen) and the nature of his gory death (a landmine). Sam found the dead man's mother, performed a ritual at the family altar, and sent the mother a monthly stipend. No nightmarish hauntings followed. They were gone.[69]

That is impossible thinking.

Ancestors and Poltergeists in the Unconscious

There are other esoteric traces of the ghost and the specter in modern thought.[70] It is difficult, for example, to avoid the observation in these anthropological situations that many such "ghosts" are on their way of becoming revered "ancestors," as the psychoanalyst Hans Loewald had it over sixty years ago now.[71] So much for the spectral turn of the 1990s. Loewald was using this language *in 1960*. No wonder his reframing has been called "a stunning re-definition of the unconscious as a crowd of ghosts."[72]

But this is not that surprising—if only we know our history of psychoanalysis. Consider the related notion in the 1950s of what the Hungarian American psychoanalyst Nandor Fodor called the "poltergeist psychosis,"

an idea that emerged from the lingering suspicion that poltergeist phenomena are ultimately about the exteriorized emotion and sexual libido of some
living, profoundly conflicted person in the room—often, but not always, an
adolescent.[73] Fodor also wrote a big book on much of the spectrum of impossible matters, from hauntings and reincarnation to sex with spirits and
wonders in the sky, all of which relied on an ontology of two intersecting
worlds and a subsequent theorization of the parapsychological as a form
of "super-nature."[74]

It is indeed astonishing how psychologically inflected so many parapsychological events are, how they seem to involve family dynamics, sexual
trauma, emotional conflict, and psychopathology—which is to say, human
suffering. But they also involve physical objects moving without any apparent means or cause. If this is what the psychoanalyst might call an "exteriorization of emotion," that emotion is doing some pretty impressive things.
In these instances, at least, a more adequate future critical theory would
become a kind of superpsychoanalysis that would take the telekinetic ghost,
the effective curse, the dramatic possession, and the terrifying haunting
to a whole new level of seriousness. Such historical events might remain
psychological projections, but projections now of a superhuman nature.

Still, for now, we remain in the realm of the social here, even if that
psychosexual realm is throwing pots and pans around. Such a spectral
turn is about the hidden workings of culture doing its own thing, which is
quite the thing.

A History of the Impossible

There are ways to go further still. Consider the aforementioned American
historian Carlos Eire. Eire grew up, as it were, with the anomalous. In his
mind, the strangest thing about his own family history was his father, who
was deeply committed to Spiritualism and Theosophy and who claimed to
see ghosts and remember previous lives. Eire's father also claimed that his
father (who died in 1927 from an infected paper cut on his tongue caused by
an envelope he licked carelessly) had been expelled from his Jesuit school because he had accidentally caught a glimpse of some Jesuits levitating a table.[75]

The family story is not without significance since levitation enters
Eire's scholarship in major and multiple ways. Consider, as a start, Eire's
biographical study of Teresa de Ahumada y Cepeda (1515–1582), otherwise
known as the abbess and saint Teresa of Avila, a woman who literally floated
off the floor in front of visitors and fellow nuns, whose contemporaries

spoke often and explicitly of her levitation (it was patently obvious), and who even levitated with another person, John of the Cross (one of my favorite mystical writers of all time for his apophatic emphasis on *nada*, or divine nothingness).[76]

What is so extraordinary about Teresa's levitations are the oddly empirical and disarmingly honest terms in which she described them, and how she finally begged God to remove them, to just stop. He did.

In the beginning, Teresa feared the levitations. The concern was both hers and her confessors: such levitations, after all, may have been from the Devil. As such, they put her in the inquisitorial spotlight. The demonic reading is not what Teresa finally concluded, but it was certainly the standard interpretive option, and Teresa knew it. When she would compose what would eventually become her spiritual autobiography, her *Vida*, or *Life*, the future saint and doctor of the church would describe these spiritual-physical raptures in terms that the spiritual directors and inquisitors could affirm; she would describe them with humility—that is, as having little, or nothing at all, to do with her. As Eire reads the Spanish text, the *Vida* is a "forced confession" in a riven theological and political context.[77]

But what a confession it was. Here are a few examples from Eire's translations:

> Once, when we were together in a choir, and I was kneeling and about to take communion, it disturbed me greatly [to levitate], for it seemed to me a most extraordinary thing and I thought it would create quite a stir; so I ordered the nuns not to speak of it (for I had already been appointed prioress). At other times, when I have felt that the Lord was going to do the same thing to me (as happened once during a sermon . . . when some great ladies were present), I have lain on the ground and the sisters have tried to hold me down, but everyone has seen me enraptured anyway.[78]

> The effects of these raptures are truly great. One of the greatest is the manifestation of the Lord's mighty power. Since we are unable to resist His Majesty's will, either in soul or in body, and are not our own masters, we are forced to realize that there is One stronger than ourselves, however painful this truth may be, and that these favors are bestowed by Him, and that we, of ourselves, can do nothing whatsoever. This imprints in us great humility. Indeed, I confess that at first it produced a great fear in me, an awful fear.[79]

Again, such things produced real fear and doubt not only in Teresa but in her confessors. Hence their advice to the nun to greet her visions of Christ

with an obscene hand gesture called "giving the fig"—basically, in today's
English lingo, giving the finger.[80] No wonder that when she finished her
Vida all copies were destroyed except one, which was then kept locked
away until she died and her sanctity became apparent through postmortem miracles.[81] Put simply, Teresa's impossible phenomena were contested,
censored by both herself and her superiors, and only later recovered after
her death. Now, at least, she was no longer a woman deceived by the Devil.
She was in the process of becoming a saint and doctor of the church, a
recognized and revered Catholic superhuman.

How exactly does one wish these things away with so much documentation, much of it as reliable as any other historical deposit? Eire does not. In
two words that sum up his argument and function as his latest book title,
They Flew. The title phrase is adamant, and just a bit angry. This is what he
calls, in perfect sync with my argument here, a "history of the impossible."

Eire's two chapters on the Italian Franciscan friar Joseph of Cupertino
(1603–1663) are especially important for my own purposes, as the saint
played an early, if relatively hidden, role in my discipline, the history of
religions, in the thought of one of its founding figures, Mircea Eliade.

In an essay entitled "Folklore as an Instrument of Knowledge" (1937),
Eliade follows the earlier "psycho-folklore" of Andrew Lang and argues for
the empirical reality behind folkloric beliefs. Exactly as the title suggested,
Eliade took Joseph of Cupertino's commonly witnessed flights, along with
other well-attested phenomena from Indian yoga and modern parapsychology, as containing a core of historical-physical truth and went as far as to
suggest that such "folklore" should function as an "instrument of knowledge."[82] This early essay is a very clear precedent for my own impossible
thinking.[83] Eliade would camouflage the same thesis (he was convinced of
it) in his fantastic literature, most obviously in his autobiographical novel,
Youth without Youth, with which I quietly began the present book.

For his part, Eire focuses on Joseph and the saint's Italian shrieks and
whoops (*urlo! strillo! grido!*)—likened to the explosions of gun powder—
that made buildings shake, as in an earthquake, and propelled him into an
olive tree or up into a church statue perched high above an altar.[84] If such
explosive cries were not supernatural, they were "at least superhuman."[85]
For Eire and his meticulous scholarship, "God is in the details here, literally
and figuratively."[86] And those details matter, however bizarre they become,
or rather, precisely *because* they become so bizarre. As Eire writes about
Joseph: "He is sui generis . . . uniquely outrageous in his ecstatic success."[87]

It is difficult to disagree. One reads of things like the suspension of time
during Joseph's ecstatic states, the contagious nature of his flights (he

could raise into the air anyone he touched, "much like prey in the talons of a raptor"[88]), an impermeability to burning over candle flame (despite the horrified screaming of nuns), and, perhaps strangest of all, the physical freeze-frame effect of his trances, which extended *even to his clothes* in a kind of "supernatural cocoon," within which the natural effects of gravity, space, and time were simply not operable. Such wild details can point us toward a future "making sense of the impossible," if only we can pay attention to them.[89]

In short, what Eliade called for in his early essay and explored in more depth only in his fantastic fiction, Eire accomplishes in abundance (and in the full light of day) in his public scholarship. But perhaps "accomplish" is not quite the right word. Eire would be the first to acknowledge his own rational puzzlement. This is, after all, a history of the *impossible*.

The historical documentation certainly bears this out. The saint flew countless times over a span of some thirty-five years in front of peasant milkmaid and aristocratic duke alike, even the pope himself, in the midst of a Protestant Reformation that would demonize the very superhuman miracles that so defined Joseph's remarkable life, and on the cusp of a scientific modernity that would deny their very possibility. Mystical rapture and the love of God become literal flights and floatings in Joseph's life. Ecstasies ("*éstasi*") and raptures ("*ratti*") were mixed and matched with the impossible physical movements ("*moti*") for which he became so famous (flying backward, levitating with confused confreres, or floating for hours during mass in trance).

Following our earlier chapter on Kevin, it is not difficult to recognize in the life of Joseph of Cupertino the socially challenging but spiritually creative presence of someone "on the autistic spectrum." Joseph's education was very basic, to say the least. Eire calls Joseph an "idiot savant" and spends numerous paragraphs discussing the "holy fool" of the Christian tradition.[90] Joseph's learned contemporaries thought that his theological knowledge was "infused"—that is, that it did not come from learning but from God.[91] And one of his most common epithets was *Bocca Aperta*, or "open mouth." Joseph would commonly stare into space with his mouth agape.

Joseph was certainly both an embarrassment and a danger. The church basically had to hide him the last decade of his life in remote monasteries and churches, as the Inquisition kept a suspicious eye on his doings. *They were afraid of him.* But the people were not. They took tiles off the roofs to see him and tore up his already poor garments as holy relics in the streets. (I would do the same. I mean, wow, this was basically Superman's cape.) People wept, fainted, converted, and ran away screaming in terror. His miracles, which were by no means restricted to levitation, were "too numinous and ominous all at once."[92] They were also too numerous. There were *thousands* of them.

Eire notes that the base ontological assumption of this baroque Catholic culture is that humans coexist on two levels or in two dimensions, a natural one and a supernatural one. There are two worlds, or two dimensions of that order. This is actually the base assumption of the general history of religions around the planet, what I have called in a comparative mode the Human as Two but also the World as Two. Eire calls this a "binary understanding of reality."[93]

Not that such a doubleness can be explained in any adequately rational terms. Eire calls all attempts to explain the relationship between these two human worlds, including those of the mystics, "futile." This, after all, is "the thing itself, the merging of natural and supernatural, begging for comprehensibility."[94] But he will also insist that it is *precisely* this same doubleness that makes human flight an effective expression of the transcendence or ineffability of the soul. The same symbolism, of course, is pre-Copernican, where up is up and the heavens are where heaven is.[95] We know that the pre-Copernican cosmology is not so (there is no "up" in a relativistic Einsteinian universe, and there is no heaven in outer space), but such a worldview worked its own miracles, nonetheless. "Such a view produces a specific conception of what is 'real' or 'possible.' It is the matrix in which the experience of levitation is constructed."[96]

In other words, as I have argued throughout these pages, *what is possible or impossible is culturally relative*, and *nature itself is culturally conditioned*, at least within human experience. Culture, as we have come to call it, may not go all the way down, but it goes pretty darn far.[97] There is another somewhat shocking way to say this: something like human levitation and flight are certainly *enabled* and *empowered* by local religious beliefs, but they are also *limited* and *constrained* by them. Belief is necessary. Belief is also limitation.[98]

This raises the following questions: What are we actually capable of? What are the limits of the human? Are there any?

I am spending more time on Eire's books for a simple reason. Carlos Eire is an intellectual-spiritual hero of mine, for his fearless historiography of the impossible, yes, but also for his insistence on telling his own story in two memoirs and for the fact that he has been called an "enemy of the state"—that is, because he has been condemned for his adamant opposition to the authoritarian Communism of his family's homeland of Cuba.[99] Eire will have none of the common academic moral relativism that makes such political opposition rare and unlikely. Nor will he stay quiet before the historian's refusal to look at, of all things, actual history.

They Flew is a most obvious and powerful exemplar of what I am arguing for in these pages. Eire's book, it turns out, has been developing for four decades, ever since 1983 when Eire was a young man and was being given a tour of the Carmelite Convent of the Incarnation where Teresa of Avila lived. The tour guide casually spoke of Teresa levitating with John of the Cross in the parlor, or *locutorio*, where the nuns could speak to their visitors behind a metal grille. The tour guide spoke in the exact same tone that she spoke of Teresa falling in a staircase and breaking her arm, or of the refectory, of the pots and pans, as it were. The shared levitation was not a maybe. It was "just another *fact*."[100]

Eire was stunned. He still is, particularly by his historical profession, which refuses to acknowledge such historical facts under any circumstances. He, too, of course, will honor his profession, applying the famous "brackets" to his historical Latin, Spanish, and Italian sources (we do not know if this happened, but we can describe the testimony, or the fact that people believed and claimed to witness the event). He will spend much time, moreover, historically analyzing the way that another Spanish woman's honest experiences of levitation and bilocation were exaggerated and embellished into legends that fit the church's purposes. Hence a line like this from María de Ágreda, the Lady in Blue (for the color of the habit seen by the Jumano people of New Mexico and Texas), on the way her bilocations from Spain to North America were received: "They are accurate about some things . . . but other things have been added and exaggerated."[101] That could well be a motto for the entire history of religions, where fraud and fact, trick and truth, and experience and exaggeration go together again and again, and again.

María's reserve and honesty are crucial, particularly since her miracles "dwarfed those of any other saint, including the founder of her order, Francis of Assisi."[102] I mean, we are talking about over five hundred claimed bilocations to North America that involved things like the nun sensing the specifics of the weather on the other side of the planet, observing the eating and ritual customs of the Jumano people, and learning the names of places and peoples (and this is before we get to the channeling of an immense biography of the Virgin Mary, *The Mystical City of God*).[103] Like the apostle Paul, María always wondered whether she was in the body or not during such spiritual travels. In the end, we are talking about "many improbable things" (*"muchas cosas inverosímiles"*), as the Inquisition put it in another understatement.[104] María's own "considered opinion of this whole case is that it really happened," although how or why was never clear.[105]

Eire will point out that "Teresa of Avila, Joseph of Cupertino, and María de Ágreda were liminal avatars of the impossible, suspended between the divine and demoniac, their sanctity revered and questioned simultaneously, perfectly poised to play the role of tricksters acting as agents of the devil, the ultimate trickster."[106] All three made it through the gauntlet, as it were. Many others, no doubt *most* others, did not. Some, disgraced, forced to confess, and judged negatively, landed in prison. Or worse.

Eire spends three full chapters toward the end of his book describing and analyzing the floating and flying attributed to malevolent or fraudulent forces: demonic deception, credulity, unwilled possession, practiced witchcraft, trickery, and Protestant-Catholic debates about the supernatural and the preternatural (basically, the natural manipulated by demons to mimic the supernatural), emphasizing in the process just how contentious the phenomena really are, *even and especially if they are considered possible*. Making something impossible possible does not end the conversation, it turns out. It only initiates the debate, especially on a moral level. To speak in secular terms, the fantastic gives rise to critical theory and moral criticism.

To make things more complicated still (if that is possible), modern scholars—this one, included—are generally very wary of distinguishing between recognized saints who are said to experience the sacred and those devotional wannabes who report very similar, or identical, experiences but are said to have "invented" the sacred.[107] Personally, I am not at all certain that there is any real difference, as I am convinced that most all experiences of the sacred are mediated by endless psychological and social filters—"invented," if you want to say that, or "constructed," as many prefer to say today. But this does not mean that such showings are not mediating something very real. Nor does it mean that there is not sometimes real human fraud, like the painted stigmata or fake wounds of Christ that Eire discusses on one of the three judged nuns he features.[108]

I personally think that one of the worst decisions ever made in the literature is the decision not to further research mediums and psychics who were caught cheating. Cheating and actual paranormal phenomena are *not* mutually exclusive and never have been. The doubled truth is difficult: fraud, deception, and revelation are all often woven imperceptibly together into a tight mix that can only frustrate the purist of any type, including the orthodox gatekeeper, the responsible historian, and the contemporary skeptic. Many times over, the trickster is *inherent* in the experience of the phenomenon, not some social or psychological add-on. Accordingly, such events do not follow orthodox lines, *whatever* those lines happen to be.

Sometimes, moreover, even the trickster can be tricked. I confess that one of my favorite descriptions in Eire's latest book is the way the nuns of the Spanish Franciscan convent of Santa Clara used Sor Luisa, or Sister Luisa. They would have her lie beneath the bed of a dying nun to act as a decoy and deflect the demonic assaults to her own body instead of that of the dying sister. Eire comments wryly: "As one might expect, Sor Luisa became something of an expert demonologist as a result of these encounters and also a successful exorcist."[109]

Why does that make me laugh?

I should add that such matters of deception, including endless mischief around beds, are especially obvious today, particularly in the abduction and contact literatures, where levitation and bilocation are common, and things are seldom, if ever, morally clear. One even finds in the abduction literature what one finds in the German folk devilry of Martin Luther, although there is no inherent connection to deception here: the phenomenon of three knocks heard in a haunted space.[110] The comparisons between the modern and premodern accounts get *that* specific. Indeed, if I were to name one feature of these contemporary literatures that cries out for more attention, I would immediately name deception and disinformation, and I do not mean that of the US intelligence organizations and military branches. I do not assume that such inherent misdirections are *not* in our present best interests. Or that they are.

Misdirections aside (for a moment), there is a phenomenon that is both very physical and very spiritual at work in our histories and in many of our lives. The bottom line is this: people *did* claim to witness "impossible" things, and repeatedly, including the endless levitations of María de Ágreda, during which she was poked and prodded and she even floated around, weightless as a feather, so that others could better gawk at her levitations through a window in her cell door or in public spaces. And, once again, it is not just then. I have been personally moved by numerous academic colleagues who, for example, have witnessed a table levitating in graduate school or concluded that spirits are real, or who refuse to go back to this or that psychedelically induced altered state *precisely* to avoid encountering terrifying entities. Once was enough, or more than enough. Their fear and caution are palpable, ethically fraught, and ontologically traumatic. These events alter, redirect, and focus entire lives.

Why are such historical facts ignored now, boxed off as something ridiculous? Why are some of the most astonishing superhuman beings in history essentially forgotten today? Almost anything can be discussed in history, it turns out, *except this*. Hence the necessity of Eire's most recent book. Hence the power of his title, *They Flew*, without the usual question mark. Eire even adds an exclamation point in the very last line of his epilogue: "They flew!"[111]

In this book, as well, I seek to push things far beyond the usual academic limitations—and past the religious ones. Take those brackets off. Take the question mark away. Add the stunned exclamation point. Hear the ecstatic and terrified screams, acknowledge the antigravitational flights, and feel the building physically shake around you.

Theorize *that*.

Coinherence in God

I want to end this final chapter with a theological conviction inspired by reading the medieval historian Barbara Newman.[112] Newman's subtle work suggests that, ultimately, the impossible, even in its most negative and destructive forms, implies the positive and ultimately cosmic existence of a shared Mind, of what people call "God." I think the same is true of the thread that runs throughout this entire book, from the supernormal and imaginal writings of Frederic Myers to the latest insectoid abduction experiences and the related telepathic phenomena. There can be no communication "from a distance" (*tele-*), much less a shared emotional entanglement (*-pathos*), if something fundamental is not shared, if there is not, frankly, love. The point seems obvious enough: a subjectivity cannot possess another subjectivity *unless they share the same basic nature*.[113] Even demonic possession implies, after all, something shared. The demon is "human," too.[114]

Newman offers any number of helpful interventions here, including her central notions of the "permeable self" and what she calls "coinherence"—that is, the medieval Christian conviction that persons coinhere or mutually participate in one another, after the model of the Trinity and the three persons of God sharing one divine nature, or the person of Christ who shares both human and divine natures. Theology matters. Theology makes possible what a "person" was or could become in the Christian medieval world.

Newman does not restrict herself to medieval sources, nor does she ignore the difficulties that such a relational model presents to modern ethical notions of the bounded self and "individual rights" (since the individual coinheres in other individuals). In her introduction, she cites contemporary authors, like the Canadian philosopher Charles Taylor in *A Secular Age* on how the modern self is more "buffered" than the premodern self, which was more "porous." In her conclusion, Newman turns to the feminist thinker Carter Heyward's most challenging book, *When Boundaries Betray Us*. Heyward shared with her therapist a series of abuses and rapes that, in her own mind, were those of other unknown people. All the secular therapist can do is render these descriptions of abuse and rape as fantasies, as unreal. That, after all, is the psychoanalytic theory of the fantastic. Heyward would not accept this: It "was not a lie, not a falsehood, not a fiction that I concocted and entered for a while. It was an actual life experience of a person, or persons, being raped and beaten." In Newman's terms, "Heyward came to believe that her psychic pain had opened her to experience the still greater pain of unknown others."[115] This, of course, is precisely what I have called the impossible, which is possible.

But Newman also goes further still, radicalizing the idea of coinherence or the nonexistence of actual boundaries with ancient biblical convictions, like the mystical theology of the Gospel of John: "I am *in* the Father and the Father is in me" and "You will know that I am in my Father, and you in me, and I in you."[116] She also cites the Pauline teaching that all are members of the same mystical body (1 Corinthians 12:12–27).

Newman looks at some increasingly extreme examples of this coinherence in phenomena like ancient and medieval pedagogical ideas (in which the teachers share something of themselves with the student); the exchange of hearts in medieval mystical literature (which Newman provocatively compares in her conclusion to modern heart transplants, which often impossibly—impossibly, anyway, for the materialist paradigm—bring with them the memories and emotions of the heart donor); the coinherences of conception, gestation, and childbirth (which we now know involve the sharing of genetic material) and medieval notions of how the mother's thoughts might affect the unborn child; and the experienced realities of mystical union and demonic possession—that is, the mutual sharing of two or more personalities in a single body.

Newman employs Platonic and medieval notions of a greater "universal Man who dwells within each individual"[117] and modern words like *telepathy* and *psi* as powerful examples of the same coinherence of selves. She also calls on the contemporary anthropologist T. M. Luhrmann and her "theory of mind" to challenge the assumptions of the modern reader.[118] There are different theories of mind, not just the atomistic individual of modern Western society, and many of these models make good sense of common human experiences of connectedness and mental transmission or spiritual possession. Put a bit differently, there are theories of mind in the world that render entirely plausible certain phenomena that secular critics consider "fringe" and will not think about, phenomena that Newman obviously wants to think and theorize about.

Newman is more sympathetic to a reductive rather than a productive theory of the brain-mind relation; that is, she is more than willing to see the brain as a reducer, receiver, or transmitter of mind, not as its producer, as we have it today in the materialist neuroscience.[119] Hers is an open-ended comparative project that does not conflate what we know in the sciences at the moment with what in fact is so and is always showing up in our historical deposits. Newman also works with a theory of the imagination that does not reduce the miracle story to a theological fiction but sees it, in modern terms, as an instance of the paranormal—that is, as something perfectly natural but not yet adequately understood.[120]

In the end, Newman's proposed "filter" theory of mind and "paranormal" model of the imagination is much more resonant with her medieval sources, which simply make no sense, and *cannot* make sense, in the materialist models that are currently dominant. To invoke Luhrmann, clearly these scientific theories are little more than local, and so relative, theories of mind—useful for a while but hardly omniscient or universal.

There is a final easy adjective that critics and commentators have used to describe the late Friedrich Nietzsche. That adjective is "insane." This, after all, is the man who entitled one of his later books *The Antichrist* (that is, the Devil) and another *Ecce Homo*, "Behold the Man," an unambiguous New Testament allusion to Christ (that is, God) in the Gospel of John. Such proclamations may well be irreverent, but they are hardly irrelevant. God and the Devil.

There is a legend that Nietzsche finally collapsed into madness while embracing the neck of a horse in a desperate attempt to keep it from being beaten by its owner. The legend is a legend. We do not know if it is true. But it speaks to a counterreception of Nietzsche that is certainly closer to the vision I have taught again in these pages. The superhuman is the man (both men, actually—though the cruel man does not know it yet), but it is also the horse. It is cross-species compassion. It is life itself. It is the only God.

And so am I. And so are you.

How to Think Impossibly

> Physics and mathematics are to be held responsible to a large extent for the return of interest in mystical ways of knowledge. . . . We are entitled to believe that our mind space has amazing properties, the most remarkable of which is that it is not limited to three dimensions.
>
> IOAN COULIANO, *Out of This World*

I have set out a philosophical framework for how to think impossibly across our present order of knowledge. I understand that this is a future potential way of thinking and not a present actualized one. How to think impossibly seems perfectly natural and obvious to me, but I recognize that it is not at all clear, much less sensible, to many others. So I will do my best to summarize the method in the following final pages.

There is a most important caveat to make first, however. It is this: Thinking impossibly is not possible for everyone—and not for superficial or accidental reasons but for neurodiverse or spiritual reasons. Thinking impossibly is not a widely available skill that anyone can learn, like, say, riding a bike or constructing a syllogism. Nor is it, frankly, egalitarian. Impossible thought is a rather obvious form of intellectual esotericism or academic Gnosticism. As such, it is a way of thinking that requires either the personal experience of actual *gnosis* (of the fundamental unity of the human and the cosmos) or a deep hermeneutical sympathy for the gnostic realizations of others.

Here is the good news, though—a caveat to the caveat. Although many people may not possess direct gnosis (I certainly do not), many *do* possess hermeneutical sympathy, and for a much more democratic reason: the humanities themselves are defined by textual deposits that are riven with esoteric traces of the gnostic experiences and subsequent mystical thoughts of their authors. To study the humanities is to be inspired by these very gnostic traces. It is to be struck by a kind of lightning, again and again and again.

This is why the humanities are the superhumanities. They are guided, really defined, by the altered states of consciousness, embodiment, and

knowledge of their authors in broad social, artistic, and intellectual movements from Renaissance humanism, Romanticism, and German idealism to
psychoanalysis and phenomenology to more recent movements like feminist theory, queer theory, Black critical theory, postcolonial theory, and
contemporary ecocritical literature. Even the present reigning postmodern
deconstructions of the humanities, most associated with the figures of
Jacques Derrida and Michel Foucault, possess such esoteric traces (consider Derrida's essay entitled simply "Telepathy" or his call to think-with
ghosts in *Specters of Marx*). Further back still, the humanities can be esoterically traced to the still astonishing Friedrich Nietzsche, a godless mystic
who taught us how to deconstruct each and every truth claim through his
genealogical method but also ceaselessly claimed various ecstatic states of
deification and ecstatic identity with an evolving superspecies in the final
decade of his conscious life.

Thunderstruck, indeed. The lightning bolt was Nietzsche's primary sign.
It was painted on the back of Zora Neale Hurston in a hoodoo initiation
ritual. Maybe it should be ours as well.

When I write of the intrinsically "spiritual" nature of the superhumanities, then, I intend lightning strikes. I mean it, but in the German sense of
geistig, an adjective derived from the noun *Geist* ("Mind" or "Spirit") that
can mean "intellectual" or "spiritual" and, as such, acts as an esoteric trace
of German idealism and, behind that, German mystical thought going back
to figures like Jakob Böhme and Meister Eckhart. Within this same series
of gnostic traces, the Germans refer to the *Geisteswissenschaften*—that is,
the "sciences of the mind (or spirit)" for what the Americans, more banally,
call the humanities. When Nietzsche, for example, called himself the "most
spiritual among us" (and he did), he was certainly not being spiritual in the
contemporary American sense of the term. He was being *geistig* in the German sense—very, very *geistig*. So, too, I would argue, is impossible thinking.

With that important caveat and qualification in mind (this is not for
everyone, but the humanities, available to so many more, are really the superhumanities), I want to proceed now. I will return to my own philosophical framework of impossible thinking (dual-aspect monism) and summarize
the five intellectual-spiritual practices that most define the practice, again
in my own mind (others would say other things—that's the point, that's
the conversation). I will conclude this conclusion with a few words about
what this all might imply for a paradoxical understanding of religious belief.
Finally, I will present another case of precognition in the epilogue, very
similarly to how I began this book in the prologue.

I present the latter precog story to force, one last time, the central idea of *How to Think Impossibly*—namely, that such thought is based on a set of extraordinary experiences that have everything to do with the physical world of space, time, matter, and energy *as well as* the inner world of human subjectivity. In the end, there simply is no ultimate difference. The Human may be Two, but the World is One. That is the final gnosis of impossible thinking, whether such a gnostic awakening is directly realized or indirectly interpreted and thought. And so we will come to this last precognitive story in the epilogue, at which point the loop of the book closes, the snake bites its own tail, and we end as we began. The shape of impossible thinking is circular.

"Profound, Implicit, Totally Perfect"

The secret substructure of this book is the philosophical perspective of dual-aspect monism. There are other ontological options. I do not employ this particular one as the last word. It is a philosophical framework that makes the impossible possible (so I adopt it), not a stopping place (so I am ready to set it down). I am certain that most thought is unconscious, in the sense that we are not generally aware of our own philosophical assumptions or, for that matter, the deeper origins of that thought as it emerges from these unknown realms: a nightscape temporarily lit up and revealed by sudden flashes of lightning, as we saw earlier with a collective conference.

I am very curious about my own enthusiasms for philosophical options other than dual-aspect monism. I understand these enthusiasms as fleeting glimpses in the mirrors of other authors.[1] I suppose I finally see ontological options as multidimensional—that is, as many, and ideally inclusive of one another, not singular, monolithic, or exclusive. Accordingly, I remain highly suspicious of many other options of which I am aware, mostly because they *are* absolute in a monolithic way and so exclude the human beings with whom I most want to think. Such systems can only tell me that these people are lying or are deluded, that their experiences are "coincidences" or "anecdotes," or that their visionary displays are "hallucinations" with no substance, all of which I find intellectually ridiculous and morally appalling. My method, then, is also finally an ethical one. It takes individual human beings at some of their deepest depths, and compassionately so. It listens. And it is perfectly willing to be abolished.

There are two basic principles of dual-aspect monism out of which we have so thought, imagined, and listened above:

The World Is One. Orders of knowledge as diverse as the modern sciences, the history of comparative mystical literature, contemporary social justice activisms, and the ecological movement all imply (but do not always admit) that any sustainable future ontology will need to be a monistic one.[2] Such a future order will stress connectedness, holism, unity, and sameness. It will study and embrace differences, perhaps even parallel worlds, but it will see these differences or other worlds as expressions of a more fundamental cosmic unity, social justice, ecosystem, or singular multiverse.

The Human Is Two. The common human experience of this One World is dualistic, or Two. The human being splits reality into two seemingly separate domains—a mental and a material domain. As modern thought has long recognized, these are epistemic states—that is, forms of knowledge and not things-in-themselves (in truth, you do not know objects or other subjects themselves but rather representations of objects or subjects in your mind). What connects, and probably cocreates, these two domains of the mental and the material is the human interface or cognitive and sensory system of the body-brain. Put differently, it is the human being that "splits" reality in Two and makes it appear as such. It is not so before or beyond the human knower.

In such an ontology, all experience and thought take place within a psychophysically neutral Ground, or One World, that *all* things and subjects share, from which they all emanate or emerge, and to which they return. This fundamental cosmic Ground is similarly active in scientific domains as distinct as quantum physics, evolutionary biology, and cosmology. It is ultimately unspeakable, indescribable, and unnameable for the simple reason that it is neither mental nor material and so is not an object of human speech, language, thought, or, indeed, intentional (dualistic) knowledge itself, including that of the sciences. Mathematics can capture something of its apparent structure and behavior for a human being, but only in a distant, indirect, third-person kind of way. Mathematical description or modeling can never get at *what it is*. The human body-brain simply possesses no cognitive, linguistic, or scientific resources to do that. All it can do is point.

But here is the thing. The human being can come to know directly—or, better, *become*—the Ground of all being (the One) *and* what it so fantastically expresses or excessively splits (the Two), since the human being is in fact already both. This is where dual-aspect monism, the One that becomes Two, meets comparative mystical literature and the elaborate imaginal, supernormal, and paranormal forms of the history of religions. To think the impossible, then, is to know or imagine oneself as *both* the Ground *and* that which the Ground dreams, projects, and lives.

We saw a most dramatic direct form of this gnosis with the case of Kevin and his Platonic surrealism. In William Blake's poetic terms in *Milton*, which we encountered in our first chapter, Kevin came to know that the Imagination is not a passing subjective state but "Human Existence itself." Kevin thus realized that the World is One and the Human is Two—by being *both* that One *and* that Two.

There are endless examples of how these two dimensions collapse entirely into an awakening into a One World before and beyond the epistemic dualisms of the human ego. Such accounts remind us that these awakenings are neither abstract nor without human import. They matter, and they matter a great deal. Here is one from an American neurologist, James Austin, who has seen the implications for these states for a renewed philosophy of mind and the future of neuroscience.[3] Austin was on a two-day Zen retreat in the United Kingdom. He was in a London subway station. He just "happened to look up and out into a bit of open sky above and beyond." That was enough:

> Instantly, with no transition, the entire view acquired three qualities: absolute reality, intrinsic rightness, and ultimate perfection. . . . There was no viewer. Every familiar psychic sense that "I" was viewing this scene had vanished. A fresh new awareness perceived the whole scene impersonally with the cool, clinical detachment of an anonymous mirror, not pausing to register the paradox that no "I-Me-Mine" was doing the viewing. This new awareness further transfigured the scene with a profound, implicit, totally perfect sense of absolute reality. . . . One of the unspoken messages was that this is the eternal state of affairs—things have always been this way and continue just so indefinitely. A second insight conveyed the message there is nothing more to do: this train station and the whole rest of this world were so totally complete and intrinsically valued that no further intervention was required. The third insight plumbed a deeper visceral level: there is absolutely nothing whatsoever to fear.[4]

One can frame such an experience-event in terms of Zen Buddhism, as Austin does here. Or one can invoke the more abstract categories of dual-aspect monism, as I have done in the present book. Or one can invoke the categories of theism and idealism and a doubled natural and supernatural cosmology. The historian of religions cannot dismiss or demean *any* of these responses. These are matters of interpretation that really matter. As a committed comparativist who takes ontological diversity seriously, I hardly have a final answer, nor would I want one. I strongly suspect that, in the end, there are multiple answers, all of which really matter.

Ontological Shock: A Brief History of the Expression

Toward such a plural possibility, I can think of at least five intellectual-spiritual practices, all of which can be summed up in the two-word phrase that the Harvard psychiatrist John E. Mack used to such effect to capture what he came to believe was ultimately at stake in the modern abduction phenomenon: "ontological shock."[5] The phrase is so important to how to think impossibly. The expression's history in the humanities, its radicalism, and its fundamental cosmic hope are often underappreciated, so let me explain something of each as I conclude.

The expression "ontological shock" appears multiple times in American culture, both in the past and very recently. It was used, for example, by Don Davis as a musical title (track 9) for the score of *The Matrix* (1999), a film that was itself a form of impossible thinking modeled on the ancient lessons of Plato's Cave (that true reality is not *at all* like the dark shadows flickering on the sensory cave walls).

In the summer of 2023, "ontological shock" was used in an even more public way by United States Air Force officer and former intelligence official David Grusch, whom we met briefly in chapter 2. Here is how Grusch put the stunning claim that the United States has retrieved craft of nonhuman origin: "I hope this revelation serves as an ontological shock sociologically and provides a generally uniting issue for nations of the world to re-assess their priorities."[6] I bet Grusch had read John Mack. It sure sounds like it.

Like most intellectual expressions encoding radical altered states (or reimagined nation-states), the historical roots of the phrase reach deep into the history of the humanities, really what I have called the superhumanities (the altered states of the humanities). The idea of "ontological shock" ultimately stems back to the Protestant theologian Paul Tillich in his *Systematic Theology* (1951), where the theologian uses the expression to describe an ecstatic state when "the negative side of the mystery of being—its abysmal element—is experienced."

Tillich will in fact define any "genuine manifestation of the mystery" as possessing both a subjective and objective dimension and observe that revelation comes through the altered states of "ecstasy" but that it is not reducible to the inevitable psychological expressions of these same altered states of consciousness. Both points resonate powerfully with how I have tried to think impossibly here. Tillich will also silently invoke the history of religions in the figure of Rudolf Otto to define the essential doubleness of the divine presence, which is, once again, helpful, resonant, and just plain honest:

> In revelation and in the ecstatic experience in which it is received, the on-tological shock is preserved and overcome at the same time. It is preserved in the annihilating power of the divine presence (*mysterium tremendum*) and is overcome in the elevating power of the divine presence (*mysterium fascinosum*). Ecstasy unites the experience of the abyss to which reason in all its functions is driven with the experience of the ground in which reason is grasped by the mystery of its own depth and of the depth of being generally.[7]

Still, as the references to "reason" in the above passage signal, Tillich's own definition of "ontological shock" is, in the end, not very shocking, as it preserves the lines of his own rational and finally Christian systematic theology. Tillich in fact does a great deal of border protection around his no-tion of ontological shock, even in the two pages that surround the page that announces the key expression. He tries, for example, on the page before to distinguish between the presence of the "divine Spirit" in "genuine" revela-tion events and the "overexcitement" or enthusiasms in "so-called ecstatic movements." He also tries, on the page after, to argue that genuine ecstasy does not destroy the cognitive or rational function, whereas demonic pos-session does.[8] In the end, this kind of theological conservatism is simply not what experiencers experience and John Mack was trying to get at with his own use of the same phrase.

It is unclear where Mack himself got "ontological shock" (as we have just seen, it had been circulating in the academic sphere for four decades, since 1951), but we know that he uses it as early as 1992 at the major confer-ence at MIT on alien abductions.[9] His use of the expression in an April 17, 1994, article in the *Washington Post* is as good a place as any to get a sense of what he meant by it. This piece appeared just three days before the of-ficial release of Mack's breakaway book and is clearly intended to function as an introduction to *Abduction: Human Encounters with Aliens*, which includes a rich indexical entry on "ontological shock."[10]

Mack begins the *Post* piece by writing of a "new and mysterious phe-nomenon" that "challenges our way of thinking." Nothing in his four decades of psychiatric work prepared Mack for a form of human experience and suffering that could not be explained psychiatrically and "was simply *not possible* within the framework of the Western scientific worldview." He writes of how his subjects "verge on being overwhelmed by the traumatic impact and philosophical implications of their experiences," and how these experiences "seem to shatter the notion that we are the preeminent intelligence in the cosmos."

This rational impossibility and this cosmic trauma are *precisely* what Mack means by ontological shock. In the book, he defines the suffering of his subjects as "multidimensional trauma," a phrase that was probably not meant metaphorically, since the suffering involves, well, other spatial and temporal dimensions.[11] He is especially struck by the reproductive focus of the abduction experience and the subsequent genetic or quasi-genetic engineering that appears in the multiple accounts of experimentation on the genitals and the removal of sexual fluids, even hybrid fetuses.[12] These unwanted sexual dimensions are very much a part of the severe trauma experienced and relived under hypnosis (the tears, the screaming, the growls), but they are also somehow related to a new sense of sacredness (many such abductions are "unequivocally spiritual"), a firm planetary consciousness, and a basic sadness with respect to looming ecological disaster and the explicable human path to global destruction that set in for the experiencers. We simply do not understand what is going on here.

In the same honest and unknowing spirit, Mack seriously questions whether our physics is up to the task of understanding the phenomenon but also observes that all sorts of physical effects are obvious in the abduction experiences, including independent sightings of a UFO in the sky at the same time. The experiencers themselves recognize this fundamental conflict with physics—the obvious presence of physical effects within a profoundly spiritual experience—when they speak of the "collapse of space/time." Mack's working hypothesis? That we may be witnessing "an awkward joining of two species, engineered by an intelligence we are unable to fathom, for a purpose that serves both our goals with difficulties for each."[13]

When Mack took up the phrase "ontological shock" in the early 1990s, then, he used it to put a very strong accent on what his subjects were *most* concerned about, what they *most* wanted to talk about, what they considered to be the core and center of their abduction experiences: the superreality of such experiences or events and, more to the present point, how this superreality was *completely* other or *entirely* different from the social, psychological, and physical realities they earlier experienced as the case. There was also a difficult interspecies encounter behind the phrase, although the intentions of such a galactic colonialism were complex and not entirely negative. Quite the contrary, for Mack at least, they often served existential, spiritual, and ecological purposes.

The important point to stress here is that Mack eventually landed pretty much exactly where I have landed in the present book—namely, with a firm sense that the experiencers are intellectual equals, "co-investigators in an exploration of the unknown" (compare my own "thinking-with," coined in the first few pages of the present book).[14] Also very much to my point, he concluded with a firm conviction that it was the utter and *completely disjunctive reality* of what they were experiencing that was the key to the whole shebang. This ontological shock—which he defines at MIT in 1992 as the "bleak realization that what they have experienced actually occurred and reality as they have defined it is forever altered"[15]—explains why such individuals often *wish* they were "crazy," or that this was "just a dream"; why the act of comparison is the "flip" that triggers the ontological shock (in these cases, comparison is performed by Mack himself as he relates to the person his familiarity with the various themes of the reported experience from other similar accounts); why the events have *no* psychiatric explanation; and why it is so very difficult to acquire funding for effective research (since no grant agency will recognize the truth, much less the cosmic importance, of what has actually happened). Moreover, again in perfect line with the arguments set out above, abductees often develop powerful psychic abilities. Mack highlights precognition, "profound healing capacities," and "light-related perceptions."[16]

On a philosophical level, Mack is clear that such ontological shocks follow neither the lines of our scientific objectivisms nor our humanistic subjectivisms and historicisms but call us to posit a third space of the superreal, what I have alternately called the superhuman, the fantastic, and, here in these pages, the impossible.[17] Mack, it turns out, was a childhood friend of Thomas Kuhn, the famous philosopher of science and author of *The Structure of Scientific Revolution* (1962), which gave us our notion of a "paradigm shift."[18] This is clearly what Mack was about as well—not a rearrangement of the chairs on a sinking ship, but a totally new ship or way of sailing (or flying above the water, or emerging out of it). Because of his relationship to Kuhn, he understood deeply that the binaries of real/unreal and objective/subjective are theological leftovers of a Western belief system that is now active in the sciences. This is not how reality actually works, what it really *is*.

This same radical disjunction, or not-fitting-in, is why I react so strongly, and so negatively, to phrases like "the paranormal is normal" or "the ordinary is extraordinary." Of course, there are forms of psi that interact invisibly or unconsciously—behind the scenes, as it were—with normal human physiological, psychological, and social functioning. I am thinking of James Carpenter's fine work on "first sight" in parapsychology and everyday life.[19] And, of course, there are forms of contemplative practice that sacralize the secular world. We just saw one above, with the practice and realizations of Zen Buddhism.

But this is *not* what we are trying to think-with here. This is a mechanical "fucking bug" coming out of the wall (and probably the future), a gigantic mantis with a purple cloak who shows every sign of controlling the little praying mantis insects of the biological world, and holy women and men floating and flying in Italian and Spanish churches, convents, and outdoors before thousands of people. How is *any* of that "normal" or "ordinary"? Obviously, something, or someone, is trying to get our attention. Are we listening? Or are we just reducing it all to our own banality with clever but vacuous phrases?

Please stop it.

Of course, whether one can stop the soothing reductions and counter the ontological shock depends on one's philosophical position, on one's ontology. I do not know how else to say this. I cannot say this enough. My own impossible thinking performed in these pages is *not* limited to Cartesian dualism (although it certainly honors such a split in our ordinary perceptual and cognitive experience). In the end, though, it thinks beyond, below, or above any and every epistemic dualism, which is the basis of *all* modern science and the general order of knowledge in the academy as a whole, as John Mack and his friend Thomas Kuhn so clearly saw. None of this implies that the sciences are not crucial to moving forward. If dual-aspect monism is correct, then every mental experience is also a physical one, and every physical experience is also a mental one. There can be no ultimate split since the ground of all being is not so split.

I sense this all, *intensely*, when I speak to experiencers who tell me about discarnate beings "coming into" their room and consciousness, effortlessly transcending any and all boundaries of self or mind. Such superintelligent beings become perfectly physical. And then they disappear, seemingly into nothing. They move at will in and out of this three-dimensional world, our so-called reality.

Little wonder that these experiencers often believe that this is where the future human will emerge, in this superreality that is at once physical and spiritual, that cares little—or nothing at all—for our assumed moral, psychological, and cultural boundaries. I laugh out loud when I hear military or government officials talking about "threats," "our airspace," or "our national borders." There are no boundaries or borders in this surreality. Nor, frankly, are there individuals, not at least as we conceive them today in the secular world. There are presences coinhabiting one another, communicating with each other on levels for which we only have science fiction today (and the religions of the past).

Five Ways to Be Ontologically Shocked

Can I say this more clearly still? Let me try. Let me list five ways to be so ontologically shocked, to coinhere, to transcend these boundaries and borders, to be a self that is not a self. I will capture each in a crystallizing phrase and then define the colloquialism as best I can.

1. **"Think Impossibly."** Such a practice begins with the realization that impossibility itself is a function of a particular cultural or cognitive system and, most of all, of what is excluded from that system. I have put this insight in a kind of meme in my written activism: *our conclusions are a function of our exclusions*. I have also used the metaphor of the table to discuss the same idea.[20]

The metaphor of the table is a simple one: the conclusions one reaches about a set of phenomena will be largely a function of what one places on one's table and so also of what one takes off that table. Take things off the table and you will eventually be able to explain what is left (because you just took everything off the table that you cannot explain). Put more things on the table, however, and things will begin to look considerably stranger. The pieces on the table, moreover, will mean very different things—and likely things that we simply cannot understand with our present concepts and their implicit exclusions. In so many ways, *impossible thinking is simply what happens when we do not take things off the table*.

I know well the defenses and strategies of the well-meaning person. I have witnessed trusted colleagues taking the impossible off the table in real time, right in front of my eyes. They invoke all the common strategies (hallucination, coincidence, anecdote, and so on), none of which actually work and all of which function more as rhetorical devices or, frankly, moral excuses not to think. It does not seem to bother them. But it bothers me. I want to know where thought goes if the impossible really happened in the sense that the historians mean it, as in *wie es eigentlich gewesen ist*—"how it really happened."[21]

Because it really happened.

2. "Get Weird." Impossible thinking focuses in on particular forms of shocking experience that are given or shown to the individual, that come unbidden, that are not consciously constructed or thought. Such a phenomenological approach is at the very core of what I am proposing in these pages. The method involves taking people's experiences, no matter how unbelievable they become, with the utmost care and seriousness, and then thinking out of those experiences as premonitions of new critical theory.

Again, the latter new thought is not the former strange experience, but the two are definitely related: the new thought is an abstract ordering or intellectual form of the original anomalous event. The new words that emerge are coded experiences of the original inrushings, revelations, or horrors. The new words *are* experiences.

There is a kind of hermeneutical loopiness here, a trickster-like twisting and turning that shows numerous signs of being related to the ancient Anglo Saxon "wyrd," whose connotations of twisting, becoming, and shaping one's own fate or future are all esoterically traced—but also largely forgotten—in our modern concept of the "weird."

But not entirely forgotten. I am thinking in particular of the historian of modern religions Erik Davis and what he theorizes as the "high weirdness" of the mushroomed American counterculture of the 1970s and its endless UFO visions and entity encounters. Davis is thinking especially of the psychedelic visionaries and activists Terence and Dennis McKenna, the occult writer Robert Anton Wilson, and the science-fiction author Philip K. Dick, with whom his project in fact began (Davis was one of the key figures in the massive Exegesis project, which was the effort to make public what Dick experienced and wrote about so excessively, and humorously, in his private journals).

Davis means something very specific by the weird. He means what I mean. He means a power over time. He means creativity. He means writing the paranormal writing us:

> This book is about high weirdness, a mode of culture and consciousness that reached a definite peak in the early seventies, when the writers and psychonauts whose stories I tell herein pushed hard on the boundaries of reality—and got pushed around in return. . . . The element of recursion . . .

also characterizes high weirdness. . . . Think of it as a kind of visionary skepticism, or critical gnosis. Within this questioning current, the object of weird fascination is folded back into the subject, constructing a strange loop of cultural play, recursive enigma, and extraordinary encounter that makes a raid on the real. . . . . High weirdness is equally a mode of *extraordinary experience*. . . . The phrase (or sometimes *high strangeness*) has served as an underground term of art for particularly intense and bizarre experiences—especially anomalous experiences associated with paranormal phenomena, occult practice, synchronicities, and psychedelics. Marginal and esoteric cultural narratives—particularly those wrapped up in conspiracy theories, extraterrestrials, occult forces, strange gods, and fantastic pulp fictions—intrude forcefully, uncannily, and sometimes absurdly into the texture of lived experience.[22]

Welcoming such a high weirdness onto our phenomenological table is a central practice of impossible thinking. Indeed, without this welcome, without this recursive skeptical affirmation, there can be no impossible thought.

In this same high weirdness, I must insist that *it is okay to be tricked, to be wrong*. Part of this is the very nature of critical thinking. To think critically is to be constantly corrected, and to be comfortable and at peace with that. But thinking impossibly it is also to practice a skepticism so deep that it questions the very roots of the questions. A genuine skeptic, after all, is as skeptical of scientific materialism as any other interpretive framework. Such skeptical writers possess astonishing levels of integrity, care, and reflexivity, like all serious intellectuals. Even weirder still, they might begin to question whether human reason, or rationality itself, is up to the task. This is what I mean by questioning the very roots of the questions.

There is something more fundamental in this kind of weird skepticism, then. To be involved in high weirdness is precisely a willingness to be duped and deceived, yes, but it is also to entertain the likelihood that the fraud and the fact are not always exclusive of one another. Perhaps this is one reason I have come to see that what finally sets an impossible thinker apart from a traditional historian is precisely the sense that *deception may well be an integral part of the phenomenon itself*.[23]

There are endless political misinformation campaigns, religious exaggerations, and psychological projections in the mythologies and folklores of human history, of course, but there is also something fundamentally deceptive about the superhuman or nonhuman source of the impossible itself. "Belief is the enemy," as John Keel had it, not because there is nothing to any of this but precisely because there is—and because it is not, at all, what it seems to be, what it pretends to be. Belief shuts down. So does debunking. It is better, much better, to be weirdly skeptical of both.

3. "Look up." Once one attempts a genuine phenomenology of the weird, it becomes clear that there is a "vertical" dimension to some such experiences that has traditionally been framed as "transcendent" or, if you want to be phenomenological about it, "transcendental." One allows such a verticality to stand. One does not immediately reduce it to this or that horizontal dimension—to a social process; a political system; a historical context; a gendered, racial, or sexual identity; a psychological need; a biological instinct; a linguistic grammar; a neurological default system; or whatever the reigning reduction might be. All of these are helpful and illuminating. None of them are adequate.

Of course, transcendence is relative to the subject in a dual-aspect monistic world. It is epistemically true and ontologically false. Outside the human subject, there is neither transcendence nor immanence. There is only the One, which is neither material nor mental, neither immanent nor transcendent. But there is certainly transcendence or an "up" in human experience, Copernicus or no.

And there are, of course, other modern ways to affirm the vertical dimension of human experience, including the new realities of the sciences, especially mathematics and geometry, which produce the hyperdimensional model first announced in Edwin Abbott's *Flatland* of 1884, and evolutionary biology, which produces the endless evolutionary esotericisms of the modern world, definitively beginning with Alfred Russel Wallace's model of a double evolution, at once biological and spiritual, and quickly morphing into Friedrich Nietzsche's still unassimilated atheistic vision of the coming superhumans. We can easily surmise with Ioan Couliano now, and through an entire spectrum of sciences and consequent models, that the mind is "not limited to three dimensions."[24]

Consider, as a modern example, the Scottish writer Grant Morrison, who has written autobiographically of an entire spectrum of hyperdimensional, evolutionary, and ecological mystical experiences in the language of "supergods"—that is, in Morrison's own mythical reimaginings of the contemporary superheroes. Morrison has also described in considerable detail an abduction-like experience they knew in Kathmandu, Nepal, in 1994. There fifth-dimensional extraterrestrial mercurial blobs came out of the walls and furniture to "turn around" the writer into another dimension and pop them into another star system.

Morrison explains coming back into the meat-body with a temporary real-world superpower, in this case the ability to see all things, including a physical cup, in five dimensions. Morrison could also now understand how the human body is in fact a billion-limbed form spread throughout time, and how these individual forms we call the body are simply temporal slices or sensed intersections of that much longer and stranger cosmic form. The superhuman writer finally came to realize that all life-forms on Earth are in essence branches of the same biotic tree, an amoeba-like superorganism or "infant god" feeding on itself within a singular block-universe "AllNow."[25]

4. "More Real than Real." Impossible thinking understands the imagination in a doubled way, much indebted to the Romantic revelation of Europe, as I explained in chapter 1. Accordingly, such thought understands perfectly well that there are different forms of the imagination, some of them, yes, imaginary and banal. But it also understands that there are specific occurrences in which the imagination, somehow, takes on a life of its own, becomes independent of the ego or conscious self, and intervenes in the life of the seer in essentially creative, surprising, and often "more real than real" ways. Indeed, such an imagination, or Imagination, can even manipulate physical reality—throw things around, levitate bodies, materialize stuff.

It bears repeating one more time: Impossible thought approaches imaginal appearances not as imagined fantasies that are symptomatic of various psychodynamic issues (the standard psychoanalytic answer), or false social illusions meant to distract from social revolution and the hard work of addressing economic injustice (the Marxist answer), or ontologically neutralized comparative patterns or historical descriptors (the traditional religious studies answer), or as tempting tangents that should be ignored for the sake of some religious goal or God (the orthodox religious answer). Impossible thought treats some imaginal appearances as *potential mediated revelations of the real*. This is a theory of the fantastic as the real. Hence the most common descriptor of these events: it was "more real than real."

It no doubt was.

I claim no originality here. Writers have struggled with these astonishing imaginal phenomena for millennia. In the history of Christianity alone, there have been detailed discussions of visionary and dream experience. One could cite here Augustine and his threefold classification of visions (corporeal, spiritual, and intellectual, or the *visio intellectualis*). Or William Blake and his "fourfold vision"—Blake, whose visionary illuminations and channeled poems have evoked for scholars any number of aesthetics and implications, from superhero serial comic art to the most paradoxical nondual philosophies of Hindu and Buddhist Asia.[26]

It is also of great significance that a contemporary medievalist like Barbara Newman, whom we treated briefly in chapter 6, has found it helpful to invoke the modern parapsychological category of telepathy to understand the medieval saintly reading of minds and communicate something actual of the power of prayer in medieval Christian writers. Contemplative prayer becomes a kind of ontological permeability, a practiced telepathy or coinherence through the medium of the Mind of God in this most remarkable work.[27]

These imaginal visions and their subsequent theorizations are endless. No one can really grasp what is happening from what Morrison might call a higher fifth dimension outside the mere three dimensions that the ordinary human mind can imagine and with which it thinks and speaks. And that, I suppose, is my point, my poignancy, and my fundamental hesitation. Whatever religious, poetic, psychedelic, or superhuman framework we choose, the basic idea remains more or less the same: sometimes the imagination imagines the real, "dreams the true" (*Wahrträumen*) before it happens— that is, outside space and time—as Schopenhauer had it. Sometimes, in other words, the imagination is our best access to what is really and truly out there, if always, yes, of course, in culturally mediated forms.

Impossible thinking is not thinking in only three dimensions. That is entirely possible. It is imagining in four, or five, or more.

5. "It's about Time." This brings us to our fifth and final feature of impossible thinking. This one is perhaps the most speculative, so I have put it last. Perhaps it requires more explanation than the others. Still, I think something like it is inevitable if we are to take the full phenomenological record seriously.

In terms of the empirical implications of impossible phenomena— that is, how they interact with space and time and what we normally call "history"—I continue to think that precognitive experiences are among our strongest and most promising instances of an alternative rationality. Those who know a future that has not yet happened *are* thinking impossibly. They are thinking in terms of moving through time, not space. They are actual biological time machines.

I have been especially influenced here by the work of the anthropologist Eric Wargo. Wargo elegantly employs psychoanalytic theory, dream interpretation, and the philosophical implications of quantum mechanics in order to argue that some of the very paragons of the humanities precognized their intellectual and literary ideas in a future that was already so and then worked toward that already accomplished future to realize their game-changing cultural creations. Hence Wargo's central notion of the self-fulfilling "time loop." Such time loops are perfectly possible for Wargo in a materialist universe because we live in a "glass block universe" in which the past, the present, and the future are all laid out, are already present, are already so. It is all one big Now. We saw this same idea in our prologue and in chapter 4.

I am much attracted to Wargo's model, not because I know it is correct in every specific detail but because the model is an ideal exemplum of impossible thought: it makes the impossible possible by providing a theory for how and why precognitive events are entirely natural. It also has *immense* implications for what we think of as human culture, religion, literature, and science, really for the entire order of knowledge. If Wargo is correct, everything changes, instantly, and for good reason: the deepest source of human creativity is outside time or, if you prefer, not bound by the linear temporality that we assume to be some kind of iron law (or chain).

As Wargo writes, "The reality of time loops calls into question *all* our taken-for-granted beliefs about originality in the realm of culture."[28] Consider this passage on the fundamental inadequacies of our present critical theories and what precognitive creativity portends about a different kind of trust and subsequent interpretative style:

> Suspicion is always the rule: Criticism often treats texts like crime scenes—dusting carefully for fingerprints, looking for what evidence has been concealed or destroyed, and always arriving at a story about guilt and fault, dissembling and deception (if only the artist's self-deception). But if the creative imagination is a precognitive imagination, no amount of reducing a work to causes, influences, and intentions can exhaust what is original in it, and it is wrong to see a work as a crime committed, or the artist/writer as a figure to be distrusted and held accountable. The artist, I believe, is no more guilty than Ezekiel was: They are a witness to the impossible. Consequently, I agree with J. F. Martel in his book *Reclaiming Art in the Age of Artifice* that a better guiding metaphor for how we approach art is an encounter with an alien technology. It follows from this that we should approach criticism more like the investigation of a UFO landing than like a crime scene. An encounter with something extraordinary happened (the inspiration), and it left material evidence behind (the work) that can be forensically examined. The witness is

the most important piece of evidence, and warrants a default "hermeneutics of trust" rather than of suspicion.[29]

Along similar lines: "The most shocking entailment of what I am arguing . . . is that the most ingenious works (and probably, by extension, all 'inspired' works) are really *time loops*. What this means is that they are actually *uncaused*, exactly what debunkers of divine inspiration have always feared—creation from nothing."[30] Wargo returns to his favorite metaphor of the UFO again, along with the superheroic mytheme of radioactive mutation, to express why we feel a kind of aesthetic shock reading a great canonical work of literature. His comments about the "end of this book" can be read as applying to this one as well:

> I hope that by end of this book, you will be persuaded that culture—high and low and everything [in between]—may consist of vessels of the super, time-defying backward transmissions from our personal and collective futures. Artists and writers channel and manifest the impossible. There's a core impossibility buried in their best works that affects and changes us, even if we are not consciously aware of why or how. Like an irradiated UFO landing site, a film like [Andrei Tarkovsky's] *Stalker* or a novel like [Virginia Woolf's] *Mrs. Dalloway* or a painted bison on a Paleolithic cave wall still glow with the energy of an encounter with the impossible. We are scarred, changed, perhaps evolved a little by that radiation. Yet because we lack the eyes to see it and largely the imagination to imagine it, we pass over this mutation, imagining ourselves moved and mutated only within permissible parameters of movement and mutation. That is, only slightly, only vaguely. So, let's expand our imagination and open our eyes—and put on sunglasses.[31]

The model gets stranger still, as its commitment to a glass block universe implies that our very physical forms morph through space-time, pretty much exactly like Grant Morrison's Hindu supergods, those billion-limbed human bodies spread out through the multiple decades of a life cycle. Here is Wargo in Morrison's AllNow: "I think it is our physical body, with its quadrillions of material (which just means measurable and thus information-carrying) particles, that connects us to the future and is thus the locus of our supernature. The body is actually a hyperbody, a higher-dimensional being. It is our bodies, including our nervous systems, that carry information from time point B backwards to time point A."[32]

I should add that, in its furthest and most speculative reaches at least, impossible thinking does not restrict this kind of power over time to precognition. It is willing to speculate that one of our major hurdles in understanding impossible phenomena is our insistence on thinking spatially instead of temporally. To give a concrete example, consider the UFO and contact experience. Many of these cases, which involve things like objects and beings popping in and out of existence, make absolutely no sense— until *one begins to think of them temporally instead of spatially*, of the objects and beings come from another time, not from another space. Once we can think like that, much that is puzzling falls into place and ceases to look so puzzling.[33]

It's about time.

Limitless Belief and the Skeptic's Overcoming

All of this finally implies a most complex view of religious belief. I do not generally like the category of belief, mostly because it works in the contemporary world to dismiss people who hold such convictions. It reduces them to participating in "a community of belief," which means a relative group of people who believe things that others do not and that, if the truth be told, do not exist.[34]

This is *not* how the experiencers I think-with understand what has happened to them. They are not believers. They are *experiencers*, and they are in very painful ontological shock. They did not ask for their experiences. They did not worship, meditate, and practice their way into some kind of wish fulfillment, false form of consciousness, or socially produced hallucination. As Karin Austin has it, the thing *physically* came out of the wall, out of nowhere, and sat on her chest, while she was perfectly awake. This is not religion. This is science, or superscience. This is not a belief. This is a bug. And a hyperdimensional mechanical one at that. Such events are not "beliefs." Nor are they supernatural. They are "ultra-natural."[35] AI, or artificial intelligence, from the future seems like a far better interpretation in this particular case than something spiritual, much less something unreal.

Still, I have to walk my resistance to the category of religious belief back a bit, since beliefs give access or at least mediation to specific aspects of reality that would not otherwise be available to the person and community. That seems beyond question.

But it also needs to be said: *religious beliefs are not religions*. The former beliefs, based on actual impossible experiences, are ancient. They are expressions of the way things are, as I have tried to argue above. One can believe in the nonphysical form of the soul, because one has experienced it as such, without accepting any of the teachings of a local religion, which one might well find culturally bound, or just oppressive. The latter religions, or cultural systems built up out of these primordial experiences and disciplined through myth, ritual, institution, and authority, are recent and simply not necessary, as the modern secular world shows us in abundance. Religious identity, in particular (that is, belonging to a particular community of belief, a "religion") is evolutionarily recent and unnecessary. Religious belief is *not* religion.[36]

I can think of at least three further things to say here.

The first is that belief can lead to trickery and fraud or, in Carlos Eire's always apt words (this time about seventeenth-century Spain, "*le grand siècle* for mystical theology"), "belief in miracles created a space for imposture as much as it did for hope in the impossible."[37] This, I take it, is the real concern of secular and humanist thinkers with respect to religion: belief and fraud go hand in hand. That is so.

The second thing to say is more positive; in Eire's words again, belief is the "immortal soul of the imagination" but also of "all mentalities, mindsets, worldviews, epistemic regimes, discourses, social imaginaries, and social facts"[38] (these are the catch phrases academics have baptized as somehow more real than that which they attempt to describe). In the supernatural realm, moreover, in the realm of faith, belief appears to be nothing short of "limitless."[39] Put differently, it is finally belief that creates the realities of the critical theories of the academy but also of the marvels and miracles of the historical religions, where people really do fly.

The third thing to say is that religion is arbitrary and relative. As culturally coded forms of other people's impossible experiences, as habits of thought, historical deposits, and concretions of culture, religious beliefs are precious reminders of who we are or might yet become. But can we today really sign our names to any of their elaborate mythologies or doctrinal systems? Most of them simply are not ours. Nor are they those of most people on the planet. There is the impossible as a function of a system again. Can anyone seriously doubt this comparative point?

I cannot but help recall Nietzsche again here. The philosopher made a fierce distinction between belief as "a mere phenomenon of consciousness" (that is, as a thought in the head that refers to nothing at all) and what he called the "'eternal' factuality" of the "psychological symbol redeemed from the concept of time" (that is, the religious teaching that is not taken literally but as a parable outside historical development).[40]

Nietzsche loved Jesus, or at least a particular interpretation of him. That teacher's kingdom of God, Nietzsche wrote, is "nothing one expects; it has no yesterday and no day after tomorrow, it will not come in a 'a thousand years'—it is an experience of the heart; it is everywhere, it is nowhere."[41] Sadly, it was this Jesus—that "great symbolist," whose "evangelic conception of everybody's equal right to be a child of God"—who became the very opposite in the history of Christianity; that is, he came to represent a set of exclusionary "ecclesiastic" doctrines about silly beliefs with absolutely no reference to reality as such.[42] Of course, this God must die.

Nietzsche also famously equated "great spirits" with "skeptics," since they, very much unlike the believer, do not land and are not convinced by this or that belief statement. The skeptic is in fact strong, the believer weak.[43] The skeptic "overcomes," to use a Nietzschean word, whereas the believer does not. Religious convictions are finally "lies." Indeed, the glory of the Renaissance and its *humanitas*, its humanism, lies precisely in its attempt at a "revaluation of Christian values"—that is, its unceasing skepticism and questioning, what the German philosopher in another context calls "further experimentation, a continuation of the fluid state of values, testing, choosing, criticizing values *in infinitum*."[44] To be a humanist is *not* to be a Christian. Such a Christian wants to suppress all of this, like the German monk who destroyed the Renaissance revaluation of all values, Martin Luther. Nietzsche hated that guy.

And yet there are other places in Nietzsche that clearly favor a most profound awareness that reason cannot answer the deepest questions of life, *in principle*. "There are questions in which man is not entitled to a decision about truth and untruth; all the highest questions, all the highest value problems, lie beyond human reason." Such an attitude "alone is truly philosophy."[45] There is a most radical and real skepticism, a weird one.

Nietzsche also took a position in regard to comparative religion (comparing, for example, Christianity and Buddhism) that is extremely close to the one advanced in these pages—namely, that the belief systems of the religions are so many projections of our own distorted and denied superhuman nature and future realization. The religions are fictions, projections, or imaginary constructions *that nevertheless share an esoteric ground*, which is most clearly realized in the Asian religions or in a philosopher like Plato. Listen to Nietzsche, that master of public suspicion and secret affirmation:

> At the bottom of Christianity there are some subtleties that belong to the Orient. Above all, it knows that it is a matter of complete indifference whether something is true, while it is of the utmost importance whether it is believed to be true. Truth and the *faith* that something is true: two completely

separate realms of interest—almost diametrically opposite realms—they are reached by utterly different paths. Having knowledge of this—that is almost the definition of the wise man in the Orient: the Brahmins understand this; Plato understands this; and so does every student of esoteric wisdom.[46]

In this same esoteric spirit, Nietzsche favorably compares Jesus "the sermonizer on the mount, lake and meadow" to "a Buddha on soil that is not at all Indian." The historical Jesus is like Nietzsche, a "free spirit," who would have used Sankhya (Indian philosophical) concepts had he lived among Indians, or those of Lao Tzu had he lived among the Chinese. It makes no real difference. "Of course, the accidents of environment, of language, of background determines a certain sphere of concepts," but these cultural constructions are just that: *accidents*. Nietzsche, in short, was a most radical comparativist.[47]

At the end of the day, this one anyway, the entire history of religions looks like so many temporary cinematic projections of our own superhuman nature that we must deny to ourselves—perhaps, as Nietzsche thought, because we are sick, weak, and stupid, or perhaps (much more likely, I think) because of the relative nature of the viewing subject and its inevitable experiences of the phenomenon as elusive, deceptive, and tricky. It has to be like this for us. We cannot turn around and look back at the projector projecting us, not at least as us, as characters on the screen.

That would be impossible.

The Three Bars

> So, again, one must ask: Why do so few people on earth know about Saint Joseph [of Cupertino]? Why is the evidence ignored or trivialized? Why has he been relegated to the history of the ridiculous rather than to the history of the impossible, or to the science of antigravitational forces?
>
> CARLOS M. N. EIRE, *They Flew: A History of the Impossible*

There are some practical ways to think about our reasonable resistances to all of this. There are concrete ways to understand where one honestly says, "That can't happen."[1] We all have that place. We need a very specific case with which to locate our own existential coordinates, though. We need something to which we can react *and then watch ourselves reacting*.

Happily, this is easy to do. Here is one such case to which you can react.

A man named Jim wrote me after reading one of my books. Jim is a professional copywriter. No doubt because of this, his letter is an especially eloquent but fairly typical example of the kinds of letters I receive almost every week now. Listen (and recognize that Jim does not use a lot of capitals):[2]

> I live in san francisco. I was asleep one night [in 2002] when I was awakened by the sound of a horrific car crash at the intersection. I heard the screech of brakes and then a violent crash. I bolted awake. my heart was pounding. my first thought was go outside. get the victims help. I ran out of my house but when I looked to the intersection there was nothing. no smashed cars and no evidence of an accident nor any cars at all.
>
> I was dumfounded. It didn't feel like a dream. it was much more specific. what woke me up was a distinct experience of *sound* as opposed to something "visual." my dreams are almost exclusively "visual" so it didn't make sense this was a dream. still i put myself back to bed and told myself it was probably a dream however unique it felt.
>
> the next day I went to work and followed my normal routine and went to bed around 11pm. then in the middle of the night while I slept, *I heard the exact same accident I heard the night before*. the violent screech of brakes,

the tremendous impact of metal vs. metal. again I bolted up in bed. I distinctly remember being confused: is this *last* night? why is this repeating? what's going on? after a moment trying to process all this, I heard someone outside on the street shout something like, "Oh my god! Somebody get help!"

At that point I felt this wave of adrenaline rush through my body. all at once, I realized this was in fact a different night than the previous one. and this time the accident was real. I ran outside and a car had been T-boned, rolled several times and landed in my neighbor's yard. I learned later that a young [man] in the car died.

later that night I broached the subject with myself, did I in fact have a precognition of the accident? a prosaic rationale would've been coincidence. but the odds of a searingly real dream of an accident coming the night before an actual accident, seemed too great. I'd never dreamed of car accidents before. besides, both of the "crashes" as I experienced them, were exact: same time of night, same location, same sounds.

I didn't know what happened but I knew it wasn't coincidence and was something "supernatural" or whatever it should be called.

Stories like Jim's can be approached in at least two different ways: one historical and descriptive; the other philosophical and normative. The first way is the acceptable way in the academy. It works well enough because it is safe enough. As I have already explained, it sets aside whether the experience was true or not—that is, whether it corresponded in some convincing way to the world.

Alas, this safe answer reminds me more than a little of what the quantum physicists were told for decades when the philosophical implications of quantum mechanics were fairly new and basically shouting in people's heads. The implications revolved around the role of human consciousness in physical experiments, in the ways that consciousness and intention seemed to influence or even determine how material reality behaves. Unfortunately, or fortunately, this influence also seemed to work *backward* in time. Words like "retrocausation" began to be used, to almost everyone's great embarrassment and shock. It was as if meaning and purpose, even— God forbid—teleology (the universe is actually going somewhere), were being snuck back into the picture.

There did not seem to be any end to the philosophizing around the empirical results of the quantum experiments. The experiments, however, were repeated and consistent in their results. But what did it all mean? Eventually, an answer formed. It was a nonanswer. We have seen it above.

It went like this: "Shut up and calculate." That is actually what the physicists were told for decades. That is how they got jobs. It was the Cold War, and there were bombs to build. Just do the math. Ignore the man behind the curtain.

Accordingly, the utterly mind-blowing implications of the new physics, like quantum entanglement—whereby invisible particles seemed to influence and affect one another across unimaginable distances instantly—would have to wait for decades, until it would become key to computer science and new forms of encryption. But that was then. This is now. For then, in the 1960s and '70s, such implications would be carried, virtually alone, by unemployed San Francisco Bay–area "hippie physicists" and their strange parapsychological and Asian mystical obsessions, which the quantum mechanics seemed to imply in more ways than one.[3]

Unfortunately, we can refigure the present humanist convention in the humanities around anomalous experiences in similar terms: "Shut up and describe." Which means something like this: "It is okay if other people believe crazy things. Sure, you can describe their beliefs and experiences. But don't, for God's sake don't, think about the philosophical implications of what they are saying, and do not, *do not* take their experiences as real in any way other than as purely subjective events. That, after all, might bring down the house of cards that we have built up so carefully over the decades and centuries. Instead, reduce those described experiences to something else: emotional need, psychological projection, scriptural editing, psychopathic hallucination, social structure, political power, cognitive modules, neurons, microtubules, *anything* other than *that*."

I am getting too crabby. Maybe I should not italicize all those words. Let me pull back a bit, then. To think in a more granular (and diplomatic) fashion, I like to imagine three bars that we can attempt to jump over in cases like this. I am thinking of my high school days on the track team. I was quite bad at it, but I remember the high jump pit. Mostly it was lanky teenagers trying to leap backward over a bar, until they couldn't, which was usually very soon. The bar clanked (it was still metal) and flew a lot. Bodies leaped and fell.

Same in this situation. It's okay to knock the bar off. You *will* at some point, and maybe quite soon. But we need very specific examples, like Jim's precog dream experience, to see which bar we cannot get over and, more importantly, exactly why we cannot get over it.

The first bar of my thought experiment comes down to a question: Do such impossible things happen at all, or are they really nonevents, functions perhaps of miscalculated coincidences, meaningless statistical outliers, or neurological blips of the pattern-seeking brain? Jim asks the question in terms of coincidence. Was his dream really just an unusual coincidence or matter of timing? He answers no, and so he is over the first bar.

The same question can be put in the terms of a much simpler question: Why do people believe impossible things? My own answer to the question goes something like this: People believe impossible things because impossible things happen to people. People around the world believe in a soul that is separable from the body because people around the world have experienced what look very much like their own souls or those of others separating from their bodies. People around the world believe in some form of reincarnation because small children routinely claim to remember, in detail, their previous lives and report them to their parents and family—especially, it turns out, when their previous lives ended violently, suddenly, and unjustly. People around the world practice dream divination because people around the world occasionally dream of the future before it happens. And so on.

To repeat myself, one does not need to believe any of the belief systems that build up around such extraordinary experiences to acknowledge that the experiences in fact happened. To invoke a story documented in the literature, this one of a Swedish mother, if any of us had a disturbing dream the night before a school field trip that our young child would be left lying asleep on train tracks in mortal danger and *exactly that* played out the next day as we worried all day at home, most of us—*all* of us?—would seriously entertain a belief in precognitive dreams, *whatever* our earlier education told us.[4] We would not need to believe in "God" or "prayer" or "guardian angels." We would only need to accept the facts of our own troubling experience. It's the same with Jim. Is he supposed to deny what happened to him over two consecutive nights? And then not think about it?

It is really that simple. Belief as intellectual assent to some past creedal statement is perfectly irrelevant in such contexts. It is a simple matter of empiricism and honesty—in these two cases, a mother's well-founded anxiety and two consecutive nights of a man's life. This first bar is not too difficult to get over, then. All one really needs is a certain level of hermeneutical charity and sympathy (or a child).

I think this is one reason I have so little trouble with anomalous events. At least at an early stage of inquiry, it does not matter to me what their philosophical import is or is not. What matters initially is *that they happen*, and that these historical events work with a thousand other cultural, moral, political, and psychosomatic processes to produce what we call "religion." This conclusion is of immense significance. I thought this is what the study of religion is all about—understanding how religion comes to be. Well, this is clearly a part of that answer.

Once one is over the first bar, though, the second bar inevitably appears. This second bar involves questions of consciousness or the reach of subjectivity: If these phenomena are a part of our mental world, what do they mean? Can such phenomena tell us something important about ourselves, about subjectivity itself, about the nature and reach of consciousness?

Take Jim's experience again. If we accept that such an event happened in both his dreamworld the first night and outside his window the second (and I do), then such events may well point to something very profound about the location of consciousness in time. The event suggests, after all, that consciousness, particularly when altered by the emotional gravity of nightmare and impending mortal tragedy, may not be entirely located or stuck in what we call the present. The "now" of consciousness might be much more expansive. We might not be the prisoners of space and time that we imagine ourselves to be.

Here one moves beyond sympathetic acknowledgment or pure description to a more philosophical question about the nature of the human being. One sees the anomalous experience, certainly not as a nonevent or cognitive mistake, or even as a curious experience to be described but not followed up on, but as a meaningful sign of some future form of consciousness trying to take shape. One begins to pay attention. One listens. One also refuses to shut up. One *interprets*.

The third bar is already implied in the second bar. The question goes something like this: If these "impossible" phenomena in fact take place in the mental dimension of the human (the first bar) and suggest something important, if future, about the nature and scope of subjectivity (the second bar), might they also signal to us fundamental truths about the objective physical universe? Might such phenomena, then, have something important to say about the behavior and structure of space, time, energy, and matter and their relationship to mind? Hence the physics, mathematics, and multiple dimensions of my conclusion's epigraph. Couliano was perfectly serious. So am I.[5]

The bar clanks and flies again as the jumper tumbles into the padded pit. The bar flies because such questions grate against almost every assumption we make in the university these days, with our neat divisions between the subjective or "soft" humanities and the objective or "hard" sciences. Take the precognitive dreams with which we have begun and now end this very book. These are hardly "just dreams." Each is also a *Gedankenexperiment*, a thought experiment, just begging to be thought. We have so thought. Such human experiences clearly imply something like the philosophical position of eternalism or a block cosmology in which the future already exists—a future, moreover, to which we occasionally have some rare but real access. None of this "proves" such a cosmology, of course, but it clearly acts as a very suggestive sign of the same.

In any case, this is precisely the general direction I would hope we might move now: the employment of altered states of consciousness (first bar) as generators of a philosophy of mind (second bar) toward the scientific modeling of the physical cosmos (third bar). I want us to try to jump over all three bars. Maybe we cannot. But we can try, and it seems good and useful to know where, when, and why the bar falls.

And that is how one thinks impossibly.

Not "humankind," but rather *superhumans* are the goal!

I think that all metaphysical and religious ways of thinking are the result of dissatisfaction with humans, the result of the drive for a superior, superhuman future.

FRIEDRICH NIETZSCHE, *Unpublished Fragments from the Period of "Thus Spoke Zarathustra"*

I believe that humankind is coming to a point where there's going to be a huge metamorphosis and our physical bodies—or maybe it's our invisible bodies— are going to change. . . . I think humanity is going to take an evolutionary leap and things are going to be different.

GLORIA E. ANZALDÚA, *Interviews Entrevistas*

You who play the zig-zag lightning of power over the world, with the grumbling thunder in your wake, think kind of those who walk in the dust. . . . Consider that with tolerance and patience, we godly demons may breed a noble world in a few hundred generations or so.

ZORA NEALE HURSTON, *Dust Tracks on a Road*

Acknowledgments

A Sociology of the Impossible

There is a kind of unconscious or barely conscious "sociology of the impossible" at work in these pages, by which I mean there is something that is fundamentally unspeakable or unknowable about the endless lines of influence and human relationship that produced such thought. Up here, on the social and historical plane, these thought experiments were all written over a four-year period, between the very beginning of 2019 and the middle of 2023, at the express invitation of others for specific higher educational purposes. They are also developments of my thought as expressed and performed at over one hundred speaking events in the United States, Canada, and Europe, most all of them before the pandemic. Impossible thinking is thus a kind of thinking-with on the most practical of levels: with thousands of colleagues, listeners, readers, and interlocutors at institutions of higher learning on both sides of the ocean.

The book is divided into two more or less equal parts: one on practice, and one on theory, although these divisions, obviously, are not absolute. It begins with three chapters on the imaginal nature of the near-death literature, the UFO phenomenon, and the abduction accounts that I wrote for three back-to-back conferences in the spring of 2022—at the University of Vienna, Harvard University, and the University of California, Berkeley. I begin the book with these chapters at the request of an anonymous reader, who wanted me to emphasize the personal dimension of impossible thought.

These three initial chapters are then followed up in the second half of the book with three others on theorizing the impossible. This second part begins with a chapter on temporality and hermeneutics, parts of which I wrote for a Festschrift in honor of the historian of medieval kabbalah Elliot Wolfson, whom I have known (and thought-with) for almost three decades now and consider to be one of the most important scholars of religion of my own, or any, generation (chapter 4). As a means of summarizing the arguments, I then include two more pieces: one I wrote substantially for a special anniversary journal issue on the philosophy of mind and matter, edited by the quantum theorist Harald Atmanspacher (chapter 5), and the other I begin with some paragraphs from the original conclusion of *The Superhumanities* that I decided not to publish there but am here sharing as a kind of open secret (chapter 6), now vastly expanded to try to answer an anonymous reader of this book manuscript who asked me the question I get all the time on the lecture and podcast circuit: "What about the negative paranormal, the demonic possession, the palpable sense of evil?"

I also experimented with these ideas at the Chicago Humanities Festival in the spring of 2019 and, again, at the University of Northampton that same fall for a conference on "trans- states." I crystallized my thought again in an inaugural address for the new Center for Advanced Studies in the Humanities and Social Sciences: Alternative Rationalities and Esoteric Practices from a Global Perspective at Friedrich-Alexander-Universität in Erlangen, Germany, at the very end of 2022. These explorations and crystallizations are spread throughout the book but define especially the prologue and epilogue, the beginning and the end, which are the same.

I also want to recognize specific individuals as especially influential. Foremost among them are Jeremy Vaeni and Stuart Davis, two of the comedic writers and extreme experiencers of chapter 2; Kevin, the central subject of chapter 3; "Jim" of the epilogue; and John Allison, whose experiences are recounted in chapter 4 and who also helped guide the entire project through his close reading and, frankly, astonishing bibliographic recommendations, particularly within the phenomenological tradition. John knows so much more than I do. He will not say that, but I just did. He also wrote a kind of précis or manifesto for our last Archives of the Impossible conference here at Rice University. John knows the impossible.

Also of central importance here is Karin Austin, the director of the John E. Mack Institute, one of the experiencers discussed in chapter 2, and now a close friend and colleague. Austin is the individual who is centrally responsible for bringing the John E. Mack archives to Rice University and an astonishing mind and experiencer in her own right. Austin is now helping to manage the scanning and archival process for the Mack papers and files here in Houston. Amanda Focke, the director of the Woodson Research Center with whose archival staff and leadership I work closely to host, collect, and theorize the Archives of the Impossible, and Anna Schparberg, the librarian for the Department of Religion and very much a professional partner in all of this, need to be recognized in this same context. I could not do any of this without these colleagues.

Part of those same Archives of the Impossible are my own papers and files, which are chock-full of experiencers who wrote me of their most intimate lives, completely unbidden, in email correspondence or, often enough, after conferences and speaking events. Such individuals are now friends, readers, and colleagues. I am not sure what I am trying to say here, other than I cannot separate these stories and their potential theorization in this book from these people, or these people from these stories and embedded ideas, now expressed here in these pages. It is all one life, or again, one sociology of the impossible. I guess that is what I am trying to say.

In the end, this is a book for them. It is about them. It *is* them.

Jeffrey J. Kripal
MAY 25, 2023

Notes

Prologue

1. Charles King, *Gods of the Upper Air: How a Circle of Renegade Anthropologists Reinvented Race, Sex, and Gender in the Twentieth Century* (New York: Random House, 2019). This is a powerful book about the origins of cultural anthropology "on the front lines of the greatest moral battle of our time: the struggle to prove that—despite differences of skin color, gender, ability, or custom—humanity is one undivided thing" (King, 4). Significantly, King's title is from Zora Neale Hurston, *Dust Tracks on a Road* (New York: Amistad, 2006), with which we also begin here. See note 12, below, for the secret initiatory context of the phrase "gods of the upper air."

2. Hence, this book begins exactly like my previous one, with a pedagogical moment, thinking-with another human being. My previous book is Jeffrey J. Kripal, *The Superhumanities: Historical Precedents, Moral Objections, New Realities* (Chicago: University of Chicago Press, 2022). I encountered the present scene from Hurston's memoir in a dissertation on channeling, *Black Elk Speaks*, and settler colonialism that I read for my home university: Sam Stoeltje's most remarkable "Of Ghosts and Justice: Spectral Politics in 20th-Century U.S. Literature" (PhD diss., Rice University, 2022).

3. Hurston, *Dust Tracks*, 7, ix. Hurston describes her father as having "grey-green eyes and light skin" with a body that made him stand out from the crowd, like a "stud-looking buck." He was said to be "a certain white man's son" (Hurston, 8–9). In a late chapter on religion, Hurston writes of the religious acumen of "Reverend Jno" (her father's nickname), of her own confusion around the fact that intense religious experience during revival meetings (when people had "visions" and "got religion") did not change people's moral or social habits, and of contradictions of basic Christian ideas, like that of human sinfulness, physical immortality, and the intense love of an invisible being. In her childhood, her Christian religious practice was a social requirement, not a spiritual conviction. Studying philosophy, history, and comparative religion in college would undermine the exclusive claims of Christianity further: she saw intensely how human all religion really is, how it begins in altered states, like that of Paul on the way to Damascus, but is then furthered by military violence and political force (Hurston, 215–23). She explains how she now knows a great deal about the "form" of religion "but little or nothing about the mysteries I sought as a child" (Hurston, 225). Finally, Hurston confesses that she does not pray, which she understands as a sign of weakness. Rather, she works out her "destiny" with her own mind and will (and precognitive

gifts, I would add). She is a pagan, at one with the infinite and the universe through the indestructibility of matter, and with no need of "organized creeds" that are little more than "collections of words around a wish" (Hurston, 226).

4. There are profound implications here for the imbrication of the history of technology—in this case, photography and film—and paranormal phenomena: "like clearcut stereopticon slides."

5. Hurston, *Dust Tracks*, 119.

6. Hurston, 43–44.

7. Hurston, 42.

8. Hurston, 42. The first, second, third, seventh, and twelfth visions are recounted at Hurston 69–70, 85, 97, 119, and 268.

9. My thanks to Eric Wargo for supplying most of this list.

10. Hurston, *Dust Tracks*, 268. The telepathic transcendence of race and class in the Harlem Renaissance reminds me of the ways telepathic experimentation functioned in Soviet-American diplomacy around the Esalen Institute in the 1970s. See Jeffrey J. Kripal, "Superpowers: Cold War Psychics and Citizen Diplomacy," in *Esalen: America and the Religion of No Religion* (Chicago: University of Chicago Press, 2007), 315–38.

11. Hurston, *Dust Tracks*, 144. King at least smirks. He is dismissive of the psychical research, describing such interests as "the idea that the ills of modernity could be cured by a retreat into the premodern past" (King, *Gods of the Upper Air*, 203).

12. Hurston, *Dust Tracks*, 156–57. See also this: "I have walked in storms with a crown of clouds about my head and the zig zag lightning playing through my fingers. The gods of the upper air have uncovered their faces to my eyes" (Hurston, 264). The esoteric initiatory context of this last sentence, which we know from the previous passage (Hurston, 156–57), suggests an actual encounter with these "gods of the upper air." The standard option, I am sure, is to suggest she was simply being poetic. That is the difference between the superhumanities and the conventional humanities.

13. Hurston, 41–44.

14. Hurston, 39. Hercules was the Greek mythical model for the American pop-cultural Superman.

15. Hurston, 41.

16. Hurston, 40.

Introduction

1. This was the second annual installment of a three-year cycle on "The Furthest Reaches of the Imagination," Esalen Institute, 1–6 November 2015.

2. Cristina Scarlat, *F. F. Coppola & Mircea Eliade: Youth without Youth: A View from Romania*, trans. Mihaela Mititelu (Bucharest: Editura Eikon, 2018), 33–34.

3. Quoted in Scarlat, *F. F. Coppola & Mircea Eliade*, 33.

4. Mircea Eliade, *Youth without Youth* (Chicago: University of Chicago Press, 2007), 68. It has always struck me, and Eliade knew this: Krishna begins to reveal the actual reality of things to Arjuna (including Krishna's own nature as God) in the *Bhagavad Gita* (2:10) with a sense of knowing bemusement, a gnostic grin: "as if smiling" (*prahasanniva*).

5. Eliade, *Youth without Youth*, 68.

6. I will say much more about the centuries-long inquiry into these problems and potentials in *Biological Gods: Evolution and the Coming Superhumans, 1766–2028*, the

tentatively titled second planned book in my upcoming Super Story trilogy on the hidden histories of science, science fiction, and the paranormal.

7. And I mean this "mutation" in more than metaphorical terms. I am indebted for my understanding of the fantastic in contemporary critical theory to Ramzi Fawaz, *The New Mutants: The Radical Imagination of American Comics* (New York: New York University Press, 2016), 26–28. I am perfectly aligned with Ramzi's queer theoretical emphasis on the social margins as the source of new thought and theory, although my own theory of the fantastic is more paranormally inflected. In this same paranormal spirit, the mutant mythology that is most relevant in the present evolutionary esoteric context—with its experiential precedents in the histories of animal magnetism, Spiritualism, psychical research, and Friedrich Nietzsche—is the X-Men mythos, with the figures of Professor Xavier and Magneto engaging in the debate about the human and the superhuman in graphic novel, film, and television. The motto of mutant education, by the way, is a Latin comparative logic: *mutatis mutandis*, or "all things being equal." I would put it this way: comparison *is* mutation. What that means will become more apparent, if never perhaps fully transparent, as we proceed, including in the next two notes, which are meant mostly to signal future thought, not stabilize present thinking.

8. In an earlier book, I used the literary genre of the fantastic to explore the experiential hesitations of the impossible as suspended between the natural and the supernatural in Victorian psychical research, the American paranormal, and the American-French UFO phenomenon. See especially Jeffrey J. Kripal, *Authors of the Impossible: The Paranormal and the Sacred* (Chicago: University of Chicago Press, 2010), 34–35. The textual dimensions of the paranormal—how these events take on narrative structures in both the inner psyche and the outer culture—lie partly in the fantastic nature of those stories or real-life literary productions and our subsequent inability as social egos to choose one reading option over another (reduction or revelation). Such are we told. I take up that impossible theorizing again here but push it further into philosophical speculation around the nature of the real itself, where those interpretive hesitations remain firmly in place but are now explained as epistemic features of a historically relative human subject whose constructedness and relativity help produce the apparent elusiveness, deception, and common moral confusion of the phenomenon. Such am I thought.

9. At the horizon or limit of the impossible, it is not clear to me whether this elusiveness is "epistemic"—that is, a response of the human subject and its ordering of the paradoxical experience—or "ontological"—that is, part of the agency of the intelligence before and beyond the relative social subject. This waffling will haunt this book throughout (since it haunts me). In terms of philosophy and anthropology, impossible thought is very much an inclusive and affirming answer to the long history of antirealist thinking in Western thought (there is nothing to religious experience . . . it is all appearance, fantasy, culture, history, power, the brain, etc.). Most of all, impossible thought constitutes a kind of "rough metaphysics" that takes the most extraordinary experiences of the fantastic and the human multiplicity that they reveal as the very source and origin of new thought toward a future conception of the human. See Peter Skafish, *Rough Metaphysics: The Speculative Thought and Mediumship of Jane Roberts* (Minneapolis: University of Minnesota Press, 2023), especially in lines like this: "Myths are primordial thoughts by which collectives shape nature's behavior toward them" (Skafish, 326). Skafish means that *physically*, *literally* (since the idea that physical

reality is separate from human consciousness is just another myth). This, I would suggest, is Coleridge's dream-flower and Eliade's materializing rose. This is also this book. Hence the call for a new ecological, political, and mind-matter myth, which depends in turn on our recognizing the *physically* effective nature of myth itself. Myths, Skafish writes, "have the power to introduce every imaginable possibility into the world" (we will see this take on a special meaning in the "Platonic Surrealism" of Kevin in chapter 3). It is no accident that Skafish's book focuses on one of the central mediums, Jane Roberts, of the human potential world in which I have long been embedded at the Esalen Institute in Big Sur, California. I will have much more to say about this plural ontology, this new myth, in the aforementioned Super Story trilogy. I hope it is clear that I read Skafish belatedly, which is to say not in time to fully integrate his thought into the impossible thinking modeled here.

10. Carlos Eire, *They Flew: A History of the Impossible* (New Haven: Yale University Press, 2023), 1.

11. Eire, *They Flew*, xv–xvi.

12. I think many assume that "the ancestors" (as a metaphor or as an otherworldly community) are somehow omniscient or all good. I make no such assumptions. Particularly if they are still attached to a culture, ethnicity, or religion, "the ancestors" or "the community of saints" can be just as bound and limited as any other human identity. Indeed, if we are to take seriously the profoundly negative nature of so much occult experience, we might guess that death solves little, or nothing at all. These pages will not worship the past.

13. The phrase is not my own. This is what the Harvard psychiatrist John E. Mack said through a medium, after he died. The phrase is ambiguous. It might refer to the alien or the departed soul, or both, or neither. See Ralph Blumenthal, "Alien Nation: Have Humans Been Abducted by Extraterrestrials?," *Vanity Fair*, May 10, 2013, https://www.vanityfair.com/culture/2013/05/americans-alien-abduction-science.

14. See Hussein Ali Agrama, Department of Anthropology, University of Chicago, https://anthropology.uchicago.edu/people/faculty/hussein-ali-agrama.

15. The title for this book, *How to Think Impossibly*, was suggested to me early in the process by my editor, Kyle Wagner.

16. I explored this "physics of mystics" in Jeffrey J. Kripal, *Secret Body: Erotic and Esoteric Currents in the History of Religions* (Chicago: University of Chicago Press, 2017). I will do so again more fully in the first volume of the aforementioned Super Story trilogy, currently entitled *The Physics of Mystics: Super Natural Experiences of the Cosmos and Their Special Effects, 1884–2025*.

17. Donald Hoffman, *The Case Against Reality: Why Evolution Hid the Truth from Our Eyes* (New York: W. W. Norton & Company, 2019).

18. Philip Ball, *Beyond Weird: Why Everything You Thought You Knew about Quantum Physics Is Different* (Chicago: University of Chicago Press, 2018), 15.

19. Jennifer Chu, "Light from Ancient Quasars Helps Confirm Quantum Entanglement: Results Are among the Strongest Evidence Yet for 'Spooky Action at a Distance,'" *MIT News*, August 19, 2018, https://news.mit.edu/2018/light-ancient-quasars-helps-confirm-quantum-entanglement-0820.

20. Ball, *Beyond Weird*, 19.

21. David Kaiser, *How the Hippies Saved Physics: Science, Counterculture, and the Quantum Revival* (New York: W. W. Norton & Company, 2012). I will return to this topic in the epilogue.

22. Richard Noakes, *Physics and Psychics: The Occult and the Sciences in Modern Britain* (Cambridge: Cambridge University Press, 2019).

23. It in this same stunned spirit that I will *not* conclude on a number of key philosophical questions, including what we might call the tension between being and becoming, between the One and the Many, what William James—drawing on the intellectual wells of mystical experience, psychoactive states, and psychical research with female mediums—called late in his life a "pluralistic universe" (the latter monistic word, uni-verse, is usually occluded, as is his clear perennialism in *The Varieties of Religious Experience*). Today our preference for difference and pluralism is sometimes radicalized as a multinaturalism—that is, the notion that nature behaves differently in different places and times, which I will also adopt in these pages. In short, *I do not land*—probably because I think both monism and pluralism are "true" on different levels of the real. On the specific instance of James and his "radical empiricism" (a clear precedent to my theorization of the imagination and insistence that we take seriously every form of human experience, no matter how strange or weird), please see my discussion of James in Jeffrey J. Kripal, *The Superhumanities: Historical Precedents, Moral Objections, New Realities* (Chicago: University of Chicago Press, 2022), 103–10.

24. I really do understand that mine is a different interpretation of myth that does not reduce the story to social realities, to representations or systems of disciplinary power. I am not against those models of myth, since I am very skeptical (I would say openly dismissive) of any and all local mythologies. I just think these social readings are limited and, in the cases I am most interested in, not very helpful. Basically, I think that what we have called "myth" sometimes has an actual physical effect; that is, it determines how nature actually behaves. In this sense, we are the ultimate power. Hence my embrace of multinaturalism, which is also an affirmation of a superhumanism.

25. Edmund Husserl, *Ideas: General Introduction to Pure Phenomenology*, with a new foreword by Dermot Moran, trans. W. R. Boyce Gibson (London: Routledge, 2012), xlix.

Chapter One

1. The present chapter was originally entitled "Supernormal, Paranormal, Imaginal: Experiential Origins, the Near-Death Literature, and the World of the Dead"; I wrote parts of it for "Adventures in the Imaginal: Henry Corbin in the 21st Century," Center for the Study of World Religions, Harvard Divinity School, Cambridge, MA, May 12–14, 2022. I also want to acknowledge that this first half of the book was named, at first quite unknowingly, after an earlier book that was gifted to me by Stanislav Grof, *When the Impossible Happens: Adventures in Non-ordinary Realities* (Boulder: Sounds True, 2006), which opens with a synchronistic mantid story (see chapter 2), "Praying Mantis in Manhattan" (Grof, 8–10). I remember the evening Stan inscribed this book to me. We were on a boat in the San Francisco harbor together, celebrating the eventual publication of my book *Esalen: America and the Religion of No Religion* (Chicago: University of Chicago Press, 2007). Thank you, Stan. I really do define the reality of the impossible as, most simply, "what happens."

2. Jeffrey J. Kripal, "The Passion of Louis Massignon: Sublimating the Homoerotic Gaze in *The Passion of al-Hallaj* (1922)," in *Roads of Excess, Palaces of Wisdom: Eroticism and Reflexivity in the Study of Mysticism* (Chicago: University of Chicago Press, 2001), 98–146.

3. For an effective and quick summary, see Joshua Rothman, "Is Heidegger Contaminated by Nazism?" *The New Yorker*, April 28, 2014, https://www.newyorker.com/books

/page-turner/is-heidegger-contaminated-by-nazism. The basic facts are that Heidegger joined the party in 1933 to become *Führer-rector*, or president, of Freiburg University. He gave up the rectorship just a year later and appears to have grown increasingly dubious of Nazi ideology. He remained a member until 1945, however, and later admitted, privately, that his participation had been "the biggest stupidity of my life." Still, he never publicly retracted his membership, nor did he ever speak or write of the Holocaust.

4. For some astute reflections on Frederic Spiegelberg, especially his readings of and relationship to Martin Heidegger and Spiegelberg's pantheistic notions of Being and the paradoxes of the "religion of no religion," rooted at once in his own mystical experience in a wheat field, Heidegger's philosophy of *Dasein*, and the spiritual writing and presence of the Indian spiritual guru Sri Aurobindo, see Ahmed M. Kabil, "The New Myth: Frederic Spiegelberg and the Rise of a Whole Earth, 1914-1968," *Integral Review* 8, no. 1 (July 2012), http://integral-review.org/documents/Kabil,%20Vol%208,%20No%20 1,%20CIIS%20Special%20Issue.pdf. Spiegelberg's description of a young Alan Watts as "an almost superhuman being" and the Nietzschean theme of the "Superman" in Spiegelberg's reading of both Aurobindo and Heidegger are worth flagging here.

5. Jeffrey J. Kripal, *Esalen: America and the Religion of No Religion* (Chicago: University of Chicago Press, 2007).

6. See Kripal, "Sex with the Angels: Nonlocal Mind, UFOs, and *An End to Ordinary History*" and "Realizing Darwin's Dream: The Transformation Project and *The Future of the Body*," in *Esalen*, 339-56, 404-25.

7. Jeffrey J. Kripal, *The Superhumanities: Historical Precedents, Moral Objections, New Realties* (Chicago: University of Chicago Press, 2022), 186-87.

8. Henry Corbin, *The Man of Light in Iranian Sufism*, trans. Nancy Pearson (New Lebanon, NY: Omega, 1994), 97.

9. Henry Corbin, "Towards a Chart of the Imaginal: Prelude to the Second Edition of *Corps spirituel et terre céleste de l'Iran Mazdéen à l'Iran Shiite*," in *Spiritual Body and Celestial Earth: From Mazdean Iran to Shi'ite Iran*, trans. by Nancy Pearson (Princeton, NJ: Princeton University Press, 1977), xviii.

10. Raymond Moody and Paul Perry, *Paranormal: My Pursuit of the Afterlife* (HarperOne, 2013), 107, 110.

11. Numerous writers have tried to insert a hell back into the picture, mostly through "negative near-death experiences," which indeed exist and are likely underreported. Still, it is not at all clear that such negative experiences really establish the ontological existence of a hellish afterlife either. What they certainly establish is that people experience what they fear and imagine in the immediate zones of the death process. In essence, we die into our imaginations.

12. See Moody and Perry, *Paranormal*, 111.

13. For a full discussion and theorization of Elizabeth Krohn's experiences, including a chapter on the central role of the religious imagination, see Elizabeth G. Krohn and Jeffrey J. Kripal, *Changed in a Flash: One Woman's Near-Death Experience and Why a Scholar Thinks It Empowers Us All* (Berkeley, CA: North Atlantic Books, 2018).

14. David Hufford, *The Terror That Comes in the Night: An Experience-Centered Approach to Supernatural Assault Traditions* (Philadelphia: University of Pennsylvania Press, 1989), xviii.

15. Bruce Greyson, *After: A Doctor Explores What Near-Death Experiences Reveal about Life and Beyond* (St. Martin's Essentials, 2021).

16. Bruce Greyson was a key member of the fourteen-year study of the nature of

consciousness at the Esalen Institute that produced three large volumes, all led by the neuroscientist Edward Kelly. I wrote about this project in "Mind Matters: Esalen's Sursem Group and the Ethnography of Consciousness," in *What Matters? Ethnographies of Value in a (Not So) Secular Age*, ed. Courtney Bender and Ann Taves (New York: Columbia University Press, 2012), 215–247. I cannot sufficiently acknowledge these three volumes: Edward Kelly et al., *Irreducible Mind: Toward a Psychology for the 21st Century* (Lanham, MA: Rowman & Littlefield, 2009); Edward F. Kelly, Adam Crabtree, and Paul Marshall, eds., *Beyond Physicalism: Toward Reconciliation of Science and Spirituality* (Lanham, MA: Rowman & Littlefield, 2019); Edward F. Kelly and Paul Marshall, eds., *Consciousness Unbound: Liberating Mind from the Tyranny of Materialism* (Lanham, MA: Rowman & Littlefield, 2021). I have an essay in the last volume, very much expressive of how to think impossibly: "The Future of the Human(ities): Mystical Literature, Paranormal Phenomena, and the Contemporary Politics of Knowledge" (Kelly and Marshall, 359–405).

17. For a newspaper article on one form of participation (that is, serving as a judge for an essay contest), see Ralph Blumenthal, "Can Robert Bigelow (and the Rest of Us) Survive Death?," *New York Times*, January 21, 2021, https://www.nytimes.com/2021/01/21/style/robert-bigelow-UFOs-life-after-death.html.

18. An earlier version of parts of this section originally appeared in Jeffrey J. Kripal, *Secret Body: Erotic and Esoteric Currents in the History of Religions* (Chicago: University of Chicago Press, 2017), 233–35, used with permission. I have corrected and nuanced this material significantly.

19. Tom Cheetham, *All the World an Icon: Henry Corbin and the Angelic Function of Beings* (Berkeley, CA: North Atlantic Books, 2012), 165. This is the same year, 1964, that the California counterculture definitively began. The category coincides precisely with the mystical and occult interests of these cultural explosions in the 1960s and '70s, explosions that were in no way restricted to the United States.

20. Corbin, "Towards a Chart of the Imaginal," viii.

21. I discuss this dreaming the real in Schopenhauer in Kripal, *The Superhumanities*, 121–22.

22. This same collapse of the imagination and the nonlocal physical environment is a central feature of the superhumanities, which I have tried to capture under the banner of the imaginal-empirical. See Kripal, *The Superhumanities*, 62–64.

23. For an ancient history of this very transformation from the enigmatic and esoteric symbol to the rational and plodding metaphor, see Peter T. Struck, *The Birth of the Symbol: Ancient Readers at the Limit of Their Texts* (Princeton, NJ: Princeton University Press, 2014).

24. The expression "the creative imagination" goes back as far as the 1730s and is a gradual development of the Enlightenment and Romantic writers. The imagination as commonly understood is, historically speaking, a creation of this same eighteenth century. See James Engell, *The Creative Imagination: Enlightenment to Romanticism* (Cambridge, MA: Harvard University Press, 1981), vii–viii.

25. Cheetham, *All the World an Icon*, 165. For the definitive history of Eranos, see Hans Thomas Hakl, *Eranos: An Alternative Intellectual History of the Twentieth Century* (Montreal: McGill-Queen's University Press, 2013).

26. I am borrowing the apt phrase "transcendent blackness" from Stephen Finley, *In & Out of this World: Terrestrial and Extraterrestrial Bodies in the Nation of Islam* (Durham, NC: Duke University Press, 2022), 7.

27. Corbin, *The Man of Light*.

28. Charles M. Stang, *Our Divine Double* (Cambridge, MA: Harvard University Press, 2016).

29. This is William Wordsworth describing his altered vision of nature on a mountaintop in M. H. Abrams, *Natural Supernaturalism: Tradition and Revolution in Romantic Literature* (New York: W. W. Norton, 1973), 78; see also 117.

30. Abrams, *Natural Supernaturalism*, 67, 91–92.

31. See Abrams, 223. The same can also be seen in Hegel's *Phenomenology of Spirit*, whose metaphysical structure and narrative plot—the Mind coming to know itself reflected in nature, history, religion, and eventually, absolutely, in philosophy—is the most famous philosophical expression of what Abrams calls the "High Romantic Argument" (Abrams, 225–37).

32. Abrams, 27, 37.

33. Abrams, 385.

34. Wordsworth quoted in Abrams, 21.

35. Abrams, 68.

36. Abrams, 334.

37. I cannot pursue the matter here, but Coleridge was very much what we would today call a dual-aspect monist who saw poetry and art as expressive "up here" of a fundamental shared ground, "superessential essence," or "one great Being." His views on the imagination were, accordingly, of a cosmic and prophetic nature that worked from a third space or "middle ground" to which God descends and the poet can reach. This is what he called his "Dynamic Philosophy." See James Engell, *The Creative Imagination: Enlightenment to Romanticism* (Cambridge, MA: Harvard University Press, 1981), 328–66. Coleridge's thought was even entomological (he employed insect transformation to illustrate his ideas), like Frederic Myers after him, and evolutionary, well before Darwin: what he called "the Symbol" is "the Evolver" of humanity (Engell, 347–48).

38. Engell, *The Creative Imagination*, 364 (italics in original). "Consubstantial" is theological language, often used to explain Jesus's identity with God. The ancient Greek behind the English word, from the Nicene creed, is *homoousios*, "one in being" or "same-being."

39. William Blake quoted in Engell, 244–45.

40. I really cannot stress these historical and intellectual roots enough. The study of religion is a combination of Enlightenment reason and Romantic imagination, even though the latter has been occluded in recent decades. This is one reason I focused so on the figure of William Blake as actual inspiration in two of my early books, *Roads of Excess, Palaces of Wisdom: Eroticism and Reflexivity in the Study of Mysticism* (Chicago: University of Chicago Press, 2001) and *The Serpent's Gift: Gnostic Reflections on the Study of Religion* (Chicago: University of Chicago Press, 2007). To speak in contemporary terms, I really do think the historical roots of the gnosis of the superhumanities lie here, in English Romanticism and German idealism. Hence, also, the importance of the imaginal symbol of revelation in this line of creative writing and thinking.

41. Harold Bloom, "Prometheus Rising," in *The Visionary Company: A Reading of English Romantic Poetry* (Ithaca, NY: Cornell University Press, 1971), xxiii–xxiv.

42. Engell, *The Creative Imagination*, 6–10.

43. I am relying here on personal conversations that I had with Antoine Faivre, the French historian of Western esotericism. There are important exceptions to the role of incarnation in Islam.

44. Henry Corbin, *Alone with the Alone: Creative Imagination in the Sufism of Ibn ʿArabī*,

with preface by Harold Bloom (Princeton, NJ: Princeton University Press, 1997), 42, 47–48, 68, 222–24. For the Sufi lore around teleportation (not named as such, of course), see Corbin 237–38.

45. Harold Bloom, preface to Corbin, *Alone with the Alone*, xvi.

46. Bloom, preface to Corbin, *Alone with the Alone*, x. Also see the discussion of the Bloom's own gnostic experiences in Harold Bloom, *Omens of Millennium: The Gnosis of Angels, Dreams, and Resurrection* (Riverhead Books, 1997).

47. This intentional apoliticism is itself deeply troubling to some intellectuals. See, for example, Steven M. Wasserstrom, *Religion after Religion: Gershom Scholem, Mircea Eliade, and Henry Corbin at Eranos* (Princeton, NJ: Princeton University Press, 1999).

48. Jeffrey J. Kripal, *The Serpent's Gift: Gnostic Reflections on the Study of Religion* (Chicago: University of Chicago Press, 2007).

49. I can think of at least two places I made this mistake: Jeffrey J. Kripal et al. *Comparing Religions: Coming to Terms* (Oxford: Wiley-Blackwell, 2014), 252; and Kripal, *Secret Body*, 234. Some of the present material on Frederic Myers and Théodore Flournoy was used in an earlier version in *Secret Body*, where the telepathic-insectoid pattern was also first identified (Kripal, *Secret Body* 233–35). I used some of that material again here, but it is all corrected, revised, and much developed in this and the following chapter.

50. Théodore Flournoy, *From India to the Planet Mars: A Case of Multiple Personality with Imaginary Languages* (Princeton, NJ: Princeton University Press, 1994), 9–10.

51. Sonu Shamdasani, "Encountering Hélène: *Théodore Flournoy and the Genesis of Subliminal Psychology*," introduction to Flournoy, *From India to the Planet Mars*, xliii.

52. Flournoy, 250.

53. Corbin, *Alone with the Alone*, 224.

54. I suspect that this is the source of my earlier mistake that Flournoy used the imaginal—the supernormal, which he used extensively, means more or less the same thing as the imaginal in Myers (so I assumed he had used the imaginal). The error, in short, contained a correct intuition.

55. Frederic W. H. Myers, "Automatic Writing—II," *Proceedings of the Society for Psychical Research* 3 (1885): 30.

56. Frederic W. H. Myers, *Human Personality and Its Survival of Bodily Death* (London: Longmans, Green, and Co., 1903), xxii.

57. Myers, *Human Personality*, xviii.

58. See, for example, Joseph Maxwell, *Les phénomènes psychiques: Recherches, observations, méthodes* (Paris: Félix Alcan, 1903), 15, 16, 28, 40, 57, 74, 99, 305 (my translations throughout, with the original book reference, guidance, and help from Renaud Evrard).

59. For a near-perfect expression of this third space, see Maxwell, *Les phénomènes psychiques*, 17.

60. Maxwell, 20.

61. Maxwell, 11.

62. Maxwell, 102.

63. Maxwell, 60.

64. Maxwell, 101–2.

65. Maxwell, 17; see also 47.

66. Maxwell, 65.

67. Maxwell, 98.

68. Maxwell, 299.

69. Maxwell, 207.
70. Maxwell, 26.
71. Maxwell, 299.
72. Maxwell, 301.
73. For the seeming intelligence and independence of the phenomena and their unspeci-fied relationship to the "will" ("*volonté*") of the experimenters, see Maxwell, 102–4. These passages recall the different conceptions of the "will" ("*der Wille*") of Arthur Schopenhauer and Friedrich Nietzsche. Schopenhauer, after all, linked such a Will to an impersonal cosmic or universal life-form and the human history of magic, whereas Nietzsche would relate it directly to the coming superspecies of the Übermensch, or superhumans. Nietzsche understood himself as *extending* Schopenhauer, not exactly denying him; see Friedrich Nietzsche, *Unpublished Fragments from the Period of "Thus Spoke Zarathustra" (Spring, 1884–Winter 1884/85)*, trans. Paul S. Loeb and David E. Tinsley, vol. 15 of *The Complete Works of Friedrich Nietzsche* (Stanford, CA: Stanford University Press, 2022), 73. I should add that my constant admiration for Nietzsche in this volume, and my previous one, does not mean I agree with him on every point, particularly his rejection of democracy and individual rights. It is entirely possibly to affirm a thinker without signing one's name to every thought. Is this not what the (super)humanities are all about?
74. Maxwell, *Les phénomènes psychiques*, 100.
75. Maxwell, 64.
76. On the future Newton, see Maxwell, 17. For a discussion of the electric-like phenomena, see Maxwell, 47, 51–52, 81–82, 106.
77. Renaud Evrard, "Joseph Maxwell," *Psypioneer* 5, no. 1 (January 2009): 21–27.
78. Maxwell, *Les phénomènes psychiques*, 14.
79. Evrard, "Joseph Maxwell," 23–24. See also Evrard's magisterial *Enquete sur 150 ans de parapsychologie: La légende de l'esprit* (Escalquens, France: Éditions Trajectoire, 2016), 189–92.
80. Maxwell, *Les phénomènes psychiques*, 16, 40–41. For a contemporary discussion of this key relationship between psychopathology and the paranormal, see Renaud Ev-rard, *Folie et paranormal: Vers une clinque des experiences exceptionelles* (Rennes: Presses Universitaires de Rennes, 2014). For more on madness and the religious, see chapter 6.
81. Maxwell, *Les phénomènes psychiques*, 43, 100.
82. Renaud Evrard, private correspondence with author, March 10, 2015. The modern reluctance of authors to integrate madness into the mystical is one I will foreground in chapter 6. Here it is given further support by the history of the paranormal, which demonstrates in abundance this very linkage.
83. Maxwell, *Les phénomènes psychiques*, 42–43. This passage is *so* reminiscent of Nietz-sche. See also Maxwell, 56: "The normal man is only an average (*moyenne*)." Here that argument that madness and mystical states are mutually intertwined is given a specific evolutionary esotericism.

Chapter Two

1. This chapter was originally written under the title, "Why the Aliens Don't Land: Mantis, Mystical Theology, and Social Criticism in the Contemporary Abduction Literature,"

parts of which were presented at the conference event "Adventures in the Imaginal: Henry Corbin in the 21st Century," Center for the Study of World Religions, Harvard Divinity School, Cambridge, MA, May 12–14, 2022.

2. This linkage of the UFO and the soul will become the core idea of the third volume of the aforementioned trilogy, tentatively entitled *The Technology of Eschatology: The UFO, the Soul, and the End of All Things, 1945–2028*. As I refused to do with "the ancestors" in chapter 1, I make no assumptions about the spiritual wisdom or evolutionary advance of the various alleged entities that show up around UFO or UAP encounters. Indeed, many such "extraterrestrial" phenomena are simply not what they pretend to be.

3. Joshua Cutchin, *Ecology of Souls: A New Mythology of Death & the Paranormal*, 2 vols. (Horse & Barrel Press, 2022).

4. See George P. Hansen, "George Hansen Discussion, June 12th, 2021: UFOs and the Paranormal," Joe Foster, posted June 14, 2021, YouTube video, 2:35:20, https://www.youtube.com/watch?v=9hcCh-N7A8I.

5. This section is based on Katharine Shilcutt, "Jeffrey Kripal on How to Think about the UFO Phenomenon," *Rice News*, June 30, 2021, https://news.rice.edu/news/2021/jeffrey-kripal-how-think-about-ufo-phenomenon.

6. One professional journalist behind both articles is Leslie Kean, who also researches and writes about the afterlife. Kean's two major books are *UFOs: Generals, Pilots, and Government Officials Go on the Record* (New York: Crown Archetype, 2010) and *Surviving Death: A Journalist Investigates Evidence for the Afterlife* (New York: Crown Archetype, 2017).

7. See, for example, Jack Butler, "Wisconsin Representative Mike Gallagher: UFOs Could Be 'Us from the Future,'" *National Review*, May 20, 2022, https://www.nationalreview.com/corner/wisconsin-representative-mike-gallager-ufos-could-be-us-from-the-future/.

8. Leslie Kean and Ralph Blumenthal, "Intelligence Officials Say U.S. Has Retrieved Craft of Non-human Origin," *The Debrief*, June 5, 2023, https://thedebrief.org/intelligence-officials-say-u-s-has-retrieved-non-human-craft/.

9. Jesse Michels, *American Alchemy*, "UFO Whistleblower Dave Grusch Tells Me Everything," September 10, 2023, YouTube video, 1:54:24, https://www.youtube.com/watch?app=desktop&si=6MMpq4OJaMW9Odb08&v=kRO5jOa06Qw&feature=youtu.be.

10. These Esalen Center for Theory and Research symposia occurred in 2015, 2016, 2019, and 2021. My cohosts were Diana Walsh Pasulka for the first three and Leslie Kean for the fourth. We cohosted the private seminars under the rubrics of "Beyond the Spinning" and "Future Technologies and Emergent Mythologies."

11. See Håkan Blomqvist, *Esotericism and UFO Research* (Örebro, Sweden, 2017), a free pdf of which is available at https://ufoarchives.blogspot.com.

12. The definitive history of the UFO phenomenon, which traces it back into the ancient and global world but also locates its present interpretation in the historical context of Cold War America, is Greg Eghigian, *After the Flying Saucers Came: A Global History of the UFO Phenomenon* (New York: Oxford University Press, 2024). I will touch briefly on this history, but it will nevertheless inform everything I have to say, particularly about what I call the "Cold War invasion narrative," which continues to define so much about how the phenomenon is presently (mis)interpreted.

13. Robert Hastings, *UFOs and Nukes: Extraordinary Encounters at Nuclear Weapons Sites* (Bloomington, IN: Author's House, 2008). Stunningly, and reflective of our double

model of the UFO/soul, Hastings has recently coauthored a book with his fellow researcher about how they are also both experiencers. See Robert Hastings and Dr. Bob Jacobs, *Confession: Our Hidden Alien Encounters Revealed* (self-pub., 2019).

14. For a documentary treatment of an African case (that took place on September 16, 1994, at Ariel School in Zimbabwe and involved over sixty school children from a variety of cultural, religious, and racial backgrounds), see Randall Nickerson, dir., *Ariel Phenomenon* (String Theory Films, 2022), DVD. For the Romanian material, including a most intriguing monastery mural, see Dan D. Farcas, *UFOs over Romania* (West Yorkshire, UK: Flying Disk Press, 2016).

15. Jörg Matthias Determann, *Islam, Science Fiction and Extraterrestrial Life: The Culture of Astrobiology in the Muslim World* (London: I. B. Taurus, 2021).

16. See James T. Lacatski, Colm A. Kelleher, and George Knapp, *Skinwalkers at the Pentagon: An Insiders' Account of the Secret Government UFO Program* (Henderson, NV: self-pub., 2021).

17. Colm A. Kelleher, personal conversation with the author, May 26, 2022.

18. Hussein Ali Agrama, "Secularity, Synchronicity, and Uncanny Science: Considerations and Challenges," *Zygon: Journal of Religion & Science* 56, no. 2 (June 2021): 395–415. See also Mayanthi Fernando, "Uncanny Ecologies: More-than-Natural, More-than-Human, More-than-Secular," *Comparative Studies of South Asia, Africa and the Middle East* 42, no. 3 (2022): 568–583. For a perspective from the study of religion that is deeply resonant with this approach, see also D. W. Pasulka, *American Cosmic: UFOs, Religion, Technology* (New York: Oxford University Press, 2019). I addressed this unity of science and religion in Jeffrey J. Kripal, *Authors of the Impossible: The Sacred and the Paranormal* (Chicago: University of Chicago Press, 2010). Such an epistemological fusion is fundamental to what I am calling the Super Story and the coming trilogy on the same. It is also an expression of the dual-aspect monism advanced in these pages.

19. Mitch Hedberg, "The Reason We Can't Find Bigfoot," Just for Laughs, September 17, 2014, YouTube video, 3:49, https://www.youtube.com/watch?v=ax0MGlIVjiY, at 2:19.

20. Jeffrey J. Kripal, book review of Jacques Vallee, *Forbidden Science 5: Pacific Heights, The Journals of Jacques Vallee 2000-2009* (Charlottesville, VA: Anomalist Books, 2023), *Journal of Scientific Exploration* 37, no. 3 (Fall 2023).

21. The account I give is based on email correspondence with Stuart Davis (used with permission) and this audio account of Davis's experiences: Stuart Davis, "Man Meets Mantis," Stuart Davis Media, November 6, 2018, YouTube video, 1:14:30, https://www.youtube.com/watch?v=Zi_8W0qCUH0 (also available on his *Aliens & Artists* podcast, https://www.aliensandartists.com/). Specific examples come from this account. The interested reader might also listen to Phil Ford and J. F. Martel, "Entities, with Stuart Davis," January 2, 2019, at *Weird Studies*, podcast, MP3 audio, 1:14:47, https://www.weirdstudies.com/37.

22. The language of the emotional, subtle, and causal bodies is probably indebted to the writer and philosopher Ken Wilber but is ultimately likely Theosophical in origin.

23. The anthropologist Hussein Ali Agrama often points out that these presences clearly have the ability to alter our perception of them. So why should we trust what we sense or see?

24. Davis will later learn that this is an injunction to stay true to his own deepest values, not a threat.

25. For a full study that links the ancient Greek *mania* to mental illness, prophecy, the

deities of Greek religion, the near-death experience, the combat fury of the battlefield, possession, poetic inspiration, erotic ecstasy, and philosophical thought, see Yulia Ustinova, *Divine Mania: Alteration of Consciousness in Ancient Greece* (Oxfordshire, UK: Routledge, 2018).

26. Ardy Sixkiller Clarke, *Encounters with Star People: Untold Stories of American Indians* (San Antonio, TX: Anomalist Books, 2012), 93, 95, 147. My thanks to Nicholas Collins again for this lead.

27. Sigrid Schmidt, "The Praying Mantis in Namibian Folklore," *Southern African Humanities* 31 (September 2018): 63–77.

28. Laurens van der Post, *The Heart of the Hunter: The Customs and Myths of the African Bushman* (London: Penguin, 1968). This work, and so much more, is cited and discussed in William L. Pressly, "The Praying Mantis in Surrealist Art," *The Art Bulletin* 55, no. 4 (1973): 600–15. The present section on the mantis and surrealism relies primarily on this essay. My sincere thanks to Joachim Koester for pointing me toward these surrealist connections, as well as toward the martial arts practices and meanings immediately below.

29. Douglas Farrer, "Becoming Animal in the Chinese Martial Arts," in *Living Beings: Perspectives on Interspecies Engagments*, ed. Penny Dransart (London: Bloomsbury, 2013), 146, 159.

30. Such superpowers are more than functions of folklore. Farrar tells the story of witnessing a slight seventy-nine-year-old grandmaster, Ip Kai Shui, who could not be moved by five English men during a 1993 demonstration. He was stuck to the ground, "as if he were a giant rock" (Farrer, "Becoming Animal," 154).

31. Farrer, 159.

32. Farrer, 146. Here, "becoming animal" (following the French philosopher Gilles Deleuze and the psychoanalytic political activist Félix Guattari) is "basically of another power," since the reality of such a transformation is not in the biological animal itself nor in some kind of metaphorical correspondence between the human and the animal but exists in itself, "in that which suddenly sweeps us up and makes us become . . . 'the Beast'" (Deleuze and Guattari quoted in Farrer, 147).

33. Pressly, "The Praying Mantis in Surrealist Art," 600. The original publication is Roger Caillois, "La Mante religieuse," *Minotaure* 5: 23–26, translated into English in Roger Caillois, "The Praying Mantis: From Biology to Psychoanalysis," in *The Edge of Surrealism: A Roger Caillois Reader*, ed. Claudine Frank (Durham, NC: Duke University Press, 2003), 69–81. For an analysis of Caillois's view of mythology as profound biology (with tremendous implications for evolutionary models of religious experience as powerful archaic expressions of our animal ancestors), see Rosa Eidelpes, "Roger Caillois' Biology of Myth and the Myth of Biology," *Anthropology & Materialism*, no. 2 (2014): 1–18.

34. Pressly, "The Praying Mantis in Surrealist Art," 608. Perhaps related to the present chapter, Bataille, and especially this book on death and sexuality, played a major role in my own education and earlier interest in erotic forms of mystical literature.

35. Pressly, 611.

36. Michael Luckman, *Alien Rock: The Rock 'n' Roll Extraterrestrial Connection* (New York: Pocket Books, 2005), 30–33. I personally doubt the nudity is a tangential fact, or that the encounter occurred shortly after sex. There is an erotic component here, vague for sure but also barely hidden. The account links to numerous others in which an occult entity "feeds" on or is attracted to sexual arousal.

37. Luckman, *Alien Rock*, 37–38.

38. Luckman, *Alien Rock*, 37–38. A more recent iteration of the story from Geller himself is in Bethany Minelle, "Uri Geller: On John Lennon's Alien Encounter, Crashed UFOs and Recruiting Citizens for His Micronation," September 15, 2022, *Sky News*, https://news.sky.com/story/uri-geller-on-john-lennons-alien-encounter-crashed-ufos-and-recruiting-citizens-for-his-micronation-12691294.

39. Luckman, *Alien Rock*, 186–89. Multiple textual references to dozens of rock musicians, which go *far* beyond the few I have summarized here, can be found in this same book, which I admit I find a bit breathless at times.

40. Stuart Davis, email communication with author, May 21, 2022.

41. Listen to Bill Hicks, "Gifts of Forgiveness," *Rant in E-Minor*," Bill Hicks - Topic, March 3, 2017, YouTube video, 5:09, https://youtu.be/sXwAU6LVb8Q?t=43. The famous Hicks image of the "third eye," itself drawn from Hindu mythology and mystical thought, helped inspire the song "Third Eye" by Tool on their double-entendre entitled album *Ænima* (1996). The famous Bill Hicks line about how we are all "one consciousness experiencing itself subjectively" and how "we are the imagination of ourselves" introduces the song. My profound thanks to Matthew Dillon, who is the real Bill Hicks expert here.

42. Cynthia True, *American Scream: The Bill Hicks Story* (New York: Pan Books, 2005), 106. For a cinematic recreation of some of these events, see Matt Harlock and Paul Thomas's *American: The Bill Hicks Story* (2009).

43. The quotations here and in the following paragraphs can be found in Kevin Booth with Michael Bertin, *Bill Hicks: Agent of Evolution* (New York: Harper, 2005), 210–18.

44. Shayla Love, "The Long, Strange Relationship Between Psychedelics and Telepathy," *VICE*, July 18, 2022, https://www.vice.com/en/article/z34xa5/the-long-strange-relationship-between-psychedelics-and-telepathy.

45. Booth, *Bill Hicks*, 210–18. Hicks, of course, was reading esoteric authors, especially the psychedelic bard Terence McKenna and the channeled classic *A Course in Miracles*. I am indebted to Kevin Booth and, again, Matthew Dillon for these facts. Mike Clelland also tells some stories of owl-synchronicities, including with Kevin Booth, who saw an owl outside his window while reading Clelland's writing on Hicks, which would become a part of Mike Clelland, *The Messengers: Owls, Synchronicity and the UFO Abductee* (Richard Dolan Press, 2015), 251–56.

46. Hicks, "Gifts of Forgiveness."

47. Booth and Bertin, *Bill Hicks*, 213.

48. See also Farah Yurdozu, *Love in an Alien Purgatory: The Life and Fantastic Art of David Huggins* (San Antonio, TX: Anomalist Books, 2009).

49. Whitley Strieber, *Communion: A True Story* (New York: Beech Tree Books, 1987), 157. My thanks to my PhD student Nicholas Collins for pointing this passage out to me.

50. Whitley Strieber and Anne Strieber, eds., *The Communion Letters* (New York: HarperPrism, 1997), 259, 261.

51. Anne and Whitley Strieber, *The Afterlife Revolution* (San Antonio, TX: Walker & Collier, 2017), 33. Note that Anne is listed an author here, from the other side.

52. The Anne and Whitley Strieber Collection, L888, Woodson Research Center, Rice University. My thanks to Nicholas Collins for this reference.

53. Paul Misraki, *Des signes dans le ciel: Les extraterrestres* (Paris: Éditions Labergerie, 1968). See especially chapter 7, "Fátima."

54. The multinaturalism (nature behaves differently in different bodies and historical times) and the subsequent multiple metaphysics of Indigenous peoples, especially in the Amazon, are most relevant here. I am thinking of Eduardo Viveiros de Castro, *Cannibal Metaphysics*, ed. and trans. Peter Skafish (Minneapolis: Univocal, 2014). Significantly, Skafish is the author of *Rough Metaphysics: The Speculative Thought and Mediumship of Jane Roberts* (Minneapolis: University of Minnesota Press, 2023). I am not suggesting that Viveiros de Castro's thought is informed by psychoactive plants. I am saying that the Amazonian cultures out of which he thinks and forms his multinaturalist anthropology are. It is no surprise to me, moreover, that some of these same Indigenous cultures report a clicking insectoid entity that bears an unusual closeness to the entity we are comparatively exploring in this chapter. The intertextual weaves are simply endless on this point.

55. Rick Strassman et al., *Inner Paths to Outer Space: Journeys to Alien Worlds through Psychedelics and Other Spiritual Technologies* (Rochester, VT: Park Street Press, 2008), 1, 3.

56. Strassman et al., *Inner Paths to Outer Space*, 5.

57. Strassman et al., 33.

58. Strassman et al., 65.

59. Strassman et al., 68–70 (brackets in original).

60. Strassman et al., 71.

61. Strassman et al., 75–76.

62. Strassman et al., 123.

63. Strassman et al., 150–53. The phrase "True Hallucination" is probably a reference to the book by Terence McKenna by the same title.

64. Strassman et al., opposite 153.

65. Strassman et al., 103. This was the exact conclusion of Ken Arnold, the originating "flying saucer" witness of the United States on June 24, 1947.

66. Stephen Hickman, foreword to Strassman et al., vii.

67. David Luke and Rory Spowers, eds., *DMT Entity Encounters: Dialogues on the Spirit Molecule with Ralph Mentzer, Chris Bache, Jeffrey Kripal, Whitley Strieber, Angela Voss, and Others* (Rochester, VT: Park Street Press, 2021). See especially my humorous but "aha!" conversation with Dennis McKenna on "The Soul Is a UFO" (Luke and Spowers, 238).

68. See Gerald Heard, *Is Another World Watching? The Riddle of the Flying Saucers* (New York: Bantam Books, 1953), 123–44.

69. An earlier version of the next five paragraphs appeared in Jeffrey J. Kripal, "What Jeremy Vaeni Can Teach Us (After We Stop Laughing)," foreword to Jeremy Vaeni, *I Am to Tell You This and I Am to Tell It Is Fiction* (Kynegion House, 2020), 3–7.

70. Jeremy Vaeni, *I Know Why the Aliens Don't Land!* (2003), *Urgency* (2011), *I Am to Tell You This* (2020), and *Aliens: The First and Final Disclosure* (2022). The novel is Jeremy Vaeni, *Into the End* (2013). All the books are published by Kynegion House.

71. Jeremy Vaeni, dir., *No One's Watching: An Alien Abductee's Story* (2007), DVD.

72. Jeremy Vaeni, *The Story of Toe: A Love Story for Everyone*, ill. John Randall (Kynegion House, 2020).

73. Vaeni, *I Know Why*, 362.

74. Vaeni, 370.

75. Vaeni, 371 (emphasis in original).

76. Vaeni, 372.
77. Vaeni, 373.
78. Vaeni, 376.
79. Jeremy Vaeni, *I Am to Tell You This*, 76.
80. Vaeni, 81.
81. Vaeni, 78. The alien insectoid motif appears again in Vaeni's novel, *Into the End*.
82. Vaeni, *Aliens*, 3.
83. Vaeni, 82.
84. Vaeni, 43. The mother's answer to her son's related confession that he was an abductee was sensible enough: "I don't believe you're being abducted by aliens, Jer. I think you read too many comics books."
85. Vaeni, 215–16.
86. Vaeni, 252.
87. Vaeni, 149.
88. Vaeni, 150.
89. It is not entirely clear to me, but I think Vaeni is working with a block cosmology—that is, the universe as "one still life painting" through which we experience ourselves moving but are not (Vaeni, *Aliens*, 226–27).
90. Vaeni, *Aliens*, 56. See also Vaeni, 139, where he discusses their politics, which are lightyears ahead of ours: "Aliens aren't here in the name of human progress. They're here in the name of universal equality."
91. Vaeni, 61.
92. Vaeni, 66.
93. See Vaeni, 199.
94. Vaeni, 71. This passage buzzes with Nietzschean themes. See also Vaeni, 242.
95. Vaeni., 10–12.
96. Vaeni, 258, 259.
97. Vaeni, 200–201.
98. Vaeni, *I Am to Tell You This*, 164, 165–66. See also Vaeni, *Aliens*, 233. The ufological visions, including those of Vaeni, sometimes look exactly like science-fiction movies—that is, fake. They are also related to occult or ceremonial magic, which sometimes really works (Vaeni, *Aliens*, 235–36).
99. Vaeni, *Aliens*, 226.
100. Vaeni, 13. See also 264 for Vaeni's simple confession that he exists both inside and outside the box. Vaeni is Two.
101. Vaeni, 186. See also Vaeni, 259, for the "partitioned mind."
102. Vaeni, 2, 176.
103. Vaeni, 209, 254.
104. Both the mechanical bug drawing and a drawing of a gray being with a "little baby" is included in the later paperback edition of John E. Mack, *Passport to the Cosmos: Human Transformation and Alien Encounters* (Largo: Kunati, 2008), xiv, 2. I included the bug drawing (without knowing Karin Austin) and discussed it in the context of Spidey's iconography in *Mutants and Mystics: Science Fiction, Superhero Comics, and the Paranormal* (Chicago: University of Chicago, 2011), 77.
105. Hold onto that "autistic" comparison. It will return.
106. This quotation and those below are based on my typed notes from a class visit on April 5, 2023, at Rice University. They were then corrected and expanded upon by Austin herself.

107. This "nonsensical" act of going back to sleep, or even walking back to bed from another room, is a very common feature of encounter experiences.

108. Such an endogenous release is implied in Joshua Cutchin, *A Trojan Feast: The Food and Drink Offerings of Aliens, Faeries, and Sasquatch* (San Antonio, TX: Anomalist Books, 2015).

109. April D. DeConick, *The Gnostic New Age: How a Countercultural Spirituality Revolutionized Religion from Antiquity to Today* (New York: Columbia University Press, 2019). DeConick distinguishes this "servant spirituality" from a "gnostic spirituality" (one that affirms our identity with divinity) that was born at the same time and that has long been resisted but is currently being affirmed again in popular culture, especially science-fiction film.

Chapter Three

1. See Michael Rabe, "Sexual Imagery on the 'Phantasmagorical Castles' of Khajuraho," *International Journal of Tantric Studies* 2, no. 2 (November 1996), http://asiatica.org /ijts/vol2_no2/sexual-imagery-phantasmagorical-castles-khajuraho/. The flight in question is primarily a postmortem one, a flying to heaven. The temple, in short, was funerary in nature for Rabe. Heavenly nymphs await the newly departed king. See also Michael Rabe, "Secret Yantras and Erotic Display for Hindu Temples," in *Tantra in Practice*, ed. David Gordon White (Princeton, NJ: Princeton University Press, 2000), 434–46.

2. David Gordon White, *The Kiss of the Yogini: "Tantric Sex" in Its South Asian Contexts* (Chicago: University of Chicago Press, 2003), 211, 213.

3. White, *Kiss of the Yogini*, 218.

4. Stith Thompson, *Motif-Index of Folk-Literature: A Classification of Narrative Elements in Folktales, Ballads, Myths, Fables, Mediaeval Romances, Exempla, Fabliaux, Jest-Books, and Local Legends* (Bloomington: Indiana University Press, 1955-1958).

5. I know this case well because I know Whitley Strieber. For a summary and analysis of Strieber's *Communion* and some of the later books, see Jeffrey J. Kripal, *Mutants and Mystics: Science Fiction, Superhero Comics, and the Paranormal* (Chicago: University of Chicago Press, 2011), chap. 7. See also the book we cowrote, Whitley Strieber and Jeffrey J. Kripal, *The Super Natural: Why the Unexplained Is Real* (New York: TarcherPerigree, 2016).

6. This is what Ted Jacobs and the artist and abduction researcher Budd Hopkins told Strieber (private communication with Whitley Strieber, May 20, 2022).

7. Whitley Strieber, *Communion: A True Story* (New York: Beech Tree Books, 1987), 285. Strieber relates the triangle to any number of religious and occult precedents in these same pages and not simply to Shakta Tantra.

8. I will return to this theme of sexuality and altered states in the abduction literature in the aforementioned and future *Biological Gods*. I simply want to note here that I do not think this comparative pattern is a tangential one and that it witnesses strongly and obviously to the super sexualities of the history of religions (from Asian Tantra to Euro-American sexual magic). Theoretical models like psychoanalysis, feminist criticism, or queer theory are important here and not to be avoided or dismissed, but, to the extent that they reduce human sexuality to a biological force or social construction, these same models are not sufficient, as I have routinely and consistently argued since the 1990s. The "impossible" relationship between erotic ecstasy and human flight (whether spiritual or physical, or both) make this limitation particularly obvious.

9. Kevin, "Why the 'Paranormal' Refuses to be Understood by the Rational Mind," unpublished essay. Kevin is referring to kundalini yoga, a specific Hindu Tantric tradition with which he is most familiar.

10. Kevin, email communication with author, May 7, 2022.

11. Kevin, email communication with author, May 5, 2022.

12. This is the basic argument of my memoir-manifesto, Jeffrey J. Kripal, *Secret Body: Erotic and Esoteric Currents in the History of Religions* (Chicago: University of Chicago Press, 2017).

13. Some of Kevin's unpublished essays include "Becoming Whole," "Deconstructing Postmodernism," "Emergent Monism," "How Darwin Saved Religion," "The Human as Three," "The Imaginal, Quantum Mechanics, and Super Determinism," "Kevin's BTUFO Summoning and Encounter of the Third Kind," "Living as an Imaginal," "Magic as the Intersection between Super Determinism and Platonic Surrealism," "Ontology in Platonic Surrealism," "Parallels Between the Phenomenon and Shakti Kundalini," "The Phenomenon and the 'Problem of Angels,'" "Platonic Surrealism: A Brief Treatment," "The Relationship Between 'Kundalini' and the 'Land of the Dead,'" "Revisiting the Term 'Subtle Energies.'" "The Symbiotic Plasma Hypothesis of Religion, Spirituality and 'UFOs,'" "Teach People How to Imagine, Not What to Imagine," "The Unary Consciousness Dilemma," and "Why the Paranormal Refuses to Be Understood by the Rational Mind."

14. Kevin, email communication with author, May 20, 2022.

15. Kevin, "Unary Consciousness Dilemma."

16. Kevin, email communication with author, May 5, 2022.

17. Diane Hennacy Powell, *The ESP Enigma: The Scientific Case for Psychic Phenomena* (New York: Walker, 2009). See also Sudhir Kakar and Jeffrey J. Kripal, eds., *Seriously Strange: Thinking Anew about Psychical Experiences* (New Delhi: Penguin, 2012).

18. Donald A. Treffert, *Extraordinary People: Understanding Savant Syndrome*, rev. ed. (Lincoln, NE: iUniverse, 2006).

19. Joshua Cutchin, *Thieves in the Night: A Brief History of Supernatural Child Abductions* (San Antonio, TX: Anomalist Books, 2018), 319–27.

20. Kevin, email communication with author, May 5, 2022, with a few corrections to the original email per Kevin.

21. This being aware in the womb or very early in life is a theme I have heard in other spiritual teachers.

22. Kevin, "Why the 'Paranormal.'"

23. Kevin, "Why the 'Paranormal.'"

24. Kevin, "Unary Consciousness Dilemma."

25. Kevin, personal communication with author, September 10, 2023.

26. Kevin, personal communication with author, September 10, 2023.

27. Kevin, email communication with author, May 10, 2022, and personal communication, May 21, 2022. Such a true Soul is not the individual, supposedly immortal soul of monotheism that needs "saving." Once you believe that a particular organization, religion, or set of beliefs are necessary, "then you are well and truly fucked" (Kevin, personal communication with author, May 21, 2022).

28. Kevin, personal communication with author, May 21, 2022.

29. Kevin, personal communication with author, September 10, 2023.

30. Kevin, "Why the 'Paranormal.'"

31. This quotation is Kevin's addition to and a personal edit of an earlier draft of this chapter from an email attachment dated May 24, 2022.
32. Kevin, "Unary Consciousness Dilemma."
33. Kevin, "Unary Consciousness Dilemma."
34. Kevin, "Unary Consciousness Dilemma."
35. Kevin, "Unary Consciousness Dilemma."
36. Kevin, "Deconstructing Postmodernism."
37. Kevin, "Teach People How to Imagine."
38. The reference is to the film *Ghostbusters* (1984), where a gigantic Stay Puft Marshmallow Man appears to signal the end-times. The idea was likely created by the actor and comedian Dan Aykroyd, who cowrote the movie, and who has a long family history of Spiritualism.
39. Kevin, "Teach People How to Imagine."
40. Kevin, email communication with author, May 5, 2022. The language here is distinctly Buddhist. The Buddhist tradition often posits "strands" of identity that separate and come back together in different life-forms. None of these are permanent, however.
41. Natasha L. Mikles and Joseph P. Laycock, "Tracking the *Tulpa*: Exploring the 'Tibetan' Origins of a Contemporary Paranormal Idea," *Nova Religio* 19, no. 1 (2015): 87–97.
42. Kevin, personal communication with author, September 10, 2023.
43. See Jeffrey J. Kripal, "Better Horrors: From Terror to Communion in Whitley Strieber's *Communion* (1987)," *Social Research* 81/4 (winter 2015): 897–921.
44. Kevin, email communication with author, March 25, 2022. Kevin's point about such a "hunger strike" or realization being counter to the current of evolution and unlikely for the species as a whole was also expressed by G. I. Gurdjieff and the Harlem Renaissance collective. See especially Jon Woodson, *To Make a New Race: Gurdjieff, Toomer, and the Harlem Renaissance* (Jackson: University Press of Mississippi, 1999).
45. Dvaita Vedanta is a classical form of Hindu dualist philosophy, not to be confused with Advaita Vedanta, or nondual philosophy.
46. Kevin, personal communication with author, September 10, 2023. The idea that the "person" is in fact composed of multiple streams or strands of consciousness is a not uncommon one in the history of religions or, for that matter, modern thought. It is obvious in many Buddhist traditions—for example, where every soul or self is said to be a combination of multiple impermanent "strands." It is also probably the signature metaphysical teaching of Jane Roberts, which she called the "multidimensional self." See Peter Skafish, *Rough Metaphysics: The Speculative Thought and Mediumship of Jane Roberts* (Minneapolis: University of Minnesota Press, 2023), 29. A secular version can be had in the scientific notion of the microbiome in modern biology and physiology; that is, the idea that every human being is in fact a vast collective of microorganisms that depend on an invisible ecology for health and well-being.
47. Kevin, "Why the Paranormal."
48. Kevin, "The Phenomenon."
49. Richard Katz, *Boiling Energy: Community Healing among the Kalahari Kung* (Cambridge, MA: Harvard University Press, 1982).
50. Kevin, email communication with author, April 4, 2022. This striking comparison of kundalini yoga and the UFO is not without precedent. It is very much related to C. G. Jung's theorizing that flying saucers are self-healing circles or, to use the Tantric Sanskrit phrase, "mandalas" in the sky. Flying saucers, at once physical and psychological,

are planetary poltergeists. See C. G. Jung, *Flying Saucers: A Modern Myth of Things Seen in the Sky*, trans. R. F. C. Hall (New York: Harcourt, Brace and Company, 1959).

51. Kevin, email communication with author, May 14, 2022.
52. Kevin, email communication with author, May 5, 2022.
53. Kevin, personal communication with author, September 10, 2023.
54. Kevin, "Kevin's BTUFO."
55. This process is long observed and commented on by psychologists of creativity, including scientific creativity, and is sometimes called the "incubation" period." See Harald Atmanspacher and Dean Rickles, *Dual-Aspect Monism and the Deep Structure of Meaning* (New York: Routledge, 2022), 68–69.
56. Kevin, "Kevin's BTUFO."
57. Kevin, "Kevin's BTUFO" (with minor edits for grammatical consistency).
58. Kevin, email communication with author, May 4, 2022 (with my own very slight grammatical edits).
59. Kevin, "Creating a Term for 'Mystical UFOs,'" unpublished essay.
60. Kevin, email communication with author, May 20, 2022.
61. Kevin, personal communication with author, May 21, 2022.
62. Kevin, personal communication with author, May 24, 2022.

Chapter Four

1. Another, much shorter, version of this chapter appears as Jeffrey J. Kripal, "The Timeswerve: Reading Elliot Wolfson in a Block Universe," in *New Paths: Festschrift in Honor of Professor Elliot Wolfson*, ed. Glenn Dynner, Susannah Heschel, and Shaul Magid (West Lafayette, IN: Purdue University Press, forthcoming).
2. These are not random books. They "glow" with a fierce honesty and moral reckoning that I want to hold up and follow impossibly here and in the future. See Edward Bever, *The Realities of Witchcraft and Popular Magic in Early Modern Europe* (New York: Palgrave Macmillan, 2013); Monica Black, *A Demon-Haunted Land: Witches, Wonder Doctors, and the Ghosts of the Past in Post-WWII Germany* (New York: Metropolitan Books, 2020); Eduardo Kohn, *How Forests Think: Toward an Anthropology beyond the Human* (Berkeley: University of California Press, 2013); C. M. Mayo, *Metaphysical Odyssey into the Mexican Revolution: Francisco I. Madero and His Secret Book*, Spiritist Manual (Palo Alto: Dancing Chiva Literary Arts, 2013); Peter Skafish, *Rough Metaphysics: The Speculative Thought and Mediumship of Jane Roberts* (Minneapolis: University of Minnesota Press, 2023); and Zofia Weaver and Krzysztof Janoszka, *The Mind at Large: Clairvoyance, Psychics, Police and Life after Death: A Polish Perspective* (White Crow Books, 2022).
3. Sam Knight, *The Premonitions Bureau: A True Account of Death Foretold* (New York: Penguin Press, 2022). I want to thank Dean Kathleen Canning for gifting this particular book to me in order to affirm and support my own intellectual interests.
4. Jason A. Josephson Storm, *The Myth of Disenchantment: Magic, Modernity, and the Birth of the Human Sciences* (Chicago: University of Chicago Press, 2017); *Metamodernism: The Future of Theory* (Chicago: University of Chicago Press, 2021).
5. I am thinking of Elliot Wolfson's long engagement with Martin Heidegger, probably the major figure in the phenomenological tradition. See, for example, Elliot R. Wolfson, *Heidegger and Kabbalah: Hidden Gnosis and the Path of Poiesis* (Bloomington: Indiana

University Press, 2019). Much in line with what I have argued about the humanities (that they are really the superhumanities—that is, deeply inflected by mystical thought), Heidegger was likely influenced by Meister Eckhart. See John Caputo, *The Mystical Element in Heidegger's Thought* (Athens, OH: Ohio University Thought, 1978).

6. Fred J. Hanna, "Husserl on the Teachings of the Buddha," *The Humanistic Psychologist* 23 (Autumn 1995): 366. My sincerest thanks to John Allison for this and the references just below on Eckhart, telepathy, Colin Wilson, and Gerda Walther.

7. Edmund Husserl, *Ideas: General Introduction to Pure Phenomenology*, trans. W. R. Boyce Gibson (London: Routledge, 2012), xxiv.

8. Edmund Husserl quoted in Hanna, "Husserl," 368.

9. Hanna, "Husserl," 369–70.

10. Hanna, 371.

11. Hanna, 366.

12. Dorion Cairns, *Conversations with Husserl and Fink*, ed. the Husserl-Archives in Louvain (The Hague: Martinus Nijhoff, 1975), 94–95.

13. Cairns, *Conversations*, 91.

14. Colin Wilson, "Phenomenology as a Mystical Discipline," *Philosophy Now*, no. 56 (July/August 2006), https://philosophynow.org/issues/56/Phenomenology _as_a_Mystical_Discipline.

15. Husserl was "bitterly disappointed" by Heidegger's *Being and Time*. He considered Heidegger's work to be only about the surface elements of consciousness, "a description of human existence in its everyday character, which completely missed the whole point of the phenomenological reduction" (Dermot Moran, foreword to Husserl, *Ideas*, xvii).

16. Anthony J. Steinbock, *Phenomenology & Mysticism: The Verticality of Religious Experience* (Bloomington: Indiana University Press, 2007), 3. My thanks to Steve Crowell for this reference.

17. Steinbock, *Phenomenology & Mysticism*, 2.

18. Steinbock, 1.

19. Steinbock, 211. Steinbock affirms critical perspectives like Marxism, structuralism, and psychoanalysis, as well as writers-against-the-idol like Friedrich Nietzsche, Jacques Derrida, and Jean-Luc Marion. The phenomenological method, then, is not about some escape from historical suffering or from social and linguistic deconstruction.

20. Steinbock, 17–18.

21. Steinbock, 5.

22. Steinbock, 25.

23. I am not against such a theological conclusion (there may well be a One God behind or within the One World), but I do not think the comparativist should presume as much, partly because such a monotheism has behaved *very* badly vis-à-vis other religious modes and revelations.

24. See, for example, Olga Louchakova-Schwartz, ed., *The Problem of Religious Experience: Case Studies in Phenomenology, with Reflections and Commentaries*, 2 vols. (Cham, Switzerland: Springer, 2019); Antonio Calcagno, "Gerda Walther and the Possibility of Telepathy as an Act of Personal Social Mind," *Symposium* 26, no. 1/2 (2022): 62–75; Niamh Burns, "A Modernist Mystic: Philosophical Essence and Poetic Method in Gerda Walther (1897–1977)," *German Life and Letters* 73, no. 2 (April 2020): 246–69.

25. Husserl, *Ideas*, 14–16. Even his positive comments on Buddhism were on Theravada

Buddhism, a tradition that is not as imaginally robust as the Mahayana and Vajrayana streams. In short, Husserl admired what looked like his own philosophy.

26. See, for example, Husserl, 18–19. Husserl clearly relates geometry, the ideal mathematics that can show how an essence becomes a spatialized material object, in the ancient world to Plato and, in the modern world, to the development of even more extensive mathematical sciences like physics, which he much admires (Husserl, 20–21). These appear to be his models for his pure phenomenology of consciousness. If reality is not mathematics, it is like mathematics.

27. See especially Charles H. Long, *Ellipsis . . . The Collected Writings of Charles H. Long* (London: Bloomsbury Academic, 2018).

28. Long, "Mircea Eliade and the Imagination of Matter," in *Ellipsis*, 117. This essay is a robust defense of Mircea Eliade's comparativism and long interest in alchemy, which in this view cannot be read as a kind of protochemistry but is itself rooted more deeply in the sacralization of matter and, further back, in what Eliade called the "Graeco-oriental mysteries"—that is, in mystical thought (Long, 126). Long is being impossible.

29. Long, "Mircea Eliade and the Imagination of Matter," in *Ellipsis*, 126.

30. Long, "The Chicago School: An Academic Mode of Being," in *Ellipsis*, 96.

31. Long, "History, Religion, and the Future," in *Ellipsis*, 329.

32. Thomas J. J. Altizer, *Living the Death of God: A Theological Memoir* (Albany: State University of New York Press, 2006), 45–46. My thanks to Leigh Eric Schmidt for the extraordinary Altizer reference and story.

33. Altizer, *Living the Death of God*, 46. I happen to think that Eliade's famous "eternal return" in the comparative study of myth and ritual was an exoteric transformation of Nietzsche's much more esoteric "eternal recurrence of the same," that the two intellectuals were godless mystics who sought (and knew personally) a particular power over time. But that is another story.

34. Edmund Husserl quoted in Steinbock, *Phenomenology & Mysticism*, 28.

35. See Jeffrey J. Kripal, "The Mystical Mirror of Hermeneutics: Gazing into Elliot Wolfson's *Speculum* (1994)," in *Roads of Excess, Palaces of Wisdom: Eroticism and Reflexivity in the Study of Mysticism* (Chicago: University of Chicago Press, 2001).

36. Jeffrey J. Kripal, *Kali's Child: The Mystical and the Erotic in the Life and Teachings of Ramakrishna*, 2nd ed. (Chicago: University of Chicago Press, 1998).

37. For one summary of the thesis and a response to a critic, see Elliot R. Wolfson, "Bifurcating the Androgyne and Engendering Sin: A Zoharic Reading of Gen 1–3," in *Hidden Truths from Eden: Esoteric Readings of Genesis 1–3*, ed. Caroline Vander Stichele and Susan Scholz, (Society of Biblical Literature, 2014), 87–114.

38. Elliot R. Wolfson, *Language, Eros, Being: Kabbalistic Hermeneutics and Poetic Imagination* (New York: Fordham University Press, 2005), xiii.

39. Wolfson, *Language, Eros, Being*, 27. The final phrase is indebted to the early Mahayana Buddhist philosopher Nāgārjuna (circa 150–250 CE).

40. Wolfson, "Bifurcating the Androgyne," 94. Wolfson is quoting Gershom Scholem, who is probably quoting Meister Eckhart and his phrase for "the one beyond all distinction."

41. Elliot R. Wolfson, "*Imago Templi* and the Meeting of the Two Seas: Liturgical Time-Space and the Feminine Imaginary in Zoharic Kabbalah," *RES: Anthropology and Aesthetics*, no. 51 (Spring 2007): 121.

42. Wolfson, "*Imago Templi*," 122.

43. Quoted in Wolfson, 123.

44. Wolfson, 123.

45. Wolfson, 124.

46. Wolfson, 124.

47. Wolfson, 124.

48. Wolfson, 135.

49. Wolfson, 125.

50. Henry Corbin quoted in Wolfson, 125.

51. Wolfson, "Bifurcating the Androgyne," 88–89.

52. Wolfson, 90.

53. Wolfson, "*Imago Templi*," 126.

54. Wolfson, "Bifurcating the Androgyne," 113–14.

55. Wolfson, 114.

56. This is one of the twenty gnomons I identify and explore in Jeffrey J. Kripal, *Secret Body: Erotic and Esoteric Currents in the History of Religions* (Chicago: University of Chicago Press, 2017).

57. Elliot R. Wolfson, *A Dream Interpreted within a Dream: Oneiropoiesis and the Prism of Imagination* (New York: Zone Books, 2011). (Does anyone other than me think that the image of the sleeping man on the cover of this book looks more or less exactly like Elliot Wolfson?)

58. Ioan P. Couliano, *The Tree of Gnosis: Gnostic Mythology from Early Christianity to Modern Nihilism* (New York: HarperCollins, 1992).

59. This essay originally appeared in the defunct and especially rare journal, *Incognita* (which Couliano edited) but later was happily employed as the opening chapter of Ioan P. Couliano, *Out of This World: Otherworldly Journeys from Gilgamesh to Albert Einstein* (New York: Shambhala, 2001).

60. Couliano, *Out of This World*, 12–31; Wolfson, *Language, Eros, Being*, xv–xxxi.

61. Wolfson, *Language, Eros, Being*, xvii.

62. Wolfson, xvii.

63. For a philosophical discussion, see Huw Price, *Time's Arrow and Archimedes' Point: New Directions for the Physics of Time* (New York: Oxford University Press, 1996).

64. My understanding and use of these two phrases—"long body" and "long self"—are indebted to the work of the graphic novelists Grant Morrison and Alan Moore and the anthropologist and theorist of precognition Eric Wargo. See Eric Wargo, *Time Loops: Precognition, Retrocausation, and the Unconscious* (San Antonio, TX: Anomalist Books, 2018), and *Precognitive Dreamwork and the Long Self: Interpreting Messages from Your Future* (Rochester, VT: Inner Traditions, 2021). I return to Morrison and Wargo in my conclusion.

65. Paul Marshall, *The Living Mirror: Images of Reality in Science and Mysticism*, 2nd ed. (London: Samphire Press, 2006), viii.

66. Note in particular the subtitle of his most recent book: Paul Marshall, *The Shape of the Soul: What Mystical Experience Tells Us about Ourselves and Reality* (Lanham, MD: Rowman & Littlefield, 2019). Here is a near-perfect expression of the realist impulse of the superhumanities.

67. I discuss more of these Nietzschean ideas, much indebted to the work of the philosopher Paul Loeb, in Jeffrey J. Kripal, *The Superhumanities: Historical Precedents, Moral Objections, New Realties* (Chicago: University of Chicago Press, 2022), 84–100.

68. Michael Allen Gillespie, *Nietzsche's Final Teaching* (Chicago: University of Chicago Press, 2019).

69. Paul S. Loeb, *The Death of Nietzsche's Zarathustra* (Cambridge: Cambridge University Press, 2010), 11.

70. One thinks of Edmund Husserl's Eternity and the greatest conversion or future of humanity that he was proposing around his transcendental phenomenology. I do, anyway.

71. See Loeb, *The Death*, 26n22.

72. This is the Buddhist hermit-poet Han Shan, or Cold Mountain. He is usually placed in the eighth or ninth century of the Western calendar. I am relying for this text on Peter Kingsley's use of it in his *Reality* (Inverness, CA: The Golden Sufi Center, 2004), 203.

73. This is British parliamentarian Christopher Mayhew, friend of Aldous Huxley, in Benny Shanon, *The Antipodes of the Mind: Charting the Phenomenology of the Ayahuasca Experience* (Oxford: Oxford University Press, 2002), 227–28. Jonathan Bricklin fills in more details in Jonathan Bricklin, *The Illusion of Will, Self, and Time: William James's Reluctant Guide to Enlightenment* (Albany: State University of New York Press, 2016). He tells us that Mayhew's experience was on mescalin, which he took for a British documentary that was never aired, lest it offend the religious. Bricklin also points out that Mircea Eliade called this same account a "prodigious document" and confessed that he "trembled with joy" when he read it, since he had long worked on "the possibility of abolishing time, and of putting oneself into a trans-temporal condition" (Bricklin, 235–36). Eliade is, of course, well known for his insistence on a version of the eternal return and endlessly bashed for his "anti-history." That is what happens to someone who tries to write about time in ways beyond the positivist or postmodern order of knowledge.

74. This is the Israeli psychologist Benny Shanon on the general phenomenology of the ayahuasca experience in Benny Shanon, *The Antipodes of the Mind: Charting the Phenomenology of the Ayahuasca Experience* (New York: Oxford University Press, 2003), 47.

75. This is a near-death experience of the contemporary writer Anita Moorjani in *Dying to Be Me: My Journey from Cancer, to Near Death, to True Healing* (Carlsbad, CA: Hay House, 2014), 67.

76. This is the New York grandmother, atheist, and cancer survivor Dina Bazar after she was given psilocybin during a clinical trial at Johns Hopkins University in Brian Muraresku, *The Immortality Key: The Secret History of the Religion with No Name* (New York: St. Martin's Press, 2020), v.

77. John Allison, email message to author, March 9, 2020 (used with permission).

78. Wolfson, *Language, Eros, Being*, xv–xvi.

Chapter Five

1. This chapter is a revised version of an essay originally written for "Beyond the Spinning: Making (Better) Sense of the UFO Phenomenon," December 12–17, 2021, Center for Theory and Research, Esalen Institute, Big Sur, California, and then published as Jeffrey Kripal, "The World Is One, and the Human Is Two: Tentative Conclusions of a Working Historian of Religions," *Mind and Matter* 20, no. 1 (2022): 121–42. My sincerest thanks to Michael Murphy and Harald Atmanspacher for their support.

2. Charles H. Long, "Mircea Eliade and the Imagination of Matter," in *Ellipsis . . . The Collected Writings of Charles H. Long* (London: Bloomsbury Academic, 2018), 117.

3. The Joseph Donahue line is from a blurb on the inner flap of the hardback copy of Jeffrey J. Kripal, *Secret Body: Erotic and Esoteric Currents in the History of Religions* (Chicago: University of Chicago Press, 2017).

4. Whitley Strieber has often quoted Anne to me along these very lines, most recently in personal correspondence, September 1, 2023.

5. Alexander Regier, *Exorbitant Enlightenment: Blake, Hamman, and Anglo-German Constellations* (Cambridge: Cambridge University Press, 2019).

6. I have been inspired here by Iain McGilchrist, *The Master and His Emissary: The Divided Brain and the Making of the Western Mind* (New Haven, CT: Yale University Press, 2010).

7. This is also why, in the end, I do not think more microhistories, translations of texts, or ethnographic studies are going to get us to the bigger picture.

8. When I write lines like that, I think I may be an idealist, although the base reality of my idealism is as much matter as it is mind, which makes me look more like a dual-aspect monist. I do not want to resolve that tension. I simply want to acknowledge it.

9. Ryan Graves, "We Have a Real UFO Problem. And It's Not Balloons," *Politico*, February 28, 2023, https://www.politico.com/news/magazine/2023/02/28/ufo-uap-navy -intelligence-00084537. It is even more problematic than Graves suspects, as more science and technology, much less more military or intelligence gathering, is never going to solve this thing.

10. Consider the reflections of David Bohm and David Peat in *Science, Order, and Creativity* (Toronto: Bantam Books, 1987), 112–14. The two authors relate "selection," "collection," and "knowing" or, more literally, a "gathering in between" (*intellegere*) in their Latin roots. Put simply, what is intelligible at all is a function of what is collected and selected out of that collection and then related, and meaning itself is a psychophysical term, connecting both the mental and the material dimensions of human experience. This is partly what I mean when I write "comparison *is* mutation." Such comparative intelligence or collecting changes us since it changes our understanding of who we are and what the world is. For a discussion, see Harald Atmanspacher and Dean Rickles, *Dual-Aspect Monism and the Deep Structure of Meaning* (New York: Routledge, 2022), 135–36: "Therefore, understanding is neither anchored in mental states alone nor in physical states alone—it gathers in between."

11. The math and models continue to change. For an example, see Iain Nicolson, *Dark Side of the Universe: Dark Matter, Dark Energy, and the Fate of the Cosmos* (Baltimore: Johns Hopkins University Press, 2007).

12. Arthur C. Clarke, foreword to John Fairley and Simon Welfare, *Arthur C. Clarke's World of Strange Powers* (London: Collins, 1984), 4. At my best count, there were three such books. This was the second in the series.

13. I am indebted for the phrase "real magic" to Dean Radin, *Real Magic: Ancient Wisdom, Modern Science, and a Guide to the Secret Power of the Universe* (New York: Harmony, 2018). My friend and colleague Timothy Morton has also used the phrase in his own way, by which I have also been informed and inspired. See Timothy Morton, *Realist Magic: Objects, Ontology, Causality* (Ann Arbor, MI: Open Humanities Press, 2013).

14. C. G. Jung quoted in Roderick Main, *Breaking the Spell of Disenchantment: Mystery, Meaning and Metaphysics in the Work of C. G. Jung* (Asheville, NC: Chiron Publications, 2022), 82.

15. Harald Atmanspacher and Christopher A. Fuchs, eds., introduction to *The Pauli-Jung Conjecture and Its Impact Today* (Exeter, UK: Imprint Academic, 2014), 2.

16. Such a model is different from "compositional" forms of dual-aspect monism, which posit a ground that is not holistic and so must build up, rather like materialism and panpsychism, into physical objects and mental states. I find this too reductive. Nor does it jive with my historical materials.

17. John Wheeler, quoted in Atmanspacher and Rickles, *Dual-Aspect Monism*, 109.

18. Atmanspacher and Fuchs, *The Pauli-Jung Conjecture*, 4.

19. John Wheeler, quoted in Atmanspacher and Rickles, *Dual-Aspect Monism*, 109; see also Atmanspacher and Rickles, 120.

20. This is the deepest meaning of my endnote phrase "comparison *is* mutation."

21. C. G. Jung, quoted in Main, *Breaking the Spell*, 84.

22. The phrase "the meaning of meaning" is from Andrew Lohrey and his nonsectarian, postsecular affirmation of "the reality of an impersonal transcendent presence" in Andrew Lohrey, *Speaking of the Numinous: The Meaning of Meaning* (Falmouth, Tasmania: Rishi, 2010), 1.

23. Atmanspacher and Rickles, *Dual-Aspect Monism*, 158–59.

24. Atmanspacher and Fuchs, *The Pauli-Jung Conjecture*, 5 (italics in original).

25. Atmanspacher and Rickles, *Dual-Aspect Monism*, 66–67, 190–92. There it is again: "comparison *is* mutation," meaningful coincidence *is* evolution.

26. Atmanspacher and Rickles, 13.

27. C. G. Jung, *Mysterium Coniunctionis: An Inquiry into the Separation and Synthesis of Psychic Opposites in Alchemy*, trans. R. F. C. Hull (Princeton, NJ: Princeton University Press, 1989), 534; cited and discussed in Atmanspacher and Fuchs, *Dual-Aspect Monism*, 79. There are excellent descriptions of dual-aspect monism in terms of Christian theology, alchemy, and microphysics (quantum physics) and depth-psychology at Jung, *Mysterium*, 462, 536, and 538, respectively.

28. Jung, *Mysterium*, 462, 534–35.

29. Atmanspacher and Rickles, *Dual-Aspect Monism*, 13; see also Atmanspacher and Rickles, 169.

30. Atmanspacher and Rickles, 92.

31. Atmanspacher and Rickles, 159. Ironically, John Wheeler was adamantly opposed to anything he perceived to be parapsychological.

32. Atmanspacher and Rickles, 169. For more on Spinoza and the melting or merger of mental/material or divine/human epistemic divisions in a most radical state of nonduality, see especially Peter Sjöstedt-Hughes, "The White Sun of Substance: The Psychedelic *Amor Dei Intellectualis*," in *Philosophy and Psychedelics: Frameworks for Exceptional Experience*, ed. Christine Hauskeller and Peter Sjöstedt-Hughes (London: Bloomsbury Academic, 2023), 211–35.

33. I should add that both men knew multiple such synchronicities in their own lives, so they were not theorizing out of thin air. They knew personally of that which they were writing.

34. Atmanspacher is defining "sense" after Gottlob Frege, "*Über Sinn und Bedeutung*" (1892), "On Sense and Reference." See Harald Atmanspacher, "Why Physics Does Not Inform the Human Condition, But Its Boundaries Do," *Foundations of Science* (2023), https://doi.org/10.1007/s10699-023-09915-y. Frege intends "his notion of sense to open up a metaphysical dimension to the concept of meaning. As a relation to a domain that is 'even more objective' than the physical: the psychophysically neutral domain of reality" (Atmanspacher). Much in the spirit of my comments on mathematics above, Atmanspacher will also write of "the fantastic relationship between physics

and mathematics" (Atmanspacher), the latter Platonically understood as preexisting truths to be discovered, ontological fields of meaning that have nothing to do in themselves with social position or history. Meaning exists implicitly in the very fabric of reality.

35. Atmanspacher, "Why Physics Does Not Inform."

36. This is where comparative mystical literature comes in. Suffice it to say that the most relevant forms of this literature work to unthink thought and unbelieve belief. They recognize that reality has nothing to do with our thoughts or beliefs. This is Nietzsche too.

37. This complex thesis has been explored by authors for decades, including and especially Albert Budden, *UFOs Psychic Close Encounters: The Electromagnetic Indictment* (London: Blandford, 1995); Budden, *Electric UFOs: Fireballs, Electromagnetics and Abnormal States* (London: Blandford, 1998); Louis Proud, *Strange Electromagnetic Dimensions: The Science of the Unexplainable* (Pompton Plains, NJ: New Page Books, 2015); and David J. Halperin, *Intimate Alien: The Hidden Story of the UFO* (Stanford: Stanford University Press, 2020). These authors are very different. I do not want to collapse their theorizing into any single simplistic model (Halperin, for example, focuses much more on the psychology of death and its anxieties), but each is fully aware of a possible electromagnetic origin or environmental "trigger" of the phenomenon. I am probably most aligned with Jacques Vallee in a book like *Forbidden Science 5: The Journals of Jacques Vallee 5: Pacific Heights, 2000–2009* (San Antonio: Anomalist Books, 2023), where he argues for some superreal energetic emanations or radiation effects of the UFO phenomenon that result in the (illusory) mythical and visionary landscapes reported in the contact and abduction literatures. There is a "there, there," then, but it is finally unknowable by our present science and past religions. I find an author like Budden too reductive. I sometimes feel the same about Halperin, although I am much drawn to his psychoanalytic readings inspired by Freud and, in a less reductive mode, Jung.

38. One major complication here is that such visionary events might also represent the manipulation of human perception. This possibility takes us into the demonological discussion of the next and final chapter.

39. Peter T. Struck, *The Birth of the Symbol: Ancient Readers at the Limits of Their Texts* (Princeton, NJ: Princeton University Press, 2014).

40. I have made this same argument more fully elsewhere, in Jeffrey J. Kripal, *The Flip: Who You Really Are and Why It Matters* (London: Penguin, 2020). The subtitle of the American edition could have been used as a title for the present chapter: "Epiphanies of Mind and the Future of Knowledge."

41. The book was released in a new revised edition as Bertrand Méheust, *Science-fiction et soucoupes volantes: Une réalité mythico-physique* (Rennes, France: Terre de Brume, 2007). Significantly, Méheust wrote his MA thesis on William James and his massive Sorbonne dissertation on the history of animal magnetism and psychical research. For a fuller study of Méheust, see chapter 4 of Jeffrey J. Kripal, *Authors of the Impossible: The Sacred and the Paranormal* (Chicago: University of Chicago Press, 2010).

Chapter Six

1. Philip K. Dick, *Radio Free Albemuth* (New York: Arbor House, 1985), 85. This was the first novel Dick wrote to try to explain to himself Valis, or the cosmic Mind that beamed into him and "reprogrammed" his own mind in 1974.

2. Apparently, it also stuck with the designers of the book. If you take off the cover (or daily clothes) of the book, you will see a bright blue and glistening red presence—a secret Superman.

3. This reluctance to speak of God, even to eliminate the role of God, is in fact a consistent feature of Romantic thought, with which we began in chapter 1. See M. H. Abrams, *Natural Supernaturalism: Tradition and Revolution in Romantic Literature* (New York: W. W. Norton, 1973), 91. Here the "unaided mind of man" is emphasized, not a "Redeemer" (Abrams, 120).

4. See, for example, Anthony B. Pinn, *Writing God's Obituary: How a Good Methodist Became a Better Atheist* (New York: Prometheus, 2014).

5. The sexual and gendered implications of this panentheistic theology are queer and kinky beyond imagination. What we really have in this view is one Being endlessly pleasuring itself, reproducing itself in literally countless ways over eons and eons of space-time.

6. Bernardo Kastrup writes playfully but seriously about this notion of mind-at-large (the phrase is taken from Aldous Huxley) and its human personality "alters" within a vast associative identity disorder throughout his growing body of work. For one of dozens of examples, see Bernardo Kastrup, *Brief Peeks Beyond: Critical Essays on Metaphysics, Neuroscience, Free Will, Skepticism and Culture* (Winchester, UK: Iff Books, 2015), 18–19.

7. See below for more on this mutual illumination of the altered states of madness and mystical experience.

8. See Nikita Petrov, "Soviet Man in Inner Cosmos," *Medium*, October 13, 2015, https://medium.com/@nikita.petrov/soviet-man-in-inner-cosmos-e4867e62bd83.

9. Martin W. Ball, *Being Infinite: An Entheogenic Odyssey into the Limitless Eternal* (Ashland, OR: Kyandra Publishing, 2014), 6. Ball's books are chock-full of such statements.

10. Martin W. Ball, *Being Human: An Entheological Guide to God, Evolution and the Fractal Energetic Nature of Reality* (Ashland, OR: Kyandra Publishing, 2019), 36 (italics in original).

11. Bill Hicks, "The War on Drugs," track 3 on *Dangerous*, Invasion Records, 1990. My thanks to Matthew Dillon for this information and guidance.

12. Michael Allen Gillespie, *Nietzsche's Final Teaching* (Chicago: University of Chicago Press, 2019), 133, 197. The line is from a letter to Jacob Burkhardt dated January 5, 1889—just a few weeks before Nietzsche collapsed.

13. William Blake, quoted in Henry Crabb Robinson, *Diary, Reminiscences, and Correspondence of Henry Crabb Robinson*, ed. Thomas Sadler (Boston: Houghton, Mifflin and Company, 1898), 2:25.

14. Sharon Hewitt Rawlette, *The Source and Significance of Coincidences: A Hard Look at the Astonishing Evidence* (self-pub., 2019), 10.

15. Rawlette, *Source and Significance of Coincidences*, 13–14.

16. Rawlette, 13.

17. Rawlette, 580.

18. Gananath Obeyesekere, *The Awakened Ones: Phenomenology of Visionary Experience* (New York: Columbia University Press, 2012), 29.

19. Friedrich Nietzsche, *The Antichrist*, in *The Portable Nietzsche*, ed. and trans. Walter Kaufmann (New York: Penguin Books, 1976), 645 (italics in original).

20. Nietzsche, *Antichrist*, 606–7 (italics in original).

21. Nietzsche, 634.

22. For Feuerbach, see Jeffrey J. Kripal, "Restoring the Adam of Light," in *The Serpent's Gift: Gnostic Reflections on the Study of Religion* (Chicago: University of Chicago Press, 2007), 59-89. For Kafka, see Franz Kafka *The Aphorisms*, with a foreword by Daniel Frank (New York: Schocken, 2015), especially no. 50. For Bloom's academic Gnosticism, see Harold Bloom, *Omens of the Millennium: The Gnosis of Angels, Dreams, and Resurrection* (New York: Riverhead Books, 1997), or his discussion of Kafka and "Indestructability" in *The Western Canon: The Books and Schools of the Ages* (New York: Riverhead Books, 1995), 416-30. Berger's place in this lineage is less central. See Peter Berger, *The Sacred Canopy: Elements of a Sociological Theory of Religion* (Garden City, NY: Anchor Books, 1969), 180. My thanks to Rabbi James Ponet for the Kafka and second Bloom references and for all he has contributed to my thinking.

23. David Gordon White, *Daemons Are Forever: Currents and Exchanges in the Eurasian Pandemonium* (Chicago: University of Chicago Press, 2020). Other scholarly literatures have arrived at resonant positions. Consider the Jewish and Christian traditions, the "watchers," and sexual intercourse between the "sons of God" and the "daughters of men," a kind of divine-human hybridity that was believed to produce a race of "giants" (Genesis 6:1-4). This was the matrix out of which the "fallen angels" of later medieval and modern lore, and so much supersexual intercourse between invisible beings and humans, developed. See, for example, Angela Kim Harkins, Kelley Coblentz Bautch, and John C. Endres, S. J., eds., *The Watchers in Jewish and Christian Traditions* (Minneapolis: Fortress Press, 2014). Also, early Christian speculations on the human divinity of Christ were angelomorphic and dependent upon earlier Jewish notions of God made manifest in human form in the "Angel of the Lord" traditions. See Charles A. Gieschen, *Angelomorphic Christology: Antecedents and Early Evidence* (Waco, TX: Baylor University Press, 2017). I mention all this scholarship because these various superhuman dimensions are of immense, if entirely unexplored, relevance for the modern abduction materials.

24. Kocku von Stuckrad, *A Cultural History of the Soul: Europe and North America from 1870 to the Present* (New York: Columbia University Press, 2022); Marc de Launay, *Nietzsche and Race*, trans. Sylvia Gorelick (Chicago: University of Chicago Press, 2023).

25. Owen Davies, *A Supernatural War: Magic, Divination, and Faith During the First World War* (New York: Oxford University Press, 2018).

26. The case is often little better in the Asian traditions, where the condemnation of magical powers and acts is depressingly similar.

27. For one page-turning account of this early battle, see Carlos Eire, *They Flew: A History of the Impossible* (New Haven, CT: Yale University Press, 2023), 36-40.

28. Eire, *They Flew*, 297-98. For more on the history of the category of superstition in Catholic and Protestant intellectuals, particularly around the "hocus-pocus" of the eucharist, see Eire, 300-302. For magic, especially divination or seeing of the future (our own chapter 4), see Eire, 302-5.

29. Walter Stephens, *Demon Lovers: Witchcraft, Sex, and the Crisis of Belief* (Chicago: University of Chicago Press, 2002); Eire, *They Flew*, 318.

30. Eire, *They Flew*, 299-307.

31. Eire, 312.

32. See Eire, 314-15, for two early cases of levitation and some most striking images.

33. For a powerful summary of the argument, see Roderick Main, *Breaking the Spell of Disenchantment: Mystery, Meaning and Metaphysics in the Work of C. G. Jung* (Asheville, NC: Chiron Publications, 2022), especially 107-9.

34. Eire, *They Flew*, 58.

35. Eire, 63.

36. Eire, 19.

37. Eire, 25.

38. See, for example, Stephen E. Braude, *The Limits of Influence: Psychokinesis and the Philosophy of Science* (New York: Routledge & Kegan Paul, 1986); *Crimes of Reason: On Mind, Nature, and the Paranormal* (Lanham, MD: Rowman & Littlefield, 2014); *Dangerous Pursuits: Mediumship, Mind, and Music* (San Antonio, TX: Anomalist Books, 2020).

39. Darren W. Ritson, *Poltergeist Parallels and Contagion*, rev. ed. (White Crow Books, 2021).

40. Moshe Sluhovsky, *Believe Not Every Spirit: Possession, Mysticism, & Discernment in Early Modern Catholicism* (Chicago: University of Chicago Press, 2007).

41. Sharon Hewitt Rawlette has reminded us of the same truth—that psychokinetic powers can be used for negative or harmful ends. See Rawlette, *Source and Significance of Coincidences*, 317–57. Of particular note in Rawlette's work are Guy Lyon Playfair, *The Flying Cow: Exploring the Psychic World of Brazil* (Guildford, UK: White Crow Books, 2011); Edith Fiore, *The Unquiet Dead: A Psychologist Treats Spirit Possession* (New York: Ballantine Books, 1987); and Joe Fisher, *The Siren Call of Hungry Ghosts: A Riveting Investigation into Channeling and Spirit Guides* (New York: Paraview Press, 2001).

42. See, for example, Terry Lovelace, *Incident at Devil's Den: A True Story* (self-pub., 2018).

43. Dave Paulides, dir., *411: The Missing UFO Connection* (Dave Paulides Productions, 2022).

44. See, for example, Ahmad Greene-Hayes, "Discredited Knowledges and Black Religious Ways of Knowing," *J19: The Journal of Nineteenth-Century Americanists* 9, no. 1 (Spring 2021): 41–49.

45. Geraldine Heng, *The Invention of Race in the European Middle Ages* (Cambridge: Cambridge University Press, 2018), especially chap. 4, "Color: Epidermal Race, Fantasmatic Race: Blackness and Africa in the Racial Sensorium."

46. Wim M. J. van Binsbergen, *Vicarious Reflections: African Explorations in Empirically Grounded Intercultural Philosophy* (Haarlem, Netherlands: Shikanda, 2015), 552. I am indebted to Nicholas Collins for pointing this key passage out to me.

47. Van Binsbergen, *Vicarious Reflections*, 551–52. We will see these different cultural models of the person pursued and developed further into a comparative theory of mind in the anthropology of T. M. Luhrmann, below.

48. Van Binsbergen, *Vicarious Reflections*, 552–53.

49. Charles H. Long, "The Humanities and 'Other' Humans," in *Ellipsis . . . The Collected Writings of Charles H. Long* (London: Bloomsbury Academic, 2018), 327. Everything is in the ellipsis of the title . . . Real thought is of the future. It is not yet accomplished. It is something impossible on the way to being possible.

50. Long, "Humanities and 'Other' Humans," 328. I cannot help but observe that the modern UFO contact experience looks more than a little like a gigantic "cargo cult." Some in those communities will even argue that the presence is technological, not magical, in nature—a technology, moreover, that will someday be our own. That is cargo cult thinking.

51. Sylvia Wynter, "Unsettling the Coloniality of Being/Power/Truth/Freedom: Towards the Human, After Man, Its Overrepresentation—an Argument," *CR: The New Centennial Review* 3, no. 3 (Fall 2003): 257–337.

52. See Stephen C. Finley, *In & Out of This World: Material and Extraterrestrial Bodies in the Nation of Islam* (Durham, NC: Duke University Press, 2022), 10.

53. Biko Gray, personal communication with author, January 14, 2023 (used with permission with slight editorial emendations). I also want to acknowledge the Esalen Institute: Stephen C. Finley and Biko Gray have led for two years (at the time of writing) an annual series of symposia at Esalen called "Black Superhumanisms."

54. See John E. Mack, MD, *Abduction: Human Encounters with Aliens* (New York: Charles Scribner's Sons, 1994); and *Passport to the Cosmos: Human Transformation and Alien Encounters* (New York: Crown, 1999). For the Indigenous stories, see John Hart, "Abductions and Indigenous Peoples' Encounters: John Mack, Phillip Deere, David Sohappy, Ardy Sixkiller Clarke," in *Third Displacement: Cosmobiology, Cosmolocality, Cosmosoceioecology* (Eugene, OR: Cascade Books, 2020), 139–91.

55. Octavia E. Butler, *Mind of My Mind* (New York: Grand Central Publishing 2020).

56. Octavia E. Butler, *Parable of the Talents* (New York: Seven Stories Press, 2016), 31, 38.

57. For a recent exemplar of mad studies that goes into significant detail about the history of the study of religion and why it is mistaken to avoid the topics of madness and psychosis as superreligious states, see especially Richard Saville-Smith, *Acute Religious Experiences: Madness, Psychosis and Religious Studies* (London: Bloomsbury Academic, 2023).

58. For an exceptionally full analysis of madness and various psychiatric, mystical, psychedelic, and philosophical states as implicated in one another and mutually illuminative, see especially Wouter Kusters, *A Philosophy of Madness: The Experience of Psychotic Thinking*, trans. Nancy Forest-Flier (Cambridge, MA: MIT Press, 2020). Kusters's work is *so* original. He changes the game by changing the rules of that game—that is, by making the mad person a companion, a genuine mystical thinker, and a philosopher of the biggest questions of all, which he understands as spiritual questions and relates directly to the history of religions (Kusters, xxii). I was especially moved by his descriptions of the most extreme reaches of "mad time" (what he calls "crystal time"), which is directly relevant to the Nietzschean eternal recurrence of the same, by "mad monism" and "mad plurality," and by his critical commentary on the Plotinian One and Eckhart's Nothing within the *via mystica psychotica*. His rereading of major historians of religions, like Mircea Eliade and R. C. Zaehner, were also eye-openers for me. They helped me understand, for example, the Nietzschean overtones of Eliade's reading of Indian yoga as a "reversal of all human values" and the cosmicization of the human being within a "macro-anthropic" superhuman that is in turn transcended within a complete and total ascetic withdrawal from the cosmos (Kusters, 98–112, 288–99, 421–23, 426–34). Kusters's definition of madness, moreover, is resonant with my own present intellectual convictions: "the socially awkward expression of a desire for infinity in a world that defines itself as finite" (Kusters, xxii). I very recently came to this impossible book and so cannot properly integrate it into this one.

59. See Philip K. Dick, "Drugs, Hallucinations, and the Quest for Reality" (1964) and "Schizophrenia & *The Book of Changes*" (1965), in *The Shifting Realities of Philip K. Dick: Selected Literary and Philosophical Writings*, ed. Lawrence Sutin (New York: Vintage Books, 1995), 167–74, 175–82.

60. The editors highlight this very phrase, no doubt because they saw it as illuminative of Philip K. Dick's genius. See Pamela Jackson and Jonathan Lethem, eds., *The Exegesis*

of Philip K. Dick, Erik Davis, annotation editor (Boston: Houghton Mifflin Harcourt, 2011), xx.

61. María del Pilar Blanco and Esther Peeren, eds., *The Spectralities Reader: Ghosts and Haunting in Contemporary Cultural Theory* (New York: Bloomsbury Academic, 2013).

62. Martha Lincoln and Bruce Lincoln, "Toward a Critical Hauntology: Bare Afterlife and the Ghosts of Ba Chúc," *Comparative Studies in Society and History* 57, no. 1 (2015): 191-220.

63. Lincoln and Lincoln, "Toward a Critical Hauntology," 194, 195.

64. Lincoln and Lincoln, 195.

65. Lincoln and Lincoln, 196.

66. Lincoln and Lincoln, 196.

67. I am reminded in this context of Mai Lan Gustafsson, *War and Shadows: The Haunting of Vietnam* (Ithaca: Cornell University Press, 2010). The Lincolns write about Gustafsson admirably as someone who refuses the hauntology that would keep the ghost as only a memory or metaphor. Indeed, in the words of the Lincolns, Gustafsson employs an "unnervingly naturalistic idiom" to write of her cases (Lincoln and Lincoln, "Toward a Critical Hauntology,"198).

68. Lincoln and Lincoln, "Toward a Critical Hauntology," 200.

69. The original case study is at Gustafsson, *War and Shadows*, 47-48, 90.

70. I am indebted to John Boyle for the notion of the "esoteric trace" in modern thought, especially psychoanalysis; see John Boyle, "'Before and After Science': Esoteric Traces in the Formation of Psychoanalysis" (doctoral thesis, University of Exeter, 2023).

71. Hans Loewald, "On the Therapeutic Action of Psycho-Analysis," *The International Journal of Psychoanalysis* 41 (1960): 29, quoted and discussed in Boyle, "Before and After Science," 16-17.

72. Donna M. Orange, "Review of *Ghosts in the Consulting Room: Echoes of Trauma in Psychoanalysis* and *Demons in the Consulting Room: Echoes of Genocide, Slavery and Extreme Trauma in Psychoanalytic Practice*," *Psychoanalysis, Self and Context* 12, no. 1 (2017): 92. My thanks to John Boyle for this reference.

73. For a historical discussion of this thesis, see Christopher Laursen, "The Poltergeist at the Intersection of the Spirit and the Material: Some Historical and Contemporary Observations," in *Super Religion*, ed. Jeffrey J. Kripal, (Farmington Hills, MI: Macmillan, 2017), 311-326.

74. Nandor Fodor, *Between Two Worlds* (West Nyack, NY: Parker Publishing Company, 1964).

75. Carlos Eire, personal email communication with author, March 10, 2023.

76. Carlos Eire, *The Life of Saint Teresa of Avila: A Biography* (Princeton, NJ: Princeton University Press, 2019), 89-95.

77. Eire, *Life of Saint Teresa*, 73.

78. Translated in Eire, 89-90.

79. Translated in Eire, 94.

80. Eire, 77.

81. Eire, 78.

82. Mircea Eliade, "Folklore as an Instrument of Knowledge," in *Mircea Eliade: A Critical Reader*, trans. Mac Linscott Ricketts, ed. Bryan Rennie (London: Equinox, 2006).

83. Another exception to the general neglect of the superhuman is Michael Grosso, *The Man Who Could Fly: St. Joseph of Copertino and the Mystery of Levitation* (Lanham,

MD: Rowman & Littlefield, 2016). I featured Grosso in Jeffrey J. Kripal et al. *Comparing Religions: Coming to Terms* (Oxford: Wiley-Blackwell, 2014).

84. Eire, *They Flew*, 145–46, 165.

85. Eire, 165.

86. Eire, 139.

87. Eire, 165.

88. Eire, 164. In other words, those Joseph took with him did not choose this.

89. Eire, 160. Central here is the film-like suspension of time in the trances.

90. Eire, 156–59.

91. Eire, 158.

92. Eire, 155.

93. Eire, 142.

94. Eire, 145. The subtle allusion, and challenge, to Immanuel Kant, who insisted we can never come to know directly the thing-in-itself, is telling. I think Kant was correct with respect to our sense-based natural knowing but incorrect with respect to our direct mystical experiences. Hence the impossibility of this thinking, of this attempt to relate the human and the superhuman, in the contemporary academy.

95. Eire, 142–43.

96. Eire, 143.

97. Eire makes these very points in Eire, 139–40.

98. I cannot help observing here that Joseph could hear nothing in his ecstatic trances, *except the voices of his religious superiors*, who often would tell him to "come down" (Eire, 144–45). He did.

99. See Carlos Eire, *Waiting for Snow in Havana: Confessions of a Cuban Boy* (New York: Free Press, 2004); and *Learning to Die in Miami: Confessions of a Refugee Boy* (New York: Free Press, 2010).

100. Eire, *They Flew*, xii (italics in original). For further discussion of this event, see Eire, 93.

101. Eire, 211.

102. Eire, 222.

103. Eire discusses this text and its troubled reception in Eire, 236–51. He explains the arguments of this "fifth gospel of sorts"—namely, that María's miracles extend those promised in the New Testament and the continuing role of the Spanish nun's miracles in the debate around the coredemptress status of the Virgin Mary herself. I would only add that there is a kind of implicit superhumanism embedded in this very Catholic devotionalism. Mary's "divinized human body," after all, *is* the "Mystical City of God" that gave birth to the incarnate God (Eire, 239). Unsurprisingly, the work was considered blasphemous, presumptuous, heretical, and sacrilegious by many, so much "legendary nonsense" (Eire, 248). And so it goes. We still lack a convincing way to read such "private revelations," of which the history of religions down to this very day is chock-full.

104. Eire, 225.

105. Eire, 230.

106. Eire, 255–56.

107. Eire, 261.

108. Eire, 276–77.

109. Eire, 282.

110. Eire, 292.

111. Eire, 378.

112. Barbara Newman, *The Permeable Self: Five Medieval Relationships* (Philadelphia: University of Pennsylvania Press, 2021).

113. I will explore this much more in *Biological Gods*, the second volume of the Super Story trilogy, but I feel the same way about totemic experiences with animal-spirits or even "nonhuman" intelligences. This is why I want to call *all* such altered states "human," not in the reductive psychological or anthropocentric sense but in the sense that human beings have a very special access to cosmic consciousness, and consciousness is consciousness *no matter in what organic form it is manifesting*. Is this "posthuman"? I doubt it. It seems more "superhuman" to me, in the sense that consciousness is cosmic and not local, egoic, personal, or even species-specific.

114. I am also inspired here by the paranormal investigator Paul Eno and his conclusion: "Our own superlives are a Unity across the worlds. By extension, in the cosmic logic of the multiverse, we literally are each other. Even as illustrated by the horrific possession phenomenon, here is the Unity as well, for better or worse." See Paul Eno, *Dancing Past the Graveyard: Poltergeists, Parasites, Parallel Worlds, and God* (Atglen, PA: REDfeather, 2019), 207.

115. Newman, *Permeable Self*, 282–83.

116. John 14:11, 14:20, quoted in Newman, 5 (italics hers).

117. Newman, 269.

118. Newman, 270–74; T. M. Luhrmann et al., "Toward an Anthropological Theory of Mind," *Journal of the Finnish Anthropological Society* 36, no. 4 (Winter 2011): 1–69.

119. Hence Newman's discussion of Frederic Myers, William James, and the contemporary neuroscientist Edward Kelly and his colleagues on this very idea (Newman, *Permeable Self*, 276–77).

120. Newman, 277.

Conclusion

1. I am thinking in particular of the analytic idealism of the computer engineer and philosopher Bernardo Kastrup, the realist idealism of the neuroscientist Edward F. Kelly (indebted to Frederic Myers and William James), the neo-Leibnizian mysticism of the scholar of religion Paul Marshall, the paranormally inflected idealism of Sharon Hewitt Rawlette, and the Spinoza-shaped nondual psychedelic philosophy of the philosopher Peter Sjöstedt-Hughes. I have intentionally referenced all these individuals and texts above because they each teach us, in different ways, how to think impossibly.

2. See Jason Ananda Josephson Storm, *Metamodernism: The Future of Theory* (Chicago: University of Chicago Press, 2021), 20: "We will never solve anthropogenic climate change, structural racism, or patriarchal hegemony if we are incapable of thinking in terms of totalities."

3. James H. Austin, *Zen and the Brain: Toward an Understanding of Meditation and Consciousness* (Cambridge, MA: MIT Press, 1999).

4. James Austin in Richard P. Boyle, *Realizing Awakened Consciousness: Interviews with Buddhist Teachers and a New Perspective on the Mind* (New York: Columbia University Press, 2015), 296–97.

5. It is no accident at all that John Mack's papers now sit in our Archives of the Impossible. The entire archival project is meant to do exactly what he most intended: ontologically shock those who would engage it adequately.

6. Leslie Kean and Ralph Blumenthal, "Intelligence Officials Say U.S. Has Retrieved Craft of Non-human Origin," *The Debrief*, June 5, 2023, https://thedebrief.org /intelligence-officials-say-u-s-has-retrieved-non-human-craft/.

7. Paul Tillich, *Systematic Theology*, 3 vols. (Chicago: University of Chicago Press, 1951), 1:113.

8. Tillich, *Systematic Theology*, 1:112–14.

9. Andrea Pritchard et al., eds., *Alien Discussions: Proceedings of the Abduction Study Conference held at MIT, Cambridge, MA* (Cambridge, MA: North Cambridge Press, 1994). See especially John E. Mack, "Ideas and Techniques for Therapists: Ideas for Helping Abductees" (Pritchard et al., 478–84). Rather astonishingly, a full treatment of this conference event can be had in C. D. B. Bryan, *Close Encounters of the Fourth Kind: A Reporter's Notebook on Alien Abduction, UFOs, and the Conference at M.I.T.* (New York: Penguin Books, 1996). I know all the following details from conversations with my colleagues of the John E. Mack Institute, on whose board of directors I sit. My thanks especially to Will Bueché for his astonishing archival research and knowledge of the history of Mack's use of "ontological shock."

10. John E. Mack, *Abduction: Human Encounters with Aliens* (New York: Charles Scribner's Sons, 1994).

11. Mack, *Abduction*, 397.

12. I do think this obvious focus on reproductive sexuality, genetics, and alien-human hybridity is more than significant but never also what it seems. I will explore these complexities in *Biological Gods*, the second volume of the aforementioned Super Story trilogy.

13. John E. Mack, "Alien Reckoning: Many Americans Claim They've Been Abducted by Extraterrestrials. A Once-Skeptical Harvard Psychiatrist Believes Them," *The Washington Post*, Sunday, April 17, 1994.

14. Mack, "Ideas and Techniques for Therapists," 480. He stresses this theme of the cocreation of knowledge in *Abduction*, for example, 391, 399. He even describes this "intersubjective unfolding of the investigator-abductee interaction" as obtaining information "non-dualistically" (Mack, 25).

15. Mack, "Ideas and Techniques for Therapists," 480–81.

16. Mack, "Ideas and Techniques for Therapists," 484.

17. I believe that this deep philosophical alignment partly stems back to our shared Esalen-based influences. John Mack began engaging the Esalen Institute in the late 1980s, first as an antinuclear activist, then, in the 1990s, as a major UFO researcher. The figures of Stanislav and Christina Grof, who lived at Esalen for almost a decade and a half and whose psychoanalytic, psychedelic, and breathwork teachings are key components of the history of the place, were especially influential on Mack's transformation from a political activist to a paradigm shifter. Their work entered Mack's own therapeutic practice in the forms of Holotropic Breathwork and body-work involving the idea of memories stored in the muscle tissue of the body (Mack, "Ideas and Techniques for Therapists," 483). Indeed, Mack's biographer will go so far as to quote the psychiatrist on how everything goes back to the Grofs: "They put a hole in my psyche and the UFOs flew in." See Ralph Blumenthal, *The Believer: Alien Encounters, Hard Science, and the Passion of John Mack* (Albuquerque: University of New Mexico Press, 2021), 8. There are many such holes in my own psyche about now. It's a frickin' sieve.

18. Mack, *Abduction*, 20.

19. James C. Carpenter, *First Sight: ESP and Parapsychology in Everyday Life* (Lanham, MD: Rowman & Littlefield, 2015).

20. I believe I adopted this metaphor from Wouter Hanegraaff and his historical argument that Western esotericism is "rejected knowledge." See Wouter J. Hanegraaff, *Esotericism in the Academy: Rejected Knowledge in Western Culture* (Cambridge: Cambridge University Press, 2014).

21. This famous phrase is from Leopold von Ranke (1795–1886).

22. Erik Davis, *High Weirdness: Drugs, Esoterica, and Visionary Experience in the Seventies* (Cambridge, MA: MIT Press, 2019), 1, 5. For another theorization of the weird around "ontological flooding," or the idea that a plethora of ontologies follow in the wake of impossible phenomena, see Jack Hunter, ed., *Deep Weird: The Varieties of High Strangeness Experience* (Guildford, Surrey: August Night Press, 2023).

23. I remain inspired and duly cautioned by George P. Hansen, *The Trickster and the Paranormal* (Philadelphia, PA: Xlibris, 2001). I find Hansen's use of the social sciences, structuralism, and cultural anthropology particularly powerful toward his double thesis that psi is both very real and deeply deceptive.

24. Ioan P. Couliano, *Out of this World: Otherworldly Journeys from Gilgamesh to Albert Einstein* (Boston: Shambhala, 1991), 3–4.

25. Grant Morrison, *Supergods: What Masked Vigilantes, Miraculous Mutants, and a Sun God from Smallville Can Teach Us about Being Human* (New York: Spiegel & Grau, 2011), 261–63, 272–78. This book is very important to me. I reviewed *Supergods* in "How We Got to Super: Grant Morrison's Visionary Gnosticism," *Religion Dispatches*, September 8, 2011, https://religiondispatches.org/how-we-got-to-super-grant-morrisons-visionary-gnosticism/.

26. Jesse Keskiaho, *Dreams and Visions in the Early Middle Ages: The Reception and Use of Patristic Ideas, 400–900* (Cambridge: Cambridge University Press, 2015); Kathryn S. Freeman, *Blake's Nostos: Fragmentation and Nondualism in "The Four Zoas"* (Albany, NY: SUNY Press, 1997); David Weir, *Brahma in the West: William Blake and the Oriental Renaissance* (Albany, NY: SUNY Press, 2003).

27. Barbara Newman, *The Permeable Self: Five Medieval Relationships* (Philadelphia: University of Pennsylvania Press, 2021).

28. Eric Wargo, "Fancy's Monster: Being and Bootstrapping in Mary Shelley's Frankenstein" (unpublished manuscript, May 15, 2023), Microsoft Word file. My gratitude to Eric for sharing this and other works in progress with me; all used with permission.

29. Eric Wargo, "The End of the Bookstore" (unpublished manuscript, May 15, 2023), Microsoft Word file. The UFO is often interpreted as a time machine coming from the future, so it works especially well here as a metaphor for Wargo's time loops.

30. Eric Wargo, "From Nowhere: Artists, Writers, and the Precognitive Imagination" (unpublished manuscript, May 15, 2023), Microsoft Word file (italics in original).

31. Wargo, "From Nowhere."

32. Wargo, "From Nowhere."

33. I have been much informed here by the work of the physical anthropologist Michael P. Masters in *The Extratempestrial Model* (Full Circle Press, 2022) and *Identified Flying Objects: A Multidisciplinary Scientific Approach to the UFO Phenomenon* (Masters Creative, 2019).

34. In a similar spirit, as much as I align myself with the scholars of new religions, and I really do, I do not like the phrase "new religious movements," as it implies these same

"communities of belief." Both phrases are also quite boring and do not speak *at all* to the astonished WTF sense of the fantastic that is so obvious in these contexts. Why do people routinely gather around gifted, charismatic individuals as God? Because they are. The mutant teacher has realized this fact in a unique and, often enough, physically transmissive way. The very strong inclination of those outside these small communities of the stunned is to reduce the radioactive superhumans and those that experience them as such (sometimes quite accurately) to social processes, psychological projection, or moral scandal; this is an external and very limited view, although, of course, such readings might also be true enough on their own levels. Charismatic energies and supersexualities *do* go together, and often neither can be socially bound, particularly when they manifest together. It is a both-and again, not an either-or. See chapter 6.

35. Rafael Antunes Almeida, "UFOs, Ufologists, and Digital Media in Brazil," in *Believing in Bits: Digital Media and the Supernatural*, ed. Simone Natale and D. W. Pasulka (New York: Oxford University Press, 2020), 182.

36. I am influenced here by the evolutionary psychologists, anthropologists, and historians, particularly Agustín Fuentes, *Why We Believe: Evolution and the Human Way of Being* (New Haven: Yale University Press, 2019). Fuentes, following Kant, calls the impossible "transcendence."

37. Carlos Eire, *They Flew: A History of the Impossible* (New Haven: Yale University Press, 2023), 258, 260.

38. Eire, *They Flew*, 378.

39. Eire, 378.

40. Friedrich Nietzsche, *The Antichrist*, in *The Portable Nietzsche*, ed. and trans. Walter Kaufmann (New York: Penguin Books, 1976), 613, 607. Nietzsche's criticism of Christian belief is complex. Psychologically speaking, he sees it as a symptom of revenge and resentment of the rabble against the noble and cultured.

41. Nietzsche, *Antichrist*, 608. Hence the subtitle of what Nietzsche considered his most important book, *Thus Spoke Zarathustra: A Book for All and None*.

42. Nietzsche, *Antichrist*, 615.

43. Nietzsche, 638.

44. Nietzsche, 653, 656, 644.

45. Nietzsche 641.

46. Nietzsche, 590–91.

47. Nietzsche, 604–5.

Epilogue

1. I first explored the model of the three bars in Jeffrey J. Kripal, "The Super Natural: Powers and Superpowers in the Modern World," Chicago Humanities Festival, recorded April 27, 2019, posted May 15, 2019, YouTube video, 55:57, https://www.youtube.com/watch?v=_2rOFo8X_rs. I refined them again for the 2019 "Trans- States: The Art of Revelation" conference at the University of Northampton; see Jeffrey J. Kripal, "The Flip: Recalibrating the Humanities and Sciences around Extraordinary Experience," Trans- States, recorded September 14, 2019, posted November 28, 2019, YouTube video, 1:11:03, https://www.youtube.com/watch?v=cNt2KoNw0ck.

2. The letter is also used with permission. I have preserved Jim's capitalization practices

and, at his request, changed a number of details to preserve the privacy of the individuals involved.

3. David Kaiser, *How the Hippies Saved Physics: Science, Counterculture, and the Quantum Revival* (New York: W. W. Norton & Co., 2012). I have a personal relationship to Kaiser's sources, subjects, and ideas. Many of his hippie physicists, after all, were associated in some way with the Esalen Institute in Big Sur, California, whose history I have written in Jeffrey J. Kripal, *Esalen: America and the Religion of No Religion* (Chicago: University of Chicago Press, 2007). In particular, I know reasonably well the inimitable Nick Herbert, with whom I have had numerous bracing (and hilarious) conversations about everything from Joseph Campbell to esoteric Islam and nondual philosophy to Roman Catholicism and LSD. See Herbert's website, "Quantum Tantra," http://quantumtantra.blogspot.com/.

4. See Phil Zuckerman, *Society without God: What the Least Religious Nation Can Teach Us about Contentment* (New York: New York University Press, 2008), 147–48.

5. This book is a kind of intellectual-spiritual debt that I owe Ioan Couliano. He was not my mentor in any official capacity, but I was formally examined by him, and I have definitely been thinking within the science fictional, hyperdimensional, and paranormal currents in which he was writing and imagining in the late 1980s and early 1990s, right before he was murdered in a bathroom stall in May 1991. I hope I have fulfilled this debt, Ioan.

Index

Is indexing the impossible possible? Maybe. Here is one attempt. I have particularly tried to trace and track "soul," "UFO," "time," and "belief," as in the subtitle of this book. As for "everything else," well, it is everything else.

Abbott, Edwin, 230
abnormal, 44, 50
Abrahams, Brad, 72
Abrams, M. H., 33–34, 258n31, 278n3
abduction, 10, 41, 52–53, 58, 62, 64, 67–78, 69, 72, 74–76, 77, 80, 81, 83, 84–89, 93–96, 101, 141, 158–59, 172, 175, 177, 194–95, 199, 202, 212, 213, 222–24, 230, 267n8, 279n23
Africa, 25, 56, 66, 94, 113, 130, 195–200, 262n14
Afrofuturism, 200
agency, x, 4, 32, 48, 51, 55, 67, 94–96, 150, 192–93, 194–95, 253n9
Agrama, Hussein Ali, 11, 59, 262n23
Allison, John, 149–50
Altizer, Thomas J. J., 130
anamnesis, 18
Angelou, Maya, x
anthropology, ix, 11, 50, 123, 203–4, 214–15, 251n1, 253n9, 265n54, 286n23
Anzaldúa, Gloria E., 245
'Arabī, Ibn, 37, 76, 137
Archives of the Impossible, 10, 53, 56, 72–73, 96–97, 99, 158, 199, 248–49
Aristotle, 32
Asperger's syndrome, 92
atheism, 130, 181, 230, 278n3
Atmanspacher, Harald, 13, 163–70, 174

Augustine, 191, 194, 231
Aurobindo, 256n4
Austin, James, 221–22
Austin, Karin, 84–89, 90–91, 235, 266n104
authoritarianism, 39, 186, 190, 210
autism, 85, 92, 97, 99, 100–101, 102, 107, 119, 208. *See also* Kevin
autopoiesis, 166
Awareness, 106, 107, 111
ayahuasca. *See* DMT

Ball, Martin W., 182
Ball, Philip, 13–15
Barnes, Lori, 73
Basso, Michele, 143
Bataille, Georges, 68
belief: and abduction, 80; and the Blakean "flip," 183; and "bracketing," 129, 241; and demonization of paranormal phenomena, 157; and dual-aspect monism, 170, 174, 178; and fraud, 236; humanistic, 178; as the enemy, 153, 229–30, 268n27; and the experience-source hypothesis, 28–30, 207, 235; as irrelevant, 27–28, 242; as limiting, 156, 72, 230; literal, 42, 209; Kantian, 166; in Marxist model of the fantastic, 104; and modernity, 192; and monotheism, 180, 191; and multinaturalism, 61; as (necessary) trick, 50; and

belief (*cont.*)
   new religious movements, 286–87n34;
   and Friedrich Nietzsche, 188, 236–38,
   277n36, 287n40; as paradox in my
   thought, 155–57, 218; and the paranormal, 46; physical effects of, 61; and
   Platonic surrealism, 107; as problem,
   153, 170; and psycho-folklore, 207; scientistic beliefs, 178, 225; and the skeptic's
   overcoming, 235–38; and the superhuman, 156; as true projection, 188–89;
   and the UFO, 55, 61, 71, 72
Besant, Annie, 109
Bharati, Agehananda, 131
big bang, 15
Bigelow, Robert, 58, 257n17
bilocation, 5, 192, 210, 212
Binsbergen, Wim van, 94, 195–96, 199
bi-unity, 33
Blackburn, Kirsten, 62
Black Panther, 198
blackness, transcendent, 33, 197–99
Blake, William, 35–36, 183, 221, 231, 258n40
block universe, 144–49, 231, 233–34, 244,
   266n89
Blomqvist, Håkan, 56
Bloom, Harold, 35, 38, 189, 279n22
Blumenthal, Ralph, 55, 254n13, 285n17
blurriness, 8, 59
Boas, Franz, ix, 17
Böhme, Jacob, 151
Bohr, Niels, 13–14
Braude, Stephen E., 194
Brentano, Franz, 169
Breton, André
Bryant, Larry, 158
Buddhism, 10, 62–63, 66, 77, 80, 109–11,
   125, 135, 138, 148, 177, 186, 221, 226, 232,
   237–38, 269n40, 269n46, 271n25
Bueché, Will, 285n9
Butler, Ocatavia E., 200

Caillois, Roger, 66–67
Campbell, Joseph, 32, 288n3
capitalism, 55
Carpenter, Thomas, 226
cessation of miracles, 193
Cheetham, Tom, 31–32, 257n19

Christology, 36–37
clairaudience, 37, 48
clairvoyance, 37, 42, 48, 177, 198
Clarke, Ardy Sixkiller, 66
Clarke, Arthur C., 160–61, 199
class, xi, 16, 24, 252n10
clown, 85
coinherence, 213–15
Cold War, 10, 57, 60, 241, 261n12
Coleridge, Samuel Taylor, xi, 1, 2, 34–35,
   36, 69, 89, 253n9, 258n37
collage, 33, 39
Collins, Nicholas, 73
colonialism, 8–9, 18, 24–25, 38, 60, 81, 84,
   90, 129, 181, 196–97, 203, 218, 225
comparison, 10, 12–13, 15, 24–25, 28, 29, 32,
   51, 53, 55–56, 69, 73, 76–77, 90, 92–93,
   100–101, 102, 110, 118–19, 124, 126, 128–30,
   132, 134, 136, 138, 140, 141, 144, 152–53,
   156, 161–62, 163, 165, 196, 209, 212, 215,
   220–22, 225, 231, 236, 237–38, 251n3,
   253n7, 265n54, 267n8, 269n50, 271n23,
   275n10, 277n36
computer science, 61, 99–101, 241
consciousness, 35, 61, 105–6, 124, 140, 151,
   158, 176–77, 184–85, 243
conspiracy, 24, 39, 80, 183, 229
Coppola, Francis Ford, 2
Corbin, Henry, 23, 24–27, 30–33, 35–38, 40–
   42, 43, 53, 114–15, 131, 132, 137–39
Couliano, Ioan, 142–43, 217, 230, 243, 288n5
Coulthart, Ross, 55
creativity, xi, 32, 35–36, 38, 40, 42, 62–63,
   65, 69, 115, 126, 202, 208, 228–29, 232–35,
   258n40, 270n55
curse, 130
Cutchin, Joshua, 53, 100–101

Dali, Salvador, 68
Darwin, Charles, 14, 52–53, 101, 160
David-Neel, Alexandra, 109
Davis, Don, 222
Davis, Erik, 228–29
Davis, Stuart, 62–66, 69–70, 77, 91, 98, 248
deception, 47, 82, 192, 207, 211–12, 229–30,
   233–34, 236, 253n8, 261n2
DeConick, April D., 91, 267n109
defense mechanism, 78

deification, 70, 77–78, 84, 96, 134, 154, 182, 183, 188–89, 218

demonology, 77, 112, 114, 189–90, 192, 195, 212–13, 277n38

Derrida, Jacques, 169, 173, 202–4, 218

Descartes, 108

Devil: as human, or superhuman, 47; and levitation, 206; and the near-death experience, 27–28; and Friedrich Nietzsche, 187, 215; and paranormal phenomena, 112, 194; pretense of being, 47; and racism, 195; sexual intercourse with, 192; as trickster, 211; and the UFO, 194–95; as us, 112, 191. See also demonology

Dick, Philip K., xi, 74, 110, 179, 201–2, 228, 277n1

Dillon, Matthew, 264n41, 265n45

divination, x, 50, 66, 175–76, 195–96, 242

DMT (Dimethyltryptamine), 73–76

docetism, 36–37

Dōgen, 76, 77

Donahue, Joseph, 153

Doniger, Wendy, 2

Dorn, Gerhard, 163

double, 2, 29, 33, 154–55, 175

Douglass, Frederick, 198

dual-aspect monism: before religion, 170; of Samuel Taylor Coleridge, 258n37; definition of, 162–63; and the imagination, 172–74; and immanence/transcendence, 167, 230; and Kevin, 98, 105–6; and meaning, 162–70; as my own philosophical framework, 13, 84, 136–37, 140, 178, 218–21, 226; and mystical language, 167, 169, 177; as mythico-physical, 174–78; as nonduality, 163; not Kantian, 167; and Wolfgang Pauli, 152; and psychical phenomena, 173; as supernature, 163; and the symbol, 175–76; and synchronicity, 168. See also monism; subject/object paradox

dualitude, 33

Du Bois, W. E. B., 198

Eckhart, Meister, 76, 126, 177, 218, 270n5, 272n40, 281n58

eclecticism, 27, 37–40

Eddington, Arthur, 167

Einstein, Albert, 71, 142–46, 209

Eire, Carlos, v, 5–6, 192, 193–94, 205–13, 236, 239, 283n103

Eliade, Mircea, 1, 2–3, 19, 32, 69, 130, 196–97, 207–8, 272n28, 272n33, 274n73

Eluard, Paul, 68

embodiment, ix, 2, 24, 26, 32, 37, 66, 94, 106, 138, 139–40, 148, 153, 180, 182, 217–18

Eno, Paul, 284n114

Esalen Institute, 1, 25, 26, 55–56, 76, 252n10, 285n17, 253n9, 255n11, 256n16, 285n17, 288n3

eternity: and block universe, 144, 244; and Ioan Couliano, 142; and Mircea Eliade, 272n33, 274n73; as eternalism, 244; experience of, 149, 189, 221; and Bill Hicks, 71; and Edmund Husserl, 131, 147, 274n70; and impossible thinking, 147; and Kevin, 106, 119; and Paul Marshall, 144; and Friedrich Nietzsche, 145–46, 236, 272n33, 281n58; and Romanticism, 34–36; and Elliot Wolfson, 139, 144. See also evolution; space-time; time

evolution: and Africa, 199; and altered states, 3; and Karin Austin, 88, 90; as collective, 182; and comparison, 253n7, 276n25; and DMT, 75–76; and esotericism, 25, 43–45, 46, 50–51, 52–53, 70, 71, 75, 94, 100, 161, 202, 230–31, 245; and Sigmund Freud, 43; and kundalini, 111; and Joseph Maxwell, 46, 50–51, 75, 260n83; and meaning, 166, 185; and Friedrich Nietzsche, 3, 218, 199; and science fiction, 3, 161, 199, 200, 202, 230–31; and the supernormal, 43–45; and synchronicity, 166, 276n25; and the UFO, 70–72; and Alfred Russel Wallace, 230; and the X-Men, 253n7. See also eternity; space-time; time

Evrard, Renaud, 49–50

experience-source hypothesis, 28–30

Ezekiel, 34–35, 59, 233

fancy, 34

fantastic, the: and Africa, 195; and belief, 156, 286n34; and deception, 4; and Docetism, 36; and dual-aspect monism, 36, 98, 101; and Ramzi Fawaz, 253n7;

fantastic, the (*cont.*)
  and impossible thinking, 4, 12, 114; in lit-
    erary theory, 4; and John E. Mack, 225;
    in Marxist theory, 4; mediated by imag-
    ination, 4, 15, 128, 153, 172, 174–78, 231;
    and moral criticism, 211; and phenom-
    enology, 128; physical expressions of,
    ix, 2, 4–5, 69, 174–78; in psychoanalytic
    theory, 4, 214; and reality, 231, 253n8; as
    religion, 26; as semiotic, 6, 253n8; and
    surrealism, 42; and synchronicity, 176;
    and the UFO, 174–78
Fawaz, Ramzi, 253n7
fever, 63
Finley, Stephen C., 198–99
Flournoy, Théodore, 23, 40–43, 72, 80
Fodor, Nandor, 204–5
Fort, Charles, v, 100–101, 110–11
Fortune, Dion, 109
Fravor, David, 55
Fröbe-Kapteyn, Olga, 32
Fuchs, Christopher, 163–70, 174
Fuller, John G., 95
fundamentalism, 27–28, 39–40
fuzziness. *See* blurriness

Gadamer, Hans-Georg, 132
Gallagher, Mike, 54–55
Garcia, Jerry, 69
Geller, Uri, 69
Gilson, Etiénne, 3
God, 180–89, 278n3
Graves, Ryan, 55
Gray, Biko Mandela, 198–99
Greyson, Bruce, 30–31, 256–57n16
Grof, Stanislav, 255n1, 285n17
Grusch, David, 55, 222
Gustafsson, Mai Lan, 202

Hana, Fred, 125
Hansen, George, 53
Hartman, Saidiya, 198
hauntology, 202–4
Heard, Gerald, 76
Hedberg, Mitch, 59
Heidegger, Martin, 24, 125, 126, 146, 151,
    255n3, 256n4, 270n5
Heisenberg, Werner, 13

Herbert, Nick, 288n3
hermeneutics, 135, 142–44, 147, 152–53, 164
Heyward, Carter, 214
Hickman, Stephen, 76
Hicks, Bill, 70–72, 183
Hill, Barney, 95
Hill, Betty, 95
history of science, 159–61
Hitler, Adolf, 80
Hoffman, Donald, 14
Huggins, David, 72
humanism, 8–10, 39–40, 196–97
Hurston, Zora Neale, ix–xiii, 11, 17, 32, 147,
    251–52n3
Husserl, Edmund, 19, 124–31, 146, 147,
    274n70
hyperdimensionality, 25, 55, 56, 88, 89, 139,
    142, 230, 235, 288n5
hypnosis, 41–42, 72, 80, 89, 224

idealism, 15, 31, 34, 35, 77, 101, 106, 114, 142,
    151, 177, 182, 183–85, 218, 221, 258n40,
    275n8, 284n11
imaginal: coinage, 31; and Henry Corbin,
    25, 27–28, 30, 31–33, 36–40, 41, 114;
    cosmic, 101, 105, 120; and Docetism, 36–
    37; and eclecticism, 38; and Théodore
    Flournoy, 40; and *imagination créatrice*,
    32, 40, 42; and *imaginatio vera*, 31; and
    impossible thinking, 220–21; as insec-
    toid, 45, 52, 62–66, 77, 89–91; and Kevin,
    98–99, 101, 104, 108, 114–15, 120; and
    mundus imaginalis, 33; and Frederic
    Myers, 45; and near-death experience,
    27–31; and nonhuman or superhuman
    intelligences, 24; and the paranormal,
    128–29; and parapsychology, 37; and
    reality, 108, 231–32; Romantic origins,
    33–36; and the supernormal, 43–45; and
    the UFO, 52–53, 65; and Elliot Wolfson,
    137–38. *See also* imagination; symbol
imagination: active, 32–33; and William
    Blake, 35–36, 183; and Samuel Taylor
    Coleridge, 34, 258n37; creative, 32–33,
    38, 40, 42; and as dreamlike narrative
    and symbols, 48; as eclectic, 38; and
    the fantastic, 4, 15, 172; and the filter
    thesis, 215; furthest reaches of, 16, 173;

Marxist model, 4, 104; of matter, 129; as mediator, 4, 15, 48, 137-38, 172-74, 176; and myth, 16, 128; and the near-death experience, 256n11; physical effects of, 173, 178, 231, 232; precognitive, 232-35; and psychical or paranormal phenomena, 45, 215; and radical empiricism, 255n23; and reality, 232; as reductively understood, 30-31, 141; theory of, 1, 2, 255n23, 257n22; and the UFO, 173. *See also* imaginal; symbol

imperialism, 8-9, 66, 90-91, 129, 188, 196-197

implant, 65, 93, 173

impossible, the: and block universe, 147; denials of, 19; and the fantastic, 12; as function of social assumptions, 10; and givenness, 127-128; and Stanislav Grof, 255n1; and madness, 200-202; and magic, 191-93; and morality, 190-91; and the movie, 238; and multinaturalism, 10, 209, 253-54n9; and neuroanatomy, 101-2, 154; and neurodiversity, 92; and Friedrich Nietzsche, 218; and paranormal phenomena, 11; and precognition, 219; and quantum physics, 14; and reality, 16-17; and religion, 191; and religious ideation, 177; and the metaphor of the table, 227-28; and scale, 92; and social (in)justice, 190, 196; the sociology of, 247-49; and space-time, 243-44; and theism, 128; as theory, 159, 178; and the tyranny of clarity, 154-55; and words, 23

Jacobs, Ted, 95

James, William, 24, 41, 255n23

John of the Cross, 210

Joseph of Cupertino, 5, 207-13

Jung, C. G., 23, 32-33, 40, 59, 152, 161, 163, 165-66, 168

Kabbalah, 131-44, 146, 151

Kafka, Franz, xi, 188

Kaiser, David, 15, 288n3

Kali, 63, 96, 198

Kant, Immanuel, 166, 167, 283n94

Karimova, Jasmine, 64

Kastrup, Bernardo, 182, 278n6

Kean, Leslie, 55

Keel, John, 109-10, 153, 195, 229-30

Keightly, Bertram, 109

Kelleher, Colm, 58

Kelly, Edward, 257n16

Kevin: on autism in the history of religions, 100-101, 208; on "energy" in yogic systems, 102-5; on the Human as Three, 105-7; humor of, 98, 120; and Little Red Wagon Theory, 107-8; Platonic surrealism of, 98, 101, 107, 114, 118, 221; on sci-fi physical effects, 113-18; and the soul, 108-12; and super sexualities, 102-5; on the traumatic secret, 112, 117-18, 120; and the UFO, 113-18

King, Jay Christopher, 62

Kipling, Rudyard, xi

Klein, Donald, 72

Knight, Sam, 124

Krohn, Elizabeth, 28-30, 33

Kuhn, Thomas, 225-27

kundalini, 77, 102-5, 111-15, 118-19, 269-70n50

kung fu, 66

Kusters, Wouter, 281n58

Lang, Andrew, 207

Lem, Stanislaw, xi

Lennon, John, 68-69

levitation, 192, 205-13

lightning, xii, 3, 28-30, 218, 245, 252n12

Limbaugh, Rush, 84

Lincoln, Bruce, 202-4

Lincoln, Martha, 202-4

Loeb, Paul, 145-46, 147

Loewald, Hans, 204

Long, Charles H., 129-30, 153, 197-98

lizard man, 81

Luhrmann, Tanya, 214-15

Luther, Martin, 212

lynching, 10

Mack, John E., 84, 158, 199-200, 222-27, 248, 254n13, 285n17

madness, 200-202

magic, 50, 69, 97, 99, 104, 109, 114, 119, 123, 128, 130, 153, 160, 161-62, 174, 190-200, 260n73, 266n98, 267n8, 275n13 *magus*, 191

maleficium, 192

mantis, 52, 62–76, 80, 89, 93, 111, 158, 226

Marcel, Gabriel, 125

María de Ágreda, 210–13

Marshall, Paul, 144

Martel, J. F., 233

martial arts, 66–67

Marxist theory, 4, 104, 202–4, 218, 231,
    271n19

Mason, Charlotte Osgood, xi–xii

Mason, Rufus Osgood, xi

Massignon, Louis, 24–25, 131

masturbation, 104–5

mathematics, 13–14, 161–62, 272n26,
    276–77n34

May, Edwin, 158

McIntyre, O. O., xi

McKenna, Dennis, 228

McKenna, Terence, 52, 53, 228

meaning, 163–70, 276–77n34

Méheust, Bertrand, 178

Merleau-Ponty, Maurice, 125, 146

Metcalfe, David, 115

Mexican revolution, 123

Michels, Jesse, 55

Minkowski, Hermann, 144

miracle, 44, 46–47, 109, 160, 193, 194, 207,
    208–10, 215, 236, 283n103

monism: and Paul Eno, 284n114; and
    mystical thought, 166; neutral, 97–98,
    163, 177; and pluralism, 255n23; and
    reductionism, 105. *See also* dual-aspect
    monism

monotheism, 108, 128, 180, 194

Moody, Raymond, 27–28

Morrison, Grant, 100–101, 230–31, 232, 234,
    273n64

Müller, Élise, 40

multinaturalism, 61, 209, 253n9, 255n23,
    255n24, 265n54

multiverse, 83, 220, 284n14

mundus imaginalis, 33

Murphy, Michael, 25

mutant, 3, 6, 101, 253n7, 286n34

Myers, Frederic, 24, 41, 45, 48, 52,
    89–91

mystical theology, 76–77, 236

mysticism, 49, 104, 106–7, 131, 134, 138, 153,
    200–202

myth, 16, 61, 160, 253n9, 255n24

Nāgārjuna, 76, 135

Namibia, 66

nationalism, 39–40

Nazism, 24, 190, 255n3

near-death experience (NDE), 11, 26–27,
    30–31, 112, 157–58, 176, 177

Neoplatonism, 31, 133

New Age, 56, 72, 84

Newman, Barbara, 213–15

new religious movement (NRM),
    286–87n34

Nickerson, Randall, 199

Nietzsche, Friedrich: and *2001: A Space
    Odyssey*, 83, 199; "all names in his-
    tory," 183; and belief, 236–38; and block
    universe, 144–46; and the Buddha, 238;
    and Christianity, 187–88; and deifica-
    tion, 215, 218; and Mircea Eliade, 272n33,
    274n73, 281n58; eternal recurrence of
    the same, 60, 145–46, 147; as the Devil,
    187; as God, 179; "God is dead," 156,
    181; and history of religions school, 130;
    and madness, 215, 281n58; and mystical
    literature, 277n36; as proto-Nazi, 190; and
    science fiction, 3; and slave mentality, 91;
    and Frederic Spiegelberg, 256n4; and the
    superhumans coming, 230; and the will,
    260n73; and willing backward, 145–46,
    149; and Elliot Wolfson, 142; and X-Men,
    253n7

9/11, 132

non-human intelligence (NHI), 55

Obeyesekere, Gananath, 186

ontological shock, 222–27

ouroboros, 135

Oz effect, 115–16, 118

Palladino, Eusapia, 49–50

panentheism, 25, 278n5

Pang, May, 68

parakinesis, 47

paranormal, the: African origins, 195–200; and brain hemispheres, 101–2; and creativity, 62–63, 65–66, 74; and culture, 7, 16; and Mircea Eliade, 2–3; and madness, 66; and Joseph Maxwell, 45–51; and the mystical, 128; and Barbara Newman, 215; and nonhuman or superhuman intelligences, 24; and philosophy, 184; physical effects, 16; regressions, 51; and social (in)justice, 190, 196; as superpersonal, 157; and technology, 58, 252n4; and textuality, 253n8; and trauma, 112

parapsychology, 37, 53

Pauli, Wolfgang, 13, 59, 152, 163, 165–66, 168

phenomenology: bracketing as apophatic, 126; bracketing and the refusal of ontology, 141, 241; and comparative study of religion, 126, 129, 130–31; and consciousness itself, 124; as mystical, 125–26; and the "naturalist standpoint," 126–27; and telepathy, 125; transcendental, 125

plasma, 110–11

Plato, 28, 32, 43, 115, 125, 142, 162, 190, 201, 214, 222, 237–38, 272n26, 276n34

Plotinus, 76, 166

poltergeist, 10, 36–37, 112, 173, 204–5, 269–70n50

postcolonialism, 18, 90, 129, 197

Powell, Diane Hennacy, 100

precognition: and abduction, 225; and African divination, 195–96; in *Arrival*, 60; and Octavia E. Butler, 200; and dreaming, 11, 29, 31–32, 128; and Zora Neale Hurston, ix–xiii; and the impossible, 177, 232; and literary creativity, xi, 233–34; and Charles Long, 130; and phenomenology, 147; as philosophically important, ix–x, 17; and quantum physics, 15; and the three bars, 239–44; in Wales, 123–24

presentism, 159–61

psychedelics, 10, 53, 70–71, 74–76, 182–83, 202, 212, 228, 229, 285n17

psychoanalysis, 4, 67–68, 204–5, 232–33

psycho-folklore, 207

psychoid, 59, 161, 163

quantum physics: and backward causation, 14, 15; complementarity principle, 14; Copenhagen interpretation, 14; and dream interpretation, 142; and dual-aspect monism, 163; entanglement, 14–15, 164, 165, 241; and hermeneutics, 164; and hippie physicists, 15, 240–41; and holism, 164, 165; and idealism, 15; and the impossible, 14; and meaning, 165–67; and mystical experience, 12–13, 15, 167; nonlocality, 14–15, 163; and observation, 14, 98, 132, 164; and parapsychology, 15; and physicalism, 164; and superposition, 107

Rabe, Michael, 93

race, 8–9, 16, 26, 94–95, 196

racism, 81, 83, 112, 161, 181, 195–200

Randles, Jenny, 115

Rawlette, Sharon Hewitt, 183–85

Regier, Alexander, 155

reincarnation, 28, 40, 66–67, 242

Rickles, Dean, 13, 164, 165–69

Roberts, Jane, 123, 269n46

Robinson, Henry Crabb, 183

Romanticism, 33–36, 59, 151, 155, 183, 231, 258n40. 278n3

Sartre, Jean-Paul, 125

saucer, flying, 53, 269–70n50

Schelling, F. W. J., 151

Scholem, Gershom, 32, 139

Schopenhauer, Arthur, 31–32, 50, 260n73

Schrödinger, Erwin, 13

science fiction, 38, 42, 60, 71, 73–76, 83, 93–96, 110, 120, 227

Scott, Ridley, 110

sense, 165, 169

Serra, Raimundo Irineu, 74

sexuality: and abduction, 93–96; and the history of religions, 267n8; and magic, 191, 192; and mystical literature, 134; and religion in Kripal's early work, 98; shamanism, 92, 99, 100–101, 110, 118, 199

Shankara, 76

Shelley, Mary, xi

Shuña, Pablo Amaringo, 76

Skafish, Peter, 253n9

Smith, Hélène. *See* Müller, Élise

soul: in African religions, 196; and belief, 236; as consciousness, 61; and the double, 33; and dual-aspect monism, 140; and evolution, 118; and experience-source hypothesis, 235–36, 242; and haunting, 203; and levitation, 206; loss of, 110; and the near-death experience (NDE), 28, 29–30, 33; nonexistence of individual, 92, 104, 108–12, 118, 119–20, 166, 182, 196, 268n27, 269n46; and Plato's "myth of Er," 28; and reincarnation, 28, 92; sexualization of, 191; suppression of in post-Nazi Germany, 190; symbolism of, 209; as *tulpa*, 111; and the UFO, 53, 62, 73, 159, 254n13, 261n2, 261n13; violence of as necessary, 186–87

space-time, ix, 16, 17, 55, 60, 71, 133, 142–44, 146, 150, 170, 232–35, 243–44

Spiegelberg, Frederic, 24–26, 256n3

Spiritualism, 15, 40, 42, 196, 205, 269n38

Stang, Charles, 33

Steinbock, Anthony, 126–27

Stephens, Walter, 192

stigmata, 173, 211

Storm (X-Men), xii

Storm, Jason Josephson, 124

Strassman, Rick, 74, 75

Strieber, Anne, 53, 72–73, 153

Strieber, Whitley, 53, 72–73, 76, 83, 95, 153

subject/object paradox, 8, 9, 35–36, 58, 59, 62, 130, 167–68, 178. *See also* dual-aspect monism

subliminal, 43–44, 47, 90

superhistorical, 139

superhuman, 3–4, 6, 18, 24, 29, 37, 40, 41, 51, 53, 62, 66, 67, 93, 96, 100, 109, 146, 153, 154, 156, 179, 183, 188, 189, 193, 197, 198, 199, 202, 207, 213, 216, 225, 230, 238, 245

superhumanities: and Black religion, 198–99; and consciousness, 284n113; and Henry Corbin, 25; and Théodore Flournoy, 41; and geo-politics, 40; and Martin Heidegger, 270–71n5; and the history of religions, 153; as horizontal/vertical, 26, 30; as indirect gnosis,

217–18; and Frederic Myers, 43; and the Nicene creed, 154; as theorized in this book, 18–19; and Elliot Wolfson, 151

superhumanity, 37, 180, 189

Superman, 179–80, 252n14

supernatural, v, 29, 33, 35, 44–45, 46, 51, 181, 192, 193, 209, 211, 221, 235, 253n8

supernormal: as coined by Frederic Myers, 44–45; in Théodore Flournoy, 40–43; and Joseph Maxwell, 46; and non-human or superhuman intelligences, 24; in Anthony Steinbock, 126

surrealism, 4, 41, 67–68, 98, 107, 114

symbol: as Evolver of humanity, 258n37; as medium, 23; as mythico-physical, 174–78; and reality, 61, 137–41; as revelation, 36; as Romantic, 34; as veil, 137; as wheel or disk, 35, 36, 59, 113

synchronicity, 37, 168, 169–70, 176, 184–85

syncretism, 38

Tantra, 96, 113, 198, 267n8

Taylor, Charles, 214

telekinesis, 42, 47

telepathy, xi–xii, 37, 42, 43–44, 48, 52, 64, 67, 82, 86, 89–91, 125, 196, 213, 218, 232, 252n10

teleportation, 5

Teresa of Avila, 205–10

testosterone, 103, 105

theism, 134, 180–81, 221

Theosophy, 49, 56, 109, 205

Thompson, Stith, 94

Thor, xii

Tillich, Paul, 222–23

time: and abduction, 224; and altered states, 3; and backward influence, 15, 133, 143–44, 145, 149, 234, 240; and the block universe, 144; circular, 60, 143; and consciousness, ix, 2, 8, 9, 17, 61, 243; and creativity, 232–35; and entanglement, 14; and eternity, 34, 35, 82, 139, 142–43, 148, 177; and "experience," 12; hermeneutics as power over, 133, 142–44, 147, 228; linear, 9, 142, 233; past, present, future as one, 8, 9, 138–39; suspension during levitation,

207; travel, 54–55, 56, 61–62, 71, 232; and the UFO, 54–55, 56, 61–62, 71, 235; and the unconscious, 44. *See also* eternity; precognition; space-time

Tolkien, J. R. R., xi

transcendence, 134, 167, 230–31

trauma, 51, 190, 202

Treffert, Donald A., 100

trickster, 4

Trump, Donald, 80, 200

Truth, Sojourner, 198

Tubman, Harriet, 198

tulpa, 109–10, 112

Twain, Mark, xi

2001: A Space Odyssey, 83, 199

Übermensch, 179, 190, 256n4, 260n73

UFO: and Africa, 199–200; and ancient alien mythology, 160–61; and angelogy, 77; and California, 56; and cryptids, 109–10, 158–59, 174; and the dead, 73, 158–59; and demonology, 77, 113, 194–95; and Esalen Institute, 55–56; and ETH (extrater-restrial hypothesis), 61; as global and ancient, 56–57, 94, 96–97; and the his-tory of religions, 159–60; and humalien hybridity, 94–95; and the imagination, 172–74; as indigenous, 57; as insectoid, 62–66, 69, 74–76; as Islamic, 57; and John Lennon, 68–69; and Marian apparition, 73; and Michael Murphy, 25; as mythico-physical, 178; and the near-death experi-ence (NDE), 53, 58, 158; as New Age, 56; and nuclear bases, 54, 57; and order of knowledge, 170–72; as the paranormal materialized, 113–18, 173–74; and Tantra, 93–36, 113; and technology, 280n50; and time travel, 54–55, 61–62; as UAP (un-identified aerial phenomena), 53, 55; in US Congress, 53–62

Underhill, Evelyn, 131

Vaeni, Jeremy: and deification, 77, 78; and the ego, 79–80; and hermeneutics, 82–83; and humor, 77, 78, 98; and kundalini, 77; and nonduality, 52, 82; postcolonial readings, 90; racial read-ings, 90; as science fiction, 83

Vallee, Jacques, 61, 277n37

Vilas-Boas, Antonio, 94

Villeneuve, Denis, 60

Wahrträumen, 31, 232

Wallace, Alfred Russel, 230

war, 112, 190–91

Wargo, Eric, 232–35

Watts, Alan, 256n4

Weber, Max, 193

weird, 228–30

Wells, H. G., 60

Wheeler, John, 14, 15, 164, 167

White, David Gordon, 93

Wigner, Eugene, 162

Wilson, Colin, 126

Wilson, Robert Anton, 228

witchcraft, 50, 123, 192, 193, 211

Wojtowicz, Slawek, 76

Wolfson, Elliot: and *apophasis*, 135, 136; and comparison, 134; and Harvard Divinity School, 131–32; and paradox or reflexivity, 135–37; and phallomorphism, 134; and reality, 140, 141; on space-time and historiography, 123; and theism, 134; and the timeswerve, 142–143, 149, 150–51

Woolf, Virginia, xi, 234

Wordsworth, William, 34

Wynter, Sylvia, 197

yoga, 97, 103, 104, 113, 115, 207, 269n50, 281n58

yogini, 93

X-Men, xii, 198, 253n7

Zaehner, R. C., 131

Zen Buddhism, 80, 135, 219–22

Zimbabwe, 199–200